Chemistry of Peptides and Proteins
Volume 1

Chemistry of Peptides and Proteins

Volume 1

Proceedings of the Third USSR–FRG Symposium
Makhachkala (USSR), October 2–6, 1980

Editors
Wolfgang Voelter · Erich Wünsch
Yuri Ovchinnikov · Vadim Ivanov

Walter de Gruyter · Berlin · New York 1982

Editors:

Wolfgang Voelter, Prof. Dr. rer. nat.
Director
Department of Physical Biochemistry
Institute for Physiological Chemistry
Hoppe-Seyler-Strasse 1
D-7400 Tübingen · Germany

Erich Wünsch, Prof. Dr. rer. nat., Dr. med. h.c.
Director
Department of Peptide Chemistry
Max-Planck-Institute for Biochemistry
D-8033 Martinsried · Germany

Yuri Ovchinnikov, Prof. Dr.
Vadim Ivanov, Prof. Dr.
Director
Shemyakin Institute for Bioorganic Chemistry
USSR Academy of Sciences
Moscow · USSR

CIP-Kurztitelaufnahme der Deutschen Bibliothek

Chemistry of peptides and proteins : proceedings of the . . .
USSR–FRG symposium. – Berlin ; New York : de Gruyter

Vol. 1. Proceedings of the third USSR–FRG symposium : Makhachkala (USSR), October 2–6, 1980. – 1982.
ISBN 3-11-008604-2

Library of Congress Cataloging in Publication Data

USSR-FRG Symposium (3rd : 1980 : Makhachkala, R.S.F.S.R.)
Chemistry of peptides and proteins.

Bibliography: p.
Includes index.
1. Peptides–Congresses. 2. Proteins–Congresses. I. Voelter, W. II. Title.
QD431.A1U88 1980 547.7′5 82-14937
ISBN 3-11-008604-2

Printing: Karl Gerike, Berlin. – Binding: Dieter Mikolai, Berlin.
Printed in Germany.

PREFACE

The USSR-FRG Symposium on the Chemistry of Peptides and Proteins was conducted to reinforce between scientists of both countries the hitherto existent contact and exchange of information at the level of the European Peptide Symposium.

Based on the scientific agreement between the Union of the Soviet Socialist Republics and the Federal Republic of Germany, represented on the one hand by the Soviet Academy of Sciences and on the other by the Deutsche Forschungsgemeinschaft, the bilateral symposium series was initiated at the 1st Meeting in Dushanbe-Tadjikistan in April, 1976.

As the scientific basis of these meetings, we have attempted to combine the three main research areas in the field of chemistry of peptides and proteins, i.e. a) isolation, b) structure elucidation and c) synthesis as a discussion platform in order to reestablish the working community previously existing in chemistry of natural products. The first meeting has already confirmed the usefulness of such a programme .

Additionally, our idea was to incorporate, if required, discussions on special, closely-related topics of particular scientific interest. Consequently, at the 2nd Meeting in Grainau-Eibsee in May, 1978, membrane chemistry was added to the programme. The positive aspects of this choice were clearly confirmed at the 3rd meeting in Makhachkala in October, 1980, where a first, tentative discussion on immunological problems in connection with the chemistry of peptides and proteins took place. We believe that this widening of the scientific programme may, in future meetings, lead to a fruitful exchange of experience and knowledge in the individual sectors of this field of chemistry. The strong resonance of the scientific communications and related discussions among the participants of both countries is clearly shown by the increasing interest in these meetings and by the continuous demand for publication of the reports. Following a simple abstract booklet on the occasion of the 1st Meeting, a proceedings volume with shortened versions of the lectures was printed for the 2nd Meeting. In the present proceedings the detailed reports of the lectures and communications of the 3rd Symposium on the Chemistry of Peptides and Proteins are published in full.

The organizers of these series of symposia on Chemistry of Peptides and Proteins would like to take the opportunity of expressing their gratitude to the Soviet Academy of Sciences as well as to the Deutsche Forschungsgemeinschaft for their generous support and sponsorship.

For the Editors
E. Wünsch

CONTENTS

I. Isolation and Structure Elucidation of Peptides and Proteins

II. Peptide Syntheses. Biological Activity and Analytical Problems of Synthetic Peptides

I Isolation and Structure Elucidation of Peptides and Proteins

THE PRIMARY STRUCTURE OF DNA-DEPENDENT RNA POLYMERASE FROM *E. coli*. NUCLEOTIDE SEQUENCE OF rpoB GENE AND AMINO ACID SEQUENCE OF THE β-SUBUNIT

Valery Lipkin, Galina Monastyrskaya, Valentin Gubanov, Sergey Guryev, Oleg Chertov, Vladimir Grinkevich, Irina Makarova, Tatjana Marchenko, Irina Polovnikova, Eugene Sverdlov, Yuri Ovchinnikov

Shemyakin Institute of Bioorganic Chemistry, USSR Academy of Sciences, Moscow, USSR

Introduction

Elucidation of the transcription mechanism requires detailed knowledge of the active center's organization of RNA polymerase at the various stages of the RNA synthesis. This, in turn, can be obtained only after determining the primary and spatial structure of the enzyme.

Earlier we had established the amino acid sequence of the α-subunit of *E. coli* DNA-dependent RNA polymerase resorting solely to the ordinary methods of protein chemistry (1). In the case of the β-and β'-subunits with their much higher molecular weights (~155.000 and ~165.000, respectively), such an approach could no longer suffice.

Results

One of the approaches to the structure determination of large protein molecules is their initial cleavage into a small number of fragments, which then can be analyzed by conventional methods. The search for the conditions of limited proteolysis of the β- and β'-subunits was undertaken. Considerable obstacles were encountered in the course of these studies owing

Chemistry of Peptides and Proteins, Vol. 1

to the fact that the RNA polymerase subunits are not the native proteins. However, the conditions for limited tryptic proteolysis of the β-subunit were found. These are an enzyme/substrate ratio of 1:500, temperature $0^{o}C$, 4 hr (2). Herein there seems, what we believe, to be an optimal set of large fragments (mol. wt. 62.000, 52.000, 37.000, 24.000 and 10.000). Initial separation of the resultant hydrolysate was carried out by chromatography on Sephadex G-100 in 6 M guanidine hydrochloride. This yielded 10 fractions. Their analysis by polyacrylamide gel electrophoresis showed that all large fragments mentioned above are in the first three fractions, while the rest contain about 95 smaller peptides. 53 low molecular weight peptides were isolated from the hydrolysate. They consist of approximately 400 amino acid residues. Their sequencing was very useful for a further structure investigation (3). Isolation of the high molecular peptides proved difficult because of both the little hydrolytic specificity and the low yield of most products. So we could not use limited proteolysis as the main procedure for β sequencing.

The progress in DNA sequencing methods allowed to realize the possibility of using the genetic code to obtain information on the primary protein structure from the nucleotide sequences. However, here there are many pitfalls in the way, requiring considerable caution to avoid possible sources of error.

In the first place the mRNA can undergo processing, leading to erroneous deduction of the protein structure. This holds particularly for eukaryotic cells, wherein "splicing" has been noted. Secondly, the protein itself can be processed. Thirdly, it is often difficult to recognize in the overall DNA structure the beginning of a structural gene. Moreover, one has to bear in mind that a single error (deletion or insertion) in the DNA sequence could lead to a completely erroneous amino acid sequence of the protein.

Thus, primary structure determination of DNA cannot serve as a substitute for the direct sequencing of the protein. In view of this, we decided to utilize the methods of both protein

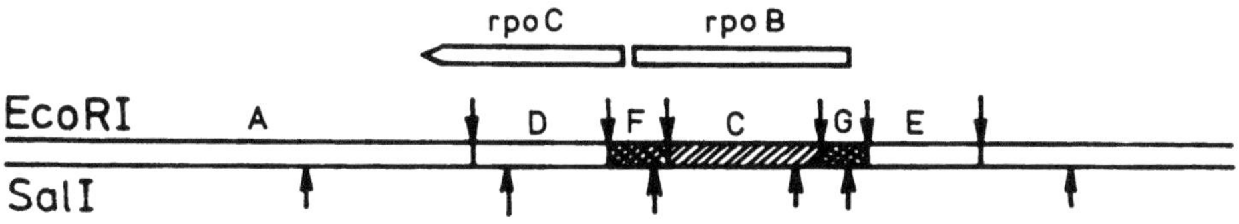

Fig. 1. EcoRI and SalI restriction cleavage map of the *E.coli* DNA region including the structural genes (rpoB and rpoC) of the β-and β'-RNA polymerase subunits.

and nucleotide chemistries, performing the parallel sequencing of the structural genes rpoB (β-subunit) and rpoC (β'-subunit) and of the corresponding proteins. Knowledge of the nucleotide sequence of the pertinent DNA segments would permit aligning of the peptide fragments from the protein analysis into an uninterrupted polypeptide chain. Such an approach provides the key to the most complicated problem in the primary structure analysis of high molecular proteins.

In Fig. 1 restriction endonucleases cleavage map of *E. coli* DNA region containing the structural genes of the β-and β'- subunits of the RNA polymerase (rpoB and rpoC correspondingly) is given. We determined the total sequence of the EcoRI-C (4), EcoRI-F (5) and EcoRI-A - HindIII fragments and partial sequence of the EcoRI-G fragment carrying the beginning of the rpoB gene (6). These fragments were obtained from DNA of $\lambda_{rif}{}^{d}47$ and $\lambda_{rif}{}^{d}18$ transducing phages, containing the *E. coli* rpoBC operon (7, 8), or corresponding plasmids by EcoRI restriction endonuclease digestion. In the case of EcoRI-A - HindIII fragment EcoRI and HindIII digestions were used.

The fragments were consecutively digested with one of the restriction endonucleases (Sau 3AI, Hinf I, Hpa II and Taq I) cleaving the DNA into relatively small blocks. The resulting subfragments were phosphorylated by means of $|\gamma-{}^{32}P|$-ATP and phage T4 polynucleotide kinase and the mixture was separated by electrophoresis on polyacrylamide gel. As a rule both

complementary chains obtained after denaturation of each subfragment and separation were analyzed. Their sequencing was performed by a modified Maxam-Gilbert procedure.
As another method of the β-subunit polypeptide chain splitting, the digestion with *Staphylococcus aureus* protease was chosen. Initial fragmentation of the hydrolysate was performed by gel filtration on biogel P-4. This yielded four fractions. Subsequent separation of the peptides was achieved by chromatography on the cation exchanger AG-50Wx4 and paper chromatography. Fraction I contained a mixture of the largest peptides. To facilitate the separation and analysis of the peptides in this fraction the mixture was additionally digested with chymotrypsin. Altogether 3 peptides were isolated from fraction IV, 73 from fraction III, 48 from fraction II and 60 from fraction I.
In order to obtain the missing fragments, the exhaustive tryptic digestion of the β-subunit was carried out after modification of the lysine residues with citraconic anhydride. The tryptic peptides were separated according to the same scheme as that used for staphylococcal peptides. After removal of the citraconic protection the high molecular weight peptides were subjected to additional tryptic cleavage at the lysine residues.
In order to determine the primary structure of the β'-subunit its cyanogen bromide cleavage was carried out. For separation of the resulting peptides we used gel filtration, paper and thin-layer chromatographies, electrophoresis in acetate cellulose slabs and butanol exstraction of hydrophobic peptides. The tryptic digestion of the β'-subunit was also carried out after modification of the lysine residues with citraconic anhydride. The amino acid sequence of the peptide fragments determined up to now covers more than 80% of the total β-subunit polypeptide chain and about 30% of the β'-subunit polypeptide chain.
During this investigation the comparison of the nucleotide and amino acid sequence was carried out continually. The search for similarities between the amino acid sequences of the

1-81	TTC CGG TCA ACA AAA TAG TGT TGC ACA AAC TGT CCG CTC AAT GGA CAG ATG GGT CGA CTT GTC AGC GAG CTG AGG AAC CCT
82-162 1-27	ATG GTT TAC TCC TAT ACC GAG AAA AAA CGT ATT CGT AAG GAT TTT GGT AAA CGT CCA CAA GTT CTG GAT GTA CCT TAT CTC Met-Val-Tyr-Ser-Tyr-Thr-Glu-Lys-Lys-Arg-Ile-Arg-Lys-Asp-Phe-Gly-Lys-Arg-Pro-Gln-Val-Leu-Asp-Val-Pro-Tyr-Leu-
163-243 28-54	CTT TCT ATC CAG CTT GAC TCG TTT CAG AAA TTT ATC GAG CAA GAT CCT GAA GGG CAG TAT GGT CTG GAA GCT GCT TTC CGT Leu-Ser-Ile-Gln-Leu-Asp-Ser-Phe-Gln-Lys-Phe-Ile-Glu-Gln-Asp-Pro-Glu-Gly-Gln-Tyr-Gly-Leu-Glu-Ala-Ala-Phe-Arg-
244-324 55-81	TCC GTA TTC CCG ATT CAG AGC TAC AGC GGT AAT TCC GAG CTG CAA TAC GTC AGC TAC CGC CTT GGC GAA CCG GTG TTT GAC Ser-Val-Phe-Pro-Ile-Gln-Ser-Tyr-Ser-Gly-Asn-Ser-Glu-Leu-Gln-Tyr-Val-Ser-Tyr-Arg-Leu-Gly-Glu-Pro-Val-Phe-Asp-
325-405 82-108	GTC CAG GAA TGT CAA ATC CGT GGC GTG ACC TAT TCC GCA CCG CTG CGC GTT AAA CTG CGT CTG GTG ATC TAT GAG CGC GAA Val-Gln-Glu-Cys-Gln-Ile-Arg-Gly-Val-Thr-Tyr-Ser-Ala-Pro-Leu-Arg-Val-Lys-Leu-Arg-Leu-Val-Ile-Tyr-Glu-Arg-Glu-
406-486 109-135	GCG CCG GAA GGC ACC GTA AAA GAC ATT AAA GAA CAA GAA GTC TAC ATG GGC GAA ATT CCG CTC ATG ACA GAC AAC GGT ACC Ala-Pro-Glu-Gly-Thr-Val-Lys-Asp-Ile-Lys-Glu-Gln-Glu-Val-Tyr-Met-Gly-Glu-Ile-Pro-Leu-Met-Thr-Asp-Asn-Gly-Thr-
487-567 136-162	TTT GTT ATC AAC GGT ACT GAG CGT GTT ATC GTT TCC CAG CTG CAC CGT AGT CCG GGC GTC TTC TTT GAC TCC GAC AAA GGT Phe-Val-Ile-Asn-Gly-Thr-Glu-Arg-Val-Ile-Val-Ser-Gln-Leu-His-Arg-Ser-Pro-Gly-Val-Phe-Phe-Asp-Ser-Asp-Lys-Gly-
568-648 163-189	AAA ACC CAC TCT TCG GGT AAA GTG CTG TAT AAC GCG CGT ATC ATC CCT TAC CGT GGT TCC TGG CTG GAC TTC GAA TTC GAT Lys-Thr-His-Ser-Ser-Gly-Lys-Val-Leu-Tyr-Asn-Ala-Arg-Ile-Ile-Pro-Tyr-Arg-Gly-Ser-Trp-Leu-Asp-Phe-Glu-Phe-Asp-
649-729 190-216	CCG AAG GAC AAC CTG TTC GTA CGT ATC GAC CGT CGC CGT AAA CTG CCT GCG ACC ATC ATT CTG CGC GCC CTG AAC TAC ACC Pro-Lys-Asp-Asn-Leu-Phe-Val-Arg-Ile-Asp-Arg-Arg-Arg-Lys-Leu-Pro-Ala-Thr-Ile-Ile-Leu-Arg-Ala-Leu-Asn-Tyr-Thr-
730-810 217-243	ACA GAG CAG ATC CTC GAC CTG TTC TTT GAA AAA GTT ATC TTT GAA ATC CGT GAT AAC AAG CTG CAG ATG GAA CTG GTG CCG Thr-Glu-Gln-Ile-Leu-Asp-Leu-Phe-Phe-Glu-Lys-Val-Ile-Phe-Glu-Ile-Arg-Asp-Asn-Lys-Leu-Gln-Met-Glu-Leu-Val-Pro-
811-891 244-270	GAA CGC CTG CGT GGT GAA ACC GCA TCT TTT GAC ATC GAA GCT AAC GGT AAA GTG TAC GTA GAA AAA GGC CGC CGT ATC ACT Glu-Arg-Leu-Arg-Gly-Glu-Thr-Ala-Ser-Phe-Asp-Ile-Glu-Ala-Asn-Gly-Lys-Val-Tyr-Val-Glu-Lys-Gly-Arg-Arg-Ile-Thr-
892-972 271-297	GCG CGC CAC ATT CGC CAG CTG GAA AAA GAC GAC GTC AAA CTG ATC GAA GTC CCG GTT GAG TAC ATC GCA GGT AAA GTG GTT Ala-Arg-His-Ile-Arg-Gln-Leu-Glu-Lys-Asp-Asp-Val-Lys-Leu-Ile-Glu-Val-Pro-Val-Glu-Tyr-Ile-Ala-Gly-Lys-Val-Val-
973-1053 298-324	GCT AAA GAC TAT ATT GAT GAG TCT ACC GGC GAG CTG ATC TGC GCA GCG AAC ATG GAG CTG AGC CTG GAT CTG CTG GCT AAG Ala-Lys-Asp-Tyr-Ile-Asp-Glu-Ser-Thr-Gly-Glu-Leu-Ile-Cys-Ala-Ala-Asn-Met-Glu-Leu-Ser-Leu-Asp-Leu-Leu-Ala-Lys-
1054-1134 325-351	CTG AGC CAG TCT GGT CAC AAG CGT ATC GAA ACG CTG TTC ACC AAC GAT CTG GAT CAC GGC CCA TAT ATC TCT GAA ACC TTA Leu-Ser-Gln-Ser-Gly-His-Lys-Arg-Ile-Glu-Thr-Leu-Phe-Thr-Asn-Asp-Leu-Asp-His-Gly-Pro-Tyr-Ile-Ser-Glu-Thr-Leu-
1135-1215 352-378	CGT GTC GAC CCA ACT AAC GAC CGT CTG AGC GCA CTG GTA GAA ATC TAC CGC ATG ATG CGC CCT GGC GAG CCG CCG ACT CGT Arg-Val-Asp-Pro-Thr-Asn-Asp-Arg-Leu-Ser-Ala-Leu-Val-Glu-Ile-Tyr-Arg-Met-Met-Arg-Pro-Gly-Glu-Pro-Pro-Thr-Arg-
1216-1296 379-405	GAA GCA GCT GAA AGC CTG TTC GAG AAC CTG TTC TTC TCC GAA GAC CGT TAT GAC TTG TCT GCG GTT GGT CGT ATG AAG TTC Glu-Ala-Ala-Glu-Ser-Leu-Phe-Glu-Asn-Leu-Phe-Phe-Ser-Glu-Asp-Arg-Tyr-Asp-Leu-Ser-Ala-Val-Gly-Arg-Met-Lys-Phe-
1297-1377 406-432	AAC CGT TCT CTG CTG CGC GAA GAA ATC GAA GGT TCC GGT ATC CTG AGC AAA GAC GAC ATC ATT GAT GTT ATG AAA AAG CTC Asn-Arg-Ser-Leu-Leu-Arg-Glu-Glu-Ile-Glu-Gly-Ser-Gly-Ile-Leu-Ser-Lys-Asp-Asp-Ile-Ile-Asp-Val-Met-Lys-Lys-Leu-
1378-1458 433-459	ATC GAT ATC CGT AAC GGT AAA GGC GAA GTC GAT GAT ATC GAC CAC CTC GGC AAC CGT CGT ATC CGT TCC GTT GGC GAA ATG Ile-Asp-Ile-Arg-Asn-Gly-Lys-Gly-Glu-Val-Asp-Asp-Ile-Asp-His-Leu-Gly-Asn-Arg-Arg-Ile-Arg-Ser-Val-Gly-Glu-Met-
1459-1539 460-486	GCG GAA AAC CAG TTC CGC GTT GGC CTG GTA CGT GTA GAG CGT GCG GTG AAA GAG CGT CTG TCT CTG GGC GAT CTG GAT ACC Ala-Glu-Asn-Gln-Phe-Arg-Val-Gly-Leu-Val-Arg-Val-Glu-Arg-Ala-Val-Lys-Glu-Arg-Leu-Ser-Leu-Gly-Asp-Leu-Asp-Thr-
1540-1620 487-513	CTG ATG CCA CAG GAT ATG ATC AAC GCC AAG CCG ATT TCC GCA GCA GTG AAA GAG TTC TTC GGT TCC AGC CAG CTG TCT CAG Leu-Met-Pro-Gln-Asp-Met-Ile-Asn-Ala-Lys-Pro-Ile-Ser-Ala-Ala-Val-Lys-Glu-Phe-Phe-Gly-Ser-Ser-Gln-Leu-Ser-Gln-
1621-1701 514-540	TTT ATG GTC CAG AAC AAC CCG CTG TCT GAG ATT ACG CAC AAA CGT CGT ATC TCC GCA CTC GGC CCA GGC GGT CTG ACC CGT Phe-Met-Val-Gln-Asn-Asn-Pro-Leu-Ser-Glu-Ile-Thr-His-Lys-Arg-Arg-Ile-Ser-Ala-Leu-Gly-Pro-Gly-Gly-Leu-Thr-Arg-
1702-1782 541-567	GAA CGT GCA GGC TTC GAA GTT CGA GAC GTA CAC CCG ACT CAC TAC GGT CGC GTA TGT CCA ATC GAA ACC CCT GAA GGT CCG Glu-Arg-Ala-Gly-Phe-Glu-Val-Arg-Asp-Val-His-Pro-Thr-His-Tyr-Gly-Arg-Val-Cys-Pro-Ile-Glu-Thr-Pro-Glu-Gly-Pro-
1783-1863 568-594	AAC ATC GGT CTG ATC AAC TCT CTG TCC GTG TAC GCA CAG ACT AAC GAA TAC GGC TTC CTT GAG ACT CCG TAT CGT AAA GTG Asn-Ile-Gly-Leu-Ile-Asn-Ser-Leu-Ser-Val-Tyr-Ala-Gln-Thr-Asn-Glu-Tyr-Gly-Phe-Leu-Glu-Thr-Pro-Tyr-Arg-Lys-Val-
1864-1944 595-621	ACC GAC GGT GTT GTA ACT GAC GAA ATT CAC TAC CTG TCT GCT ATC GAA GAA GGC AAC TAC GTT ATC GCC CAG GCG AAC TCC Thr-Asp-Gly-Val-Val-Thr-Asp-Glu-Ile-His-Tyr-Leu-Ser-Ala-Ile-Glu-Glu-Gly-Asn-Tyr-Val-Ile-Ala-Gln-Ala-Asn-Ser-
1945-2025 622-648	AAC TTG GAT GAA GAA GGC CAC TTC GTA GAA GAC CTG GTA ACT TGC CGT AGC AAA GGC GAA TCC AGC TTG TTC AGC CGC GAC Asn-Leu-Asp-Glu-Glu-Gly-His-Phe-Val-Glu-Asp-Leu-Val-Thr-Cys-Arg-Ser-Lys-Gly-Glu-Ser-Ser-Leu-Phe-Ser-Arg-Asp-
2026-2106 649-675	CAG GTT GAC TAC ATG GAC GTA TCC ACC CAG CAG GTG GTA TCC GTC GGT GCG TCC CTG ATC CCG TTC CTG GAA CAC GAT GAC Gln-Val-Asp-Tyr-Met-Asp-Val-Ser-Thr-Gln-Gln-Val-Val-Ser-Val-Gly-Ala-Ser-Leu-Ile-Pro-Phe-Leu-Glu-His-Asp-Asp-
2107-2187 676-702	GCC AAC CGT GCA TTG ATG GGT GCG AAC ATG CAA CGT CAG GCC GTT CCG ACT CTG CGC GCT GAT AAG CCG CTG GTT GGT ACT Ala-Asn-Arg-Ala-Leu-Met-Gly-Ala-Asn-Met-Gln-Arg-Gln-Ala-Val-Pro-Thr-Leu-Arg-Ala-Asp-Lys-Pro-Leu-Val-Gly-Thr-
2188-2268 703-729	GGT ATG GAA CGT GCT GTT GCC GTT GAC TCC GGT GTA ACT GCG GTA GCT AAA CGT GGT GGT GTC GTT CAG TAC GTG GAT GCT Gly-Met-Glu-Arg-Ala-Val-Ala-Val-Asp-Ser-Gly-Val-Thr-Ala-Val-Ala-Lys-Arg-Gly-Gly-Val-Val-Gln-Tyr-Val-Asp-Ala-
2269-2349 730-756	TCC CGT ATC GTT ATC AAA GTT AAC GAA GAC GAG ATG TAT CCG GGT GAA GCA GGT ATC GAC ATC TAC AAC CTG ACC AAA TAC Ser-Arg-Ile-Val-Ile-Lys-Val-Asn-Glu-Asp-Glu-Met-Tyr-Pro-Gly-Glu-Ala-Gly-Ile-Asp-Ile-Tyr-Asn-Leu-Thr-Lys-Tyr-
2350-2430 757-783	ACC CGT TCT AAC CAG AAC ACC TGT ATC AAC CAG ATG CCG TGT GTG TCT CTG GGT GAA CCG GTT GAA CGT GGC GAC GTG CTG Thr-Arg-Ser-Asn-Gln-Asn-Thr-Cys-Ile-Asn-Gln-Met-Pro-Cys-Val-Ser-Leu-Gly-Glu-Pro-Val-Glu-Arg-Gly-Asp-Val-Leu-
2431-2511 784-810	GCA GAC GGT CCG TCC ACC GAC CTC GGT GAA CTG GCG CTT GGT CAG AAC ATG CGC GTA GCG TTC ATG CCG TGG AAT GGT TAC Ala-Asp-Gly-Pro-Ser-Thr-Asp-Leu-Gly-Glu-Leu-Ala-Leu-Gly-Gln-Asn-Met-Arg-Val-Ala-Phe-Met-Pro-Trp-Asn-Gly-Tyr-
2512-2592 811-837	AAC TTC GAA GAC TCC ATC CTC GTA TCC GAG CGT GTT GTT CAG GAA GAC CGT TTC ACC ACC ATC CAC ATT CAG GAA CTG GCG Asn-Phe-Glu Asp-Ser-Ile-Leu-Val-Ser-Glu-Arg-Val-Val-Gln-Glu-Asp-Arg-Phe-Thr-Thr-Ile-His-Ile-Gln-Glu-Leu-Ala-
2593-2673 838-864	TGT GTG TCC CGT GAC ACC AAG CTG GGT CCG GAA GAG ATC ACC GCT GAC ATC CCG AAC GTG GGT GAA GCT GCG CTC TCC AAA Cys-Val-Ser-Arg-Asp-Thr-Lys-Leu-Gly-Pro-Glu-Glu-Ile-Thr-Ala-Asp-Ile-Pro-Asn-Val-Gly-Glu-Ala-Ala-Leu-Ser-Lys-
2674-2754 865-891	CTG GAT GAA TCC GGT ATC GTT TAC ATT GGT GCG GAA GTG ACC GGT GGC GAC ATT CTG GTT GGT AAG GTA ACG CCG AAA GGT Leu-Asp-Glu-Ser-Gly-Ile-Val-Tyr-Ile-Gly-Ala-Glu-Val-Thr-Gly-Gly-Asp-Ile-Leu-Val-Gly-Lys-Val-Thr-Pro-Lys-Gly-

2755-2835 GAA ACT CAG CTG ACC CCA GAA GAA AAA CTG CTG CGT GCG ATC TTC GGT GAG AAA GCC TCT GAC GTT AAA GAC TCT TCT CTG

892-918 Glu-Thr-Gln-Leu-Thr-Pro-Glu-Glu-Lys-Leu-Leu-Arg-Ala-Ile-Phe-Gly-Glu-Lys-Ala-Ser-Asp-Val-Lys-Asp-Ser-Ser-Leu-

2836-2916 CGC GTA CCA AAC GGT GTA TCC GGT ACG GTT ATC GAC GTT CAG GTC TTT ACT CGC GAT GGC GTA GAA AAA GAC AAA CGT GCG

919-945 Arg-Val-Pro-Asn-Gly-Val-Ser-Gly-Thr-Val-Ile-Asp-Val-Gln-Val-Phe-Thr-Arg-Asp-Gly-Val-Glu-Lys-Asp-Lys-Arg-Ala-

2917-2997 CTG GAA ATC GAA GAA ATG CAG CTC AAA CAG GCG AAG AAA GAC CTG TCT GAA GAA CTG CAG ATC CTC GAA GCG GGT CTG TTC

946-972 Leu-Glu-Ile-Glu-Glu-Met-Gln-Leu-Lys-Gln-Ala-Lys-Lys-Asp-Leu-Ser-Glu-Glu-Leu-Gln-Ile-Leu-Glu-Ala-Gly-Leu-Phe-

2998-3078 AGC CGT ATC CGT GCT GTG CTG GTA GCC GGT GGC GTT GAA GCT GAG AAG CTC GAC AAA CTG CCG CGC GAT CGC TGG CTG GAG

973-999 Ser-Arg-Ile-Arg-Ala-Val-Leu-Val-Ala-Gly-Gly-Val-Glu-Ala-Glu-Lys-Leu-Asp-Lys-Leu-Pro-Arg-Asp-Arg-Trp-Leu-Glu-

3079-3159 CTG GGC CTG ACA GAC GAA GAG AAA CAA AAT CAG CTG GAA CAG CTG GCT GAG CAG TAT GAC GAA CTG AAA CAC GAG TTC GAG

1000-1026 Leu-Gly-Leu-Thr-Asp-Glu-Glu-Lys-Gln-Asn-Gln-Leu-Glu-Gln-Leu-Ala-Glu-Gln-Tyr-Asp-Glu-Leu-Lys-His-Glu-Phe-Glu-

3160-3240 AAG AAA CTC GAA GCG AAA CGC CGC AAA ATC ACC C*AG GGC GAC GAT CTG GCA CCG GGC GTG CTG AAG ATT GTT AAG GTA TAT

1027-1053 Lys-Lys-Leu-Glu-Ala-Lys-Arg-Arg-Lys-Ile-Thr-Gln-Gly-Asp-Asp-Leu-Ala-Pro-Gly-Val-Leu-Lys-Ile-Val-Lys-Val-Tyr-

3241-3321 CTG GCG GTT AAA CGC CGT ATC CAG CC*T GGT GAC AAG ATG GCA GGT CGT CAC GGT AAC AAG GGT GTA ATT TCT AAG ATC AAC

1054-1080 Leu-Ala-Val-Lys-Arg-Arg-Ile-Gln-Pro-Gly-Asp-Lys-Met-Ala-Gly-Arg-His-Gly-Asn-Lys-Gly-Val-Ile-Ser-Lys-Ile-Asn-

3322-3402 CCG ATC GAA GAT ATG CCT TAC GAT GAA AAC GGT ACG CCG GTA GAC ATC GTA CTG AAC CCG CTG GGC GTA CCG TCT CGT ATG

1081-1107 Pro-Ile-Glu-Asp-Met-Pro-Tyr-Asp-Glu-Asn-Gly-Thr-Pro-Val-Asp-Ile-Val-Leu-Asn-Pro-Leu-Gly-Val-Pro-Ser-Arg-Met-

3403-3483 AAC ATC GGT CAG ATC CTC GAA ACC CAC C*TG GGT ATG GCT GCG AAA GGT ATC GGC GAC AAG ATC AAC GCC ATG CTG AAA CAG

1108-1134 Asn-Ile-Gly-Gln-Ile-Leu-Glu-Thr-His-Leu-Gly-Met-Ala-Ala-Lys-Gly-Ile-Gly-Asp-Lys-Ile-Asn-Ala-Met-Leu-Lys-Gln-

3484-3564 CAG CAA GAA GTC GCG AAA CTG CGC GAA TTC ATC CAG CGT GCG TAC GAT CTG GGC GCT GAC GTT CGT CAG AAA GTT GAC CTG

1135-1161 Gln-Gln-Glu-Val-Ala-Lys-Leu-Arg-Glu-Phe-Ile-Gln-Arg-Ala-Tyr-Asp-Leu-Gly-Ala-Asp-Val-Arg-Gln-Lys-Val-Asp-Leu-

3565-3645 AGT ACC TTC AGC GAT GAA GAA GTT ATG CGT CTG GCT GAA AAC CTG CGC AAA GGT ATG CCA ATC GCA ACG CCG GTG TTC GAC

1162-1188 Ser-Thr-Phe-Ser-Asp-Glu-Glu-Val-Met-Arg-Leu-Ala-Glu-Asn-Leu-Arg-Lys-Gly-Met-Pro-Ile-Ala-Thr-Pro-Val-Phe-Asp-

3646-3726 GGT GCG AAA GAA GCA GAA ATT AAA GAG CTG CTG AAA CTT GGC GAC CTG CCG ACT TCC GGT CAG ATC CGC CTG TAC GAT GGT

1189-1215 Gly-Ala-Lys-Glu-Ala-Glu-Ile-Lys-Glu-Leu-Leu-Lys-Leu-Gly-Asp-Leu-Pro-Thr-Ser-Gly-Gln-Ile-Arg-Leu-Tyr-Asp-Gly-

3727-3807 CGC ACT GGT GAA CAG TTC GAG CGT CCG GTA ACC GTT GGT TAC ATG TAC ATG CTG AAA CTG AAC CAC C*TG GTC GAC GAC AAG

1216-1242 Arg-Thr-Gly-Glu-Gln-Phe-Glu-Arg-Pro-Val-Thr-Val-Gly-Tyr-Met-Tyr-Met-Leu-Lys-Leu-Asn-His-Leu-Val-Asp-Asp-Lys-

3808-3888 ATG CAC GCG CGT TCC ACC GGT TCT TAC AGC C*TG GTT ACT CAG CAG CCG CTG GGT GGT AAG GCA CAG TTC GGT GGT CAG CGT

1243-1269 Met-His-Ala-Arg-Ser-Thr-Gly-Ser-Tyr-Ser-Leu-Val-Thr-Gln-Gln-Pro-Leu-Gly-Gly-Lys-Ala-Gln-Phe-Gly-Gly-Gln-Arg-

3889-3969 TTC GGG GAG ATG GAA GTG TGG GCG CTG GAA GCA TAC GGC GCA GCA TAC ACC CTG CAG GAA ATG CTC ACC GTT AAG TCT GAT

1270-1296 Phe-Gly-Glu-Met-Glu-Val-Trp-Ala-Leu-Glu-Ala-Tyr-Gly-Ala-Ala-Tyr-Thr-Leu-Gln-Glu-Met-Leu-Thr-Val-Lys-Ser-Asp-

3970-4050 GAC GTG AAC GGT CGT ACC AAG ATG TAT AAA AAC ATC GTG GAC GGC AAC CAT CAG ATG GAG CCG GGC ATG CCA GAA TCC TTC

1297-1323 Asp-Val-Asn-Gly-Arg-Thr-Lys-Met-Tyr-Lys-Asn-Ile-Val-Asp-Gly-Asn-His-Gln-Met-Glu-Pro-Gly-Met-Pro-Glu-Ser-Phe-

4051-4129 AAC GTA TTG TTG AAA GAG ATT CGT TCG CTG GGT ATC AAC ATC GAA CTG GAA GAC GAG TAA TTC TCG CTC AAA CAG GTC A

1324-1342 Asn-Val-Leu-Leu-Lys-Glu-Ile-Arg-Ser-Leu-Gly-Ile-Asn-Ile-Glu-Leu-Glu-Asp-Glu TER

4130-4210 CTG CTG TCG GGT TAA AAC CCG GCA GCG GAT TGT GCT AAC TCC GAC GGG AGC AAA TCC GTG AAA GAT TTA TTA AAG TTT CTG

1-8 Met-Lys-Asp-Leu-Leu-Lys-Phe-Leu-

4211-4291 AAA GCG CAG ACT AAA ACC GAA GAG TTT GAT GCG ATC AAA ATT GCT CTG GCT TCG CCA GAC ATG ATC CGT TCA TGG TCT TTC

9-35 Lys-Ala-Gln-Thr-Lys-Thr-Glu-Glu-Phe-Asp-Ala-Ile-Lys-Ile-Ala-Leu-Ala-Ser-Pro-Asp-Met-Ile-Arg-Ser-Trp-Ser-Phe-

4292-4372 GGT GAA GTT AAA AAG CCG GAA ACC ATC AAC TAC CGT ACG TTC AAA CCA GAA CGT GAC GGC CTT TTC TGC GCC CGT ATC TTT

36-62 Gly-Glu-Val-Lys-Lys-Pro-Glu-Thr-Ile-Asn-Tyr-Arg-Thr-Phe-Lys-Pro-Glu-Arg-Asp-Gly-Leu-Phe-Cys-Ala-Arg-Ile-Phe-

4373-4453 GGG CCG GTA AAA GAT TAC GAG TGC CTG TGC GGT AAG TAC AAG CGC CTG AAA CAC CGT GGC GTC ATC TGT GAG AAG TGC GGC

63-89 Gly-Pro-Val-Lys-Asp-Tyr-Glu-Cys-Leu-Cys-Gly-Lys-Tyr-Lys-Arg-Leu-Lys-His-Arg-Gly-Val-Ile-Cys-Glu-Lys-Cys-Gly-

4454-4534 GTT GAA GTG ACC CAG ACT AAA GTA CGC CGT GAG CGT ATG GGC CAC ATC GAA CTG GCT TCC CCG ACT GCG CAC ATC TGG TTC

90-116 Val-Glu-Val-Thr-Gln-Thr-Lys-Val-Arg-Arg-Glu-Arg-Met-Gly-His-Ile-Glu-Leu-Ala-Ser-Pro-Thr-Ala-His-Ile-Trp-Phe-

4535-4615 CTG AAA TCG CTG CCG TCC CGT ATC GGT CTG CTG CTC GAT ATG CCG CTG CGC GAT ATC GAA CGC GTA CTG TAC TTT GAA TCC

117-143 Leu-Lys-Ser-Leu-Pro-Ser-Arg-Ile-Gly-Leu-Leu-Leu-Asp-Met-Pro-Leu-Arg-Asp-Ile-Glu-Arg-Val-Leu-Tyr-Phe-Glu-Ser-

4616-4696 TAT GTG GTT ATC GAA GGC GGT ATG ACC AAC C*TG GAA CGT CAG CAG ATC CTG ACT GAA GAG CAG TAT CTG GAC GCG CTG GAA

144-170 Tyr-Val-Val-Ile-Glu-Gly-Gly-Met-Thr-Asn-Leu-Glu-Arg-Gln-Gln-Ile-Leu-Thr-Glu-Glu-Gln-Tyr-Leu-Asp-Ala-Leu-Glu-

4697-4714 GAG TTC GGT GAC GAA TTC

171-176 Glu-Phe-Gly-Asp-Glu-Phe

Fig. 2. The nucleotide sequence of the rpoBC segment, the total amino acid sequence of the β-subunit and the N-terminal amino acid sequence of the β'-subunit of *E. coli* RNA polymerase. The restriction EcoRI cleavage sites dividing the fragments EcoRI-G, EcoRI-C and EcoRI-F are situated between nucleotides 640-641 and 3508-3509. Here and in Fig. 3 the nucleotide sequence of the complementary DNA chain, equal to the sequence of mRNA, is given. The underlined amino acid sequences are those the structure of which has been determined from analysis of corresponding peptides. C* refer to 5-methylcytidine residues.

```
1-75      GAA TTC GGT CGT ACT AAA GAA AGC TAC AAA GTA CCT TAC GGT GCG GTA CTG GCG AAA GGC GAT GGC GAA CAG GTT
1-25      Glu-Phe-Gly-Arg-Thr-Lys-Glu-Ser-Tyr-Lys-Val-Pro-Tyr-Gly-Ala-Val-Leu-Ala-Lys-Gly-Asp-Gly-Glu-Gln-Val-

76-150    GCT GGC GGC GAA ACC GTT GCA AAC TGG GAC CCG CAC ACC ATG CCG GTT ATC ACC GAA GTA AGC GGT TTT GTA CGC
26-50     Ala-Gly-Gly-Glu-Thr-Val-Ala-Asn-Trp-Asp-Pro-His-Thr-Met-Pro-Val-Ile-Thr-Glu-Val-Ser-Gly-Phe-Val-Arg-

151-225   TTT ACT GAC ATG ATC GAC GGC CAG ACC ATT ACG CGT CAG ACC GAC GAA CTG ACC GGT CTG TCT TCG CTG GTG GTT
51-75     Phe-Thr-Asp-Met-Ile-Asp-Gly-Gln-Thr-Ile-Thr-Arg-Gln-Thr-Asp-Glu-Leu-Thr-Gly-Leu-Ser-Ser-Leu-Val-Val-

226-300   CTG GAT TCC GCA GAA CGT ACC GCA GGT GGT AAA GAT CTG CGT CCG GCA CTG AAA ATC GTT GAT GCT CAG GGT AAC
76-100    Leu-Asp-Ser-Ala-Glu-Arg-Thr-Ala-Gly-Gly-Lys-Asp-Leu-Arg-Pro-Ala-Leu-Lys-Ile-Val-Asp-Ala-Gln-Gly-Asn-

301-375   GAC GTT CTG ATC CCA GGT ACC GAT ATG CCA GCG CAG TAC TTC CTG CCG GGT AAA GCG ATT GTT CAG CTG GAA GAT
101-125   Asp-Val-Leu-Ile-Pro-Gly-Thr-Asp-Met-Pro-Ala-Gln-Tyr-Phe-Leu-Pro-Gly-Lys-Ala-Ile-Val-Gln-Leu-Glu-Asp-

376-450   GGC GTA CAG ATC AGC TCT GGT GAC ACC CTG GCG CGT ATT CCG CAG GAA TCC GGC GGT ACC AAG GAC ATC ACC GGT
126-150   Gly-Val-Gln-Ile-Ser-Ser-Gly-Asp-Thr-Leu-Ala-Arg-Ile-Pro-Gln-Glu-Ser-Gly-Gly-Thr-Lys-Asp-Ile-Thr-Gly-

451-525   GGT CTG CCG CGC GTT GCG GAC CTG TTC GAA GCA CGT CGT CCG AAA GAG CCG GCA ATC CTG GCT GAA ATC AGC GGT
151-175   Gly-Leu-Pro-Arg-Val-Ala-Asp-Leu-Phe-Glu-Ala-Arg-Arg-Pro-Lys-Glu-Pro-Ala-Ile-Leu-Ala-Glu-Ile-Ser-Gly-

526-600   ATC GTT TCC TTC GGT AAA GAA ACC AAA GGT AAA CGT CGT CTG GTT ATC ACC CCG GTA GAC GGT AGC GAT CCG TAC
176-200   Ile-Val-Ser-Phe-Gly-Lys-Glu-Thr-Lys-Gly-Lys-Arg-Arg-Leu-Val-Ile-Thr-Pro-Val-Asp-Gly-Ser-Asp-Pro-Tyr-

601-675   GAA GAG ATG ATT CCG AAA TGG CGT CAG CTC AAC GTG TTC GAA GGT GAA CGT GTA GAA CGT GGT GAC GTA ATT TCC
201-225   Glu-Glu-Met-Ile-Pro-Lys-Trp-Arg-Gln-Leu-Asn-Val-Phe-Glu-Gly-Glu-Arg-Val-Glu-Arg-Gly-Asp-Val-Ile-Ser-

676-750   GAC GGT CCG GAA GCG CCG CAC GAC ATT CTG CGT CTG CGT GGT GTT CAT GCT GTT ACT CGT TAC ATC GTT AAC GAA
226-250   Asp-Gly-Pro-Glu-Ala-Pro-His-Asp-Ile-Leu-Arg-Leu-Arg-Gly-Val-His-Ala-Val-Thr-Arg-Tyr-Ile-Val-Asn-Glu-

751-825   GTA CAG GAC GTA TAC CGT CTG CAG GGC GTT AAG ATT AAC GAT AAA CAC ATC GAA GTT ATC GTT CGT CAG ATG CTG
251-275   Val-Gln-Asp-Val-Tyr-Arg-Leu-Gln-Gly-Val-Lys-Ile-Asn-Asp-Lys-His-Ile-Glu-Val-Ile-Val-Arg-Gln-Met-Leu-

826-900   CGT AAA GCT ACC ATC GTT AAC GCG GGT AGC TCC GAC TTC CTG GAA GGC GAA CAG GTT GAA TAC TCT CGC GTC AAG
276-300   Arg-Lys-Ala-Thr-Ile-Val-Asn-Ala-Gly-Ser-Ser-Asp-Phe-Leu-Glu-Gly-Glu-Gln-Val-Glu-Tyr-Ser-Arg-Val-Lys-

901-975   ATC GCA AAC CGC GAA CTG GAA GCG AAC GGC AAA GTG GGT GCA ACT TAC TCC CGC GAT CTG CTG GGT ATC ACC AAA
301-325   Ile-Ala-Asn-Arg-Glu-Leu-Glu-Ala-Asn-Gly-Lys-Val-Gly-Ala-Thr-Tyr-Ser-Arg-Asp-Leu-Leu-Gly-Ile-Thr-Lys-

976-1050  GCG TCT CTG GCA ACC GAG TCC TTC ATC TCC GCG GCA TCG TTC CAG GAG ACC ACT CGC GTG CTG ACC GAA GCA GCC
326-350   Ala-Ser-Leu-Ala-Thr-Glu-Ser-Phe-Ile-Ser-Ala-Ala-Ser-Phe-Gln-Glu-Thr-Thr-Arg-Val-Leu-Thr-Glu-Ala-Ala-

1051-1125 GTT GCG GGC AAA CGC GAC GAA CTG CGC GGC CTG AAA GAG AAC GTT ATC GTG GGT CGT CTG ATC CCG GCA GGT ACC
351-375   Val-Ala-Gly-Lys-Arg-Asp-Glu-Leu-Arg-Gly-Leu-Lys-Glu-Asn-Val-Ile-Val-Gly-Arg-Leu-Ile-Pro-Ala-Gly-Thr-

1126-1200 GGT TAC GCG TAC CAC CAG CAT CGT ATG CGT CGC CGT GCT GCG GGT GAA GCT CCG GCT GCA CCG CAG GTG ACT GCA
376-400   Gly-Tyr-Ala-Tyr-His-Gln-Asp-Arg-Met-Arg-Arg-Arg-Ala-Ala-Gly-Glu-Ala-Pro-Ala-Ala-Pro-Gln-Val-Thr-Ala-

1201-1277 GAA GAC GCA TCT GCC AGC CTG GCA GAA CTG CTG AAC GCA GGT CTG GGC GGT TCT GAT AAC GAG TAA TCGTTAATCCG
401-421   Glu-Asp-Ala-Ser-Ala-Ser-Leu-Ala-Glu-Leu-Leu-Asn-Ala-Gly-Leu-Gly-Gly-Ser-Asp-Asn-Glu Ter

1278-1376 CAAATAACGTAAAAACCCGCTTCGGCGGGTTTTTTTATGGGGGGAGTTTAGGGAAAGAGCATTTGTCAGAATATTTAAGGAATTTCTGAATACTCATAA

1377-1475 TCAATGTAGAGATGACTAATATCCTGAAACTGACTGAACTAATTGAGTCAAACTCGGCAAGGATTCGATACTATTCCTGTGTAACTTTCTTAAGGAACG

1476-1574 AGAATGAAACAGGAAGTGGAAAAGTGGCGACCTTTTGGACATCCGGATGGTGATATTCGTGATTTATCATTTCTTGATGCTCATCAGGCTGTCTACGTT

1575-1673 CAGCATCATGAGGGCAAAGAGCCTTTAGAGTATCGCTTTTGGGTTACCTACTCTCTTCACTGCTTCACAAAAGATTATGAACATCAGACGAACGAAGAA

1674-1772 AAACAATCGTTAATGTACCACGCGCCTAAAGAATCTCGTCCCTTCTGCCAGCACCGTTATAACTTAGCGCGCACACACTTAAAAAGAACTATTTTGGCG

1773-1871 CTGCCAGAAAGCAACGTTATTCATGCCGGGTATGGTAGCTATGCCGTGATTGAGGTGGACTTAGACGGAGGAGATAAGGCATTTTACTTTGTTGCGTTC

1872-1970 AGGGCTTTCAGGGAAAAGAAAAAACTCCGTTTGCATGTAACTAGCGCTTATCCCATTTCTGAAAAACAGAAAGGTAAATCAGTGAAATTTTTCACCATT

1971-2069 GCCTACAACTTATTGAGAAATAAGCAGCTTCCTCAGCCCTCAAAATAACAAAACCCACCTTAAGGTGGGTTTCGCCAGAGAATTATCTCTGGTATTCAG

2070-2168 AACGCCATTACCGGACTTTGCCTTGACCTTGCGATAATCGCAGGTTGCGGGATGTCTGAATTTCTTCAGTCTGCTGCATCCTGGAAGATGAGAACATGT

2169-2267 GTTCTTATTTTCGTCTCTATCATAGTTGAGTATTTACTCTCTTACAATCAGATCTCTTTCATTGCTCAACAGGCGATGGCTTCAGACTTTGCATTACGG

2268-2306 AATTTTTAAGAAAGGCAGGGCGAAACGAGGAACAAGCTT
```

Fig. 3. The nucleotide sequence of the EcoRI-A - HindIII fragment and the C-terminal amino acid sequence of the β'-subunit of *E. coli* RNA polymerase.

peptides and the nucleotide sequences of the corresponding DNA fragments was carried out by means of a computer. A conjunction of the methods of protein and nucleotide chemistry for the combined structural investigation of a protein and DNA sharply accelerated and considerably simplified the solution of both problems and enhanced the reliability of the structural analysis. As a result the nucleotide sequences of the two segments of the rpoBC operon (4714 and 2306 base pairs) embracing the entire rpoB gene, the initial and terminal parts of the rpoC gene and the intercistronic region, together with the total amino acid sequence of the β- subunit, comprising 1342 residues (9, 10), and the N- and C-terminal sequences of the β'-subunit (176 and 421 residues) were determined (Fig. 2, 3).

References

1. Ovchinnikov, Yu.A., Lipkin, V.M., Modyanov, N.N., Chertov, O.Yu., Smirnov, Yu.V.: FEBS Letters 76, 108-111 (1977).

2. Marchenko, T.V., Modyanov, N.N., Lipkin, V.M., Ovchinnikov, Yu.A.: Bioorg. Khim. 6, 325-331 (1980).

3. Lipkin, V.M., Marchenko, T.V., Khokhryakov, V.S., Polovnikova, I.N., Potapenko, N.A., Modyanov, N.N., Ovchinnikov, Yu.A.: Bioorg. Khim. 6, 332-342 (1980).

4. Ovchinnikov, Yu.A., Sverdlov, E.D., Lipkin, V.M., Monastyrskaya, G.S., Chertov, O.Yu., Gubanov, V.V., Guryev, S.O., Modyanov, N.N., Grinkevich, V.A., Makarova, I.A., Marchenko, T.V., Polovnikova, I.N.: Bioorg. Khim. 6, 655-665 (1980).

5. Monastyrskaya, G.S., Gubanov, V.V., Guryev, S.O., Lipkin, V.M., Sverdlov, E.D.: Bioorg. Khim. 6, 1106-1109 (1980).

6. Monastyrskaya, G.S., Gubanov, V.V., Guryev, S.O., Lipkin, V.M., Sverdlov, E.D.: Bioorg. Khim. 6, 1423-1426 (1980).

7. Mindlin, S.S., Ilyina, T.S., Gorlenko, Ch.M., Hachikyan, N.A., Kovalev, Yu.N.: Genetika 12, 116-130 (1976).

8. Kirschbaum, J.B., Konrad, B.E.: J. Bacteriology 116, 517-526 (1973).

9. Ovchinnikov, Yu.A., Monastyrskaya, G.S., Gubanov, V.V., Guryev, S.O., Chertov, O.Yu., Modyanov, N.N., Crinkevich, V.A., Makarova, I.A., Marchenko, T.V., Polovnikova, I.N., Lipkin, V.M., Sverdlov, E.D.: Dokl. Akad. Nauk SSSR 253,

994-999 (1980).

10. Ovchinnikov, Yu.A., Monastyrskaya, G.S., Gubanov, V.V., Guryev, S.O., Chertov, O.Yu., Modyanov, N.N., Grinkevich, V.A., Makarova, I.A., Marchenko, T.V., Polovnikova, I.N., Lipkin, V.M., Sverdlov, E.D.: Eur. J. Biochem. 116, 621-629 (1981).

STUDY ON THE PRIMARY STRUCTURE OF THE ELONGATION FACTOR G FROM *E.coli*

Yuli B. Alakhov, Ludmila P. Motuz, Natalia V. Dovgas and Yuri A. Ovchinnikov

Institute of Protein Research, Academy of Sciences of the USSR, 142292 Poustchino, Moscow Region, USSR

Introduction

The bacterial protein-synthesizing system is known to contain three soluble proteins participating in synthesis of polypeptide chains on the ribosome at the elongation stage. One of them, the elongation factor G (EF-G), catalyzes the GTP-dependent translocation of the peptidyl-tRNA·mRNA complex from the A-site into the P-site on the ribosome. This is accompanied by the release of the deacylated tRNA from the P-site and the removal of the tRNA from the ribosome (1).

EF-G represents a protein consisting of one polypeptide chain with M_r of about 80,000. The use of traditional methods for the determination of the primary structure did not seem expedient for such a large protein. Therefore, choosing the strategy of structural analysis, we opted at the first stage for limited proteolysis which splits the protein molecule into a small number of large fragments relatively stable to further protease action. It has been shown earlier that mild tryptic hydrolysis splits the elongation factor G into five fragments (2-4), four of them (T_4-T_7) [for the fragments nomenclature see (3)] encompass the whole protein polypeptide chain. Thus, the study of the EF-G primary structure amounted to the determination of the structure of these four fragments and the search for the peptides which assemble the fragments of limited trypsinolysis into one polypeptide chain.

Results

From the data of amino acid analysis and determination of the molecular mass of peptide fragments which are the products of tryptic hydrolysis it can be concluded that limited trypsinolysis does not result in splitting off of any notable regions of the polypeptide chain (3). At the same time, elucidation of the structure of peptides obtained during the study of the EF-G cyanogen bromide cleavage permitted us to assemble all the fragments of limited trypsinolysis into one polypeptide chain. Hence it follows that there are really no insertions between the fragments of limited trypsinolysis, i.e. trypsin produces point disruptions of the EF-G polypeptide chain (4).

Fragment T_6 is N-terminal, it is followed along the chain by fragment T_7 which contains the functionally important cysteine residue. The GTP-binding site is located within this fragment (5). Fragment T_7 is followed by fragment T_4 and, finally, comes fragment T_5 representing the C-terminal part of the EF-G molecule.

The study of the structure of fragments T_5-T_7 turned out to be simple enough because of their rather low molecular mass. The molecular mass of fragment T_6 is 6500, that of T_7 - 7500 and that of T_5 - 25,000. Using traditional chemical and enzymatic methods for cleavage of the polypeptide chain, we determined completely the primary structure of fragments T_6 (4), T_7 (6,7) and T_5 which all together embrace more than 350 amino acid residues.

We have shown earlier (3) that all the fragments of limited EF-G trypsinolysis have arginine as the C-terminal amino acid residue while in the G-factor lysine is the C-terminal amino acid residue. Studying the products of cyanogen bromide cleavage of the whole EF-G molecule we isolated a peptide having the sequence analogous to that of the C-terminal cyanogen bromide peptide of fragment T_5 but differing by two amino acid residues at the C-terminus: Gly-Lys. Therefore we concluded that limited trypsinolysis of EF-G resulted in splitting off the C-terminal di-peptide.

To obtain overlapping between the fragments of limited trypsinolysis and to elucidate the structure of fragment T_4 representing the middle part of the protein polypeptide chain with M_r 41,000, we studied the products of cyanogen bromide cleavage of the whole EF-G molecule. As a result, all 24 cyanogen bromide peptides have been isolated, among which 13 peptides constituting the structure of fragment T_4. Thus, the study of the structure of fragment T_4 amounted to the elucidation of the amino acid sequence of 13 cyanogen bromide peptides and

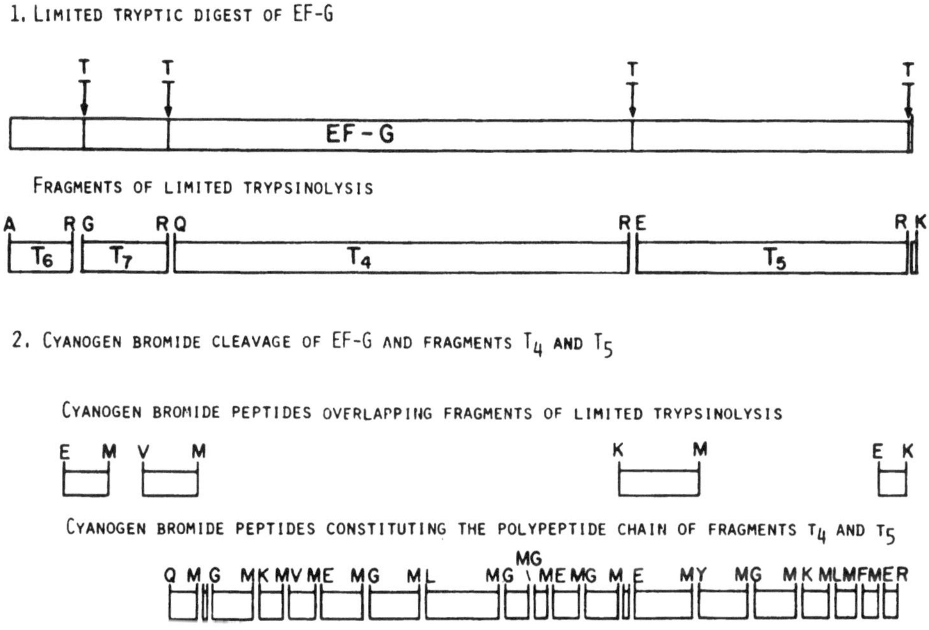

Scheme. The pathway of studying the primary structure of EF-G.
To obtain overlapping between cyanogen bromide peptides of fragments T_4 and T_5 products of their hydrolysis by trypsin and *Staphylococcus aureus* protease were studied

the search for overlapping between them. To obtain overlapping peptides, we used hydrolysis of fragment T_4 by trypsin and *Staphylococcus aureus* protease.

These data pemitted us to represent the partial structure of the elongation factor G. Its polypeptide chain contains over 700 amino acid residues.

Several regions of the obtained amino acid sequence have been recently confirmed by the study of the nucleotide sequence of the gene coding for EF-G. Thus, the study of *str* operon of *E.coli* resulted in determination of the nucleotide sequence corresponding to 92 amino acid residues from the N-terminus of EF-G (8); elucidation of the nucleotide sequence of the *tuf* A gene coding for the elongation factor Tu gave the structure of the region of the *fus* gene coding for the C-terminal part of EF-G consisting of twenty amino acid residues (9). In both cases there is a coincidence between the amino acid sequences determined directly and those resulting from studies of the DNA structure.

Since EF-G and EF-Tu perform similar functions on the ribosome, i.e. possess ribosome-dependent GTPase activity, it is interesting to compare their amino acid sequences forming the GTP-binding sites. Though the EF-G molecule is almost twice that of EF-Tu, it turned out that in both proteins these regions are located in the N-terminal part of the molecule. The cysteine residue located in close proximity to the GTP-binding site occupies a similar position in both proteins.

It is difficult at first glance to find homology between the amino acid sequences of the GTP-binding sites in both proteins. However, if we take into account several insertions or deletions the homology is 30% and if we take into account conservative replacements the homology amounts to 45% (10). These calculations have been made taking into account deletions in 10 amino acid residues in EF-Tu (position 23-32 in the figure). Laursen

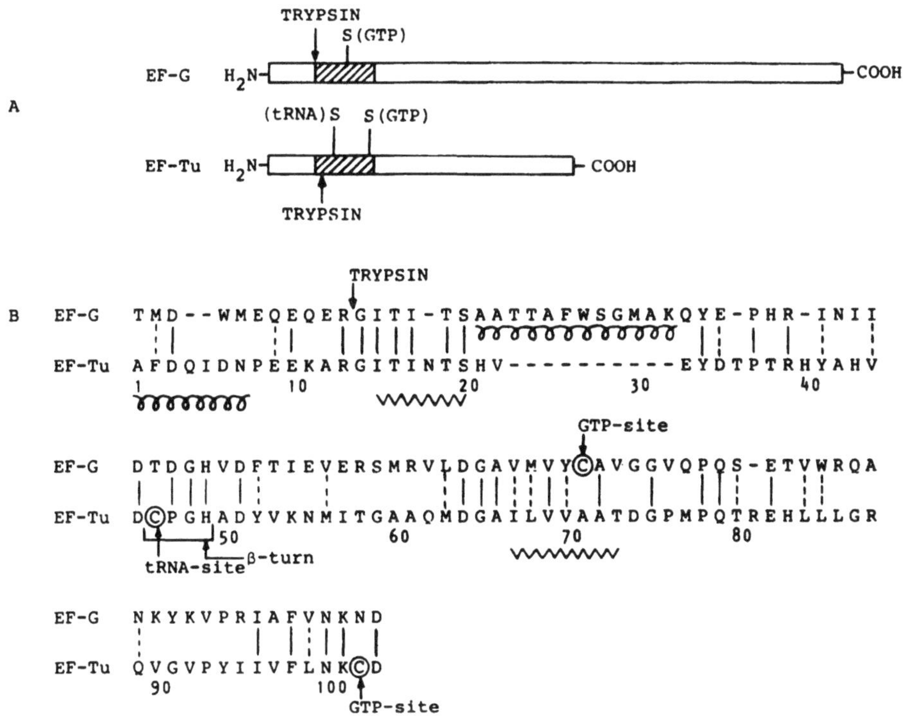

A comparison of the amino acid sequences of EF-Tu and EF-G. Solid vertical lines denote identical amino acid residues, dashed lines - conservative replacements, zigzag lines - common β-structural regions, helical line - the α-helical region. The arrow indicates the place of the trypsin attack.

and Daffy (10) have calculated conformational parameters P_α, P_β and P_t for the amino acid residues in α-helices, β-sheets and β-turns for the compared regions of both proteins. A common β-structure has been predicted for residues 15-20 and 67-73 and a β-turn for 46 to 49 residues. The main difference is the absence of the α-helix in EF-Tu (residues 21-23). There is also another interesting structural feature of the sequences compared: in both proteins there is a sequence ...-Arg-Gly-Ile-Thr-Ile-... (the region between fragments T_6 and T_7 in EF-G). Limited tryptic hydrolysis of both proteins results in disruption of the polypeptide chains at the Arg-Gly bond (position 58-59 in EF-G and EF-Tu). It is interesting to note that the sequence Arg-Gly-Ile is also present in actin (position 62-64) (11) and the Arg-Gly bond in this position is disrupted by limited trypsinolysis (12).

References

1. Nishizuka, Y., Lipmann, F.: Proc. Nat. Acad. Sci. USA 55, 212-219 (1966).
2. Skar, D. C., Rohrbach, M. S., Bodley, J. W.: Biochemistry 14, 3922-3926 (1975).
3. Alakhov, Yu. B., Motuz, L. P., Stengrevics, O. A., Vinokurov, L. M., Ovchinnikov, Yu. A.: Bioorg. Khimiya 3, 1333-1345 (1977).
4. Motuz, L. P., Bundulis, Yu. P., Alakhov, Yu. B.: Bioorg. Khimiya 5, 814-827 (1979).
5. Girshovich, A. S., Bochkareva, E. S., Pozdnyakov, V. A., Ovchinnikov, Yu. A.: FEBS Lett. 85, 283-286 (1978).
6. Alakhov, Yu. B., Motuz, L. P., Stengrevics, O. A., Ovchinnikov, Yu. A.: FEBS Lett. 85, 287-290 (1978).
7. Alakhov, Yu. B., Motuz, L. P., Stengrevics, O. A., Vinokurov, L. M.: Bioorg. Khimiya 4, 1301-1313 (1978).
8. Post, L. E., Nomura, M.: J. Biol. Chem. 255, 4660-4666 (1980).
9. Yokota, T., Sugisaki, H., Takamami, M., Kaziro, Y.: The Nucleotide Sequence of *E.coli tuf* A, preprint (1980).
10. Laursen, R. A., Daffy, L.: FEBS Lett. 92, 200-202 (1978).
11. Rosenbuch, J. P., Jacobson, G. R., Jaton, J.-C.: J. Supramol. Structure 5, 391-396 (1976).
12. Jacobson, G. R., Rosenbuch, J. P.: Proc. Nat. Acad. Sci. USA 73, 2742-2746 (1976).

INTRACELLULAR SERINE PROTEASES OF BACILLI

Alexander Ya.Strongin, Valentin M.Stepanov

Institute of Genetics & Selection of Industrial Micro-organisms, Moscow 113545, USSR

During the first hours of sporulation the level of the extracellular and intracellular protease production by bacilli cells increases substantially. Intracellular proteases are presumably involved in the sporulation-specific turnover of vegetative proteins. Their structural and functional properties are therefore of great interest. During the last years the intracellular serine proteases (ISP) isolated from several species of bacilli were extensively studied in our laboratory. In this paper, the results of these studies will be briefly reviewed with the special emphasis on the chemistry and substrate specificity of ISP. The authors with great pleasure note the personal contributions made by Drs. L.S.Izotova, Z.T.Abramov, L.A.Lyublinskaya, A.N.Markaryan, V.V.Yanonis, K.A.Shaginyan and other colleagues in this work.

Among various isolation procedures affinity chromatography of the enzyme on Sepharose 4B, containing covalently bound cyclopeptide antibiotic gramicidin S, turned to be the most effective one (1). The efficiency of gramicidin S as a ligand for ISP chromatography has to be explained by rather high content of hydrophobic residues in its structure, facilitating the ligand-enzyme interaction, and the remarkable resistance to proteolysis connected with the presence of D-amino acid and Pro residues. Using affinity chromatography, we succeeded in isolation ISP from the following species: Bacillus amyloliquefaciens A-50 and 103 (the enzymes are further named ISP-Bam A-50 and ISP-Bam 103, respectively) - the strains producing the extracellular serine proteases (subtilisins) of the BPN' type, B.subtilis 168 Marburg (ISP-Bsu 168) and B.licheniformis 1001 (ISP-Bli) which both produce subtilisins of the Carlsberg type (2-6).

All isolated ISP possessed common features which clearly distinct them from secretory subtilisins (Table 1). ISP reveal absolute requirement for Ca^{2+} ions and therefore are completely and irreversibly inactivated with EDTA and EGTA. ISP have pI values near to 4 that is much lower than those of secretory subtilisins (pI near 8). The molecular weight of ISP polypeptide chains assayed by SDS-polyacrylamide gel-electrophoresis is 31000 ± 1000 (to be compared with the value of 28000 found for subtilisins BPN' and Carlsberg). In contrast with subtilisins that contain only one polypeptide chain, the molecules of native ISP are dimeric being composed of two apparently identical subunits. It might be well that quaternary structure of ISP being rather unusual for the proteolytic enzymes is of functional importance. ISP differ from subtilisins by the immunochemical properties, e.g. purified rabbit antibodies against ISP-Bam A-50, forming a precipitate bond with the latter, did not react with subtilisins BPN' and Carlsberg, as well as with bovine trypsin and chymotrypsin. On the other hand, antibodies against subtilisins BPN' and Carlsberg did not interact specifically with ISP.

Amino acid composition of ISP showing some common features with that of subtilisins is in comparison with them about 3-fold enriched in Glu content and has substantially more Leu, Lys and Phe (Table 2). Of special interest is the presence of 1-2 half-cystine residues in ISP molecule, that is uncommon for subtilisins. Already the data on the amino acid composition did not allow to consider ISP as an intracellular precursor of secretory subtilisins and show that two distinct structural genes encode intracellular enzyme and secretory subtilisin. This was confirmed by the direct comparison of the N-terminal sequence of ISP and subtilisins (Table 3).

TABLE 1. Comparison of ISP and subtilisin properties

Property	ISP	Subtilisin
Mol.weight of polypeptide chain	31000	28000
Mol.weight of native protein	56000	28000
Isoelectric point of:		
ISP from *B.licheniformis*	3,7	
ISP from *B.amyloliquefaciens*	4,3	ca.8
ISP from *B.subtilis* Marburg	4,9	
pH-maximum of stability	6-7	7-8
pH-maximum of activity	7-10	8-10
Residual activity after treatment with:		
diisopropylfluorophosphate (1mM) or		
benzylsulfofluoride (1mM)	0	0
EDTA or EGTA (2mM)	0	ca.100%
p-hydroxymercuribenzoate (5mM)	80%	100%

TABLE 2. Amino acid composition of ISP and subtilisins. ISP abbreviations - see the text. BPN', CAR - subtilisins BPN' and Carlsberg, respectively

Residue	ISP-Bli	ISP-Bsu 168	ISP-Bam A-50	ISP-Bam 103	BPN'	CAR
Lys	12	17	20	18	11	9
His	10	10	6	6	6	5
Arg	5	6	6	6	2	4
Asp	33	36	36	36	28	28
Thr	13	13	13	13	13	19
Ser	34	23	25	25	37	32
Glu	42	38	33	29	15	12
Pro	13	13	13	13	14	9
Gly	42	40	36	34	33	35
Ala	28	26	32	30	37	41
Val	15	19	19	19	30	31
Met	-	5	5	5	5	5
Ile	11	13	13	14	13	10
Leu	19	22	25	26	15	16
Tyr	6	7	7	8	10	13
Phe	8	8	7	9	3	4
Trp	-	-	-	-	3	1
1/2Cys	1-2	1-2	1-2	1-2	-	-

The appearance of the "stugged" N-ends is characteristic for ISP. As a rule, the preparations of the enzyme show the presence of two chains that differ solely by the length.

TABLE 3. N-terminal sequences of ISP and subtilisins. BPN', CAR, AMY - subtilisins BPN', Carlsberg and Amylosacchariticus, respectively.

		-1	1									10	
ISP-*Bam*	A-50	Asn	Val	Xxx	Glu	Leu	Pro	Glu	Gly	Ile	Gln	Val	Ile
ISP-*Bsu*	168	Asn	Val	Xxx	Glu	Leu	Pro	Glx	Gly	Ile	Glx	Val	Ile
ISP-*Bli*		Xxx	Val	Xxx	Glx	Leu	Pro	Glu	Gly	Ile	Xxx	Ala	Ile/
BPN'			Ala	Gln	Ser	Val	Pro	Tyr	Gly	Val	Ser	Gln	Ile
CAR			Ala	Gln	Thr	Val	Pro	Tyr	Gly	Ile	Pro	Leu	Ile
AMY			Ala	Gln	Ser	Val	Pro	Tyr	Gly	Ile	Ser	Gln	Ile
										20			
ISP-*Bam*	A-50	Lys	Ala	Pro	Gln	Leu	Trp	Ala	Gln	Gly	Phe	Lys	Gly
ISP-*Bsu*	168	Lys	Ala	Pro	Glx	Leu	Xxx	Ala/					
BPN'		Lys	Ala	Pro	Ala	Leu	His	Ser	Gln	Gly	Tyr	Thr	Gly
CAR		Lys	Ala	Asp	Lys	Val	Gln	Ala	Gln	Gly	Phe	Lys	Gly
AMY		Lys	Ala	Pro	Ala	Leu	His	Ser	Gln	Gly	Tyr	Thr	Gly
								30					
ISP-*Bam*	A-50	Ser	Asp	Ile	Lys	Ile	Ala	Val	Leu	Asp	Thr	Gly	Ile
BPN'		Ser	Asn	Val	Lys	Val	Ala	Val	Ile	Asp	Ser	Gly	Ile
CAR		Ala	Asn	Val	Lys	Val	Ala	Val	Leu	Asp	Thr	Gly	Ile
AMY		Ser	Asn	Val	Lys	Val	Ala	Val	Ile	Asp	Ser	Gly	Ile
						40							
ISP-*Bam*	A-50	Asp	Val	Xxx	Xxx	Pro	Asn	Leu	Asp	Xxx	Xxx	Gly	Gly
BPN'		Asp	Ser	Ser	His	Pro	Asn	Leu	Lys	Val	Ala	Gly	Gly
CAR		Gln	Ala	Ser	His	Pro	Asn	Leu	Asn	Val	Val	Gly	Gly
AMY		Asp	Ser	Ser	His	Pro	Asn	Leu	Asn	Val	Arg	Gly	Gly
				50									
ISP-*Bam*	A-50	-	Xxx	Phe	Val	Ala	Gly/						
BPN'		Ala	Ser	Met	Val	Pro	Ser						
CAR		Ala	Ser	Phe	Val	Ala	Gly						
AMY		Ala	Ser	Phe	Val	Pro	Ser						

The longer chain with N-terminal Asn differs from the shorter one with N-terminal Glu with the extension Asn-Val-Xxx. The ratio of Asn/Glu chains was 1:5 for ISP-*Bam* A-50, but 3:2 for ISP-*Bsu* 168. No Glu-chain was identified in case of ISP-*Bli*. The appearance of ISP chains with the "ragged" ends might be connected with the limited proteolysis of an unknown ISP precursor. The partial hydrolysis of N-terminal part of ISP molecules by aminopeptidase(s) can not be excluded.

The N-terminal sequence of ISP is clearly homologous to that of subtilisins. The extent of divergency between ISP and subtilisins is roughly equal to that among different extracellular subtilisins. E.g., the minimal number of amino acid substitutions in the regions compared is 22 for ISP-Bam A-50 *versus* subtilisin BPN', 20 - for ISP *versus* subtilisin Amylosacchariticus and 17 - for ISP *versus* subtilisin Carlsberg. This is very close to 20 substitutions between the respective stretches of BPN' and Carlsberg subtilisins. These data clearly show the existence of at least two separate, but homologous genes coding for serine proteases in bacilli- one coding for extracellular subtilisin, and another one for intracellular protease. The presence of two structurally homologous genes within one genome might be ascribed to the subsequent duplication of an ancestral gene. The coexistence of homologous genes in bacilli genome might lead to the acceleration of their evolution by means of intragenic recombination, as well as point mutations, deletions and/or insertions (7). This might at least partially explain the substantial divergency between different members of the secretory subtilisin family. To our knowledge, the gene duplication of bacilli serine proteases was the first one confirmed by sequence data (8). More recently, the data were obtained showing that the homologous genes were quite common for bacterial genome (e.g., *tuf*A and *tuf*B genes of *E.coli*, coding for the elongation factor Tu).

The difference in the primary structure between ISP-*Bam* A-50, ISP-*Bsu* 168 and ISP-*Bli* is substantially less than that between corresponding extracellular subtilisins BPN', 168 and Carlsberg produced by *B.amyloliquefaciens* A-50, *B. subtilis* 168 Marburg and *B.licheniformis*, respectively. A rationale for this might be seen in more stringent selection constraints imposed on the structure and functions of the intracellular enzymes during the evolution.

A very characteristic pattern of ISP N-terminal sequence has to be mentioned. In the stretch of 26 residues, each fourth residue is Glu (Gln), Asp (Asn) or Lys. Taking in account the presence of two short α-helixes in this part of the subtilisin molecule, one can assume that the side chains of these residues of ISP located on the tops of the helixes form a cluster of the surface functional groups capable of binding Ca^{2+} ions shown to be indispensable for ISP, since the exact role of this cluster has to be defined. This specific distribution of Glx, Asx and Lys residues is absent in the N-terminal sequence of subtilisins.

Hence, it should be emphasized that ISP possessing specific immunological reactions, amino acid composition and primary structure form a separate group (a subfamily) of subtilisin-like enzymes and are not the additional members of the well-known family of secretory subtilisins.

Being homologous in primary structure, ISP and subtilisins, nevertheless, differ significantly with the substrate specificity, that is especially pronounced for the hydrolysis of large protein substrates. As a rule, in comparison to subtilisins ISP are more active against low molecular weight synthetic substrates, but much less active against native proteins - hemoglobin, bovine serum albumin, ribonuclease and even its' presumed natural substrates - intracellular proteins of B.amyloliquefaciens (6, 9). We undertook the direct comparison of the enzyme specificity with that of subtilisins, using native and carboxymethylated urea-denatured ribonuclease (RNase), B-chain of oxidized bovine insulin and p-nitroanilides (pNA) of benzyloxycarbonyl-peptides as the substrates, as well as organophosphorus and bororganic compounds as the inhibitors.

An evidence was obtained that ISP specificity against the peptide substrates lies within the limits of subtilisin specificity, being somewhat more restricted. As well as the subtilisins ISP splits a single peptide bond Ala_{20}-Ser_{21} of trypsin-insensitive native pancreatic RNase A, thus forming its trypsin-sensitive derivative, RNase S, that consists of the non-covalently bound S-peptide and S-protein (6). Non-specific cleavage of RNase S was much less for ISP in comparison with the subtilisin hydrolysis, hence ISP appeared to be more suitable for the RNase S production. ISP cleaves a single peptide bond Leu_{15}-Tyr_{16} in the B-chain of insulin.

Under comparable experimental conditions in contrast with ISP subtilisins cleave four additional bonds, Gln_4-His_5, Ser_9-His_{10}, Tyr_{17}-Leu_{18} and Phe_{25}-Tyr_{26} (6).

As well as the subtilisins, ISP cleave p-nitroanilides of N-protected peptide substrates with the hydrophobic residues of Leu and Phe in P_1 - Z-Gly-Gly-Leu-pNA, Z-Gly-Gly-Phe-pNA, Z-Ala-Ala-Leu-pNA, Z-Ala-Ala-Phe-pNA, Z-Ala-Leu-pNA and Z-Gly-Leu-pNA. The values of K_M, k_{cat} and k_{cat}/K_M for ISP and subtilisins BPN' and Carlsberg are represented in Table 4. The best substrate for ISP and subtilisin Carlsberg is Z-Ala-Ala-Leu-pNA, for subtilisin BPN' - Z-Ala-Ala-Phe-pNA. The specificity of subtilisin Carlsberg more closely resembles that of ISP as compared with the specificity of subtilisin BPN'. There is also a difference in the secondary substrate specificity of these enzymes. The substitution of Gly residues in P_2 and P_3 of the substrate for Ala favors about 40-fold the binding of the corresponding substrate with ISP, but only about 2-fold for both subtilisins. As compared with p-nitroanilides of tripeptides, dipeptide substrates possess much lower activity with ISP.

By studying inactivation kinetics of ISP and subtilisins BPN' and Carlsberg with the series of organophosphorus /$(C_nH_{2n+1}O)(CH_3)P(O)OC_6H_4NO_2$, where n= 2-8/ and bororganic /$H(CH_2)_n$-$B(OH)_2$, where n= 2-6/ inhibitors, A.Aaviksaar, T.Rotanova and their colleagues have shown that there was no substantial difference in topography of the active site regions of these enzymes.

Thus, catalytic centers of ISP are potentially very active, and the obtained data might only partially explain a dramatically low activity of the enzyme against a variety of the proteins. It is tempting to suppose that ISP remarkably low activity against protein substrates is due to its reluctance to attack the bulky substrates, e.g. non-denatured or partially denatured proteins. It might be connected with the existence of ISP as a dimer composed of two apparently identical subunits. In contrast with ISP, subtilisins were never

TABLE 4. Hydrolysis of p-nitroanilide substrates by ISP and subtilisins BPN' and Carlsberg.

Substrate	ISP-Bam 103			Subtilisin BPN'			Subtilisin Carlsberg		
	K_M mM	k_{cat}/K_M $M^{-1}sec^{-1}$	k_{cat} sec^{-1}	KM mM	k_{cat}/K_M $M^{-1}sec^{-1}$	k_{cat} sec^{-1}	K_M mM	k_{cat}/K_M $M^{-1}sec^{-1}$	k_{cat} sec^{-1}
Z-Gly-Gly-Leu-pNA	8	150	1.2	1.1	180	0.2	2.2	100	0.22
Z-Ala-Ala-Leu-pNA	0.18	11000	2.04	0.55	1100	0.6	1	19000	19
Z-Gly-Gly-Phe-pNA	-	160	-	0.83	360	0.3	-	50	-
Z-Ala-Ala-Phe-pNA	9	1600	14.5	0.072	12000	0.83	0.28	4500	1.26
Z-Ala-Leu-pNA	-	19	-	1.1	130	0.13	4	490	1.96
Z-Gly-Leu-pNA	-	56	-	-	-	-	-	45	-

observed as dimers. The dimerization might create steric hindrance at the active sites of both subunits thus making it difficult for the enzyme to bind and hydrolyze the bulky native globular proteins, but not the relatively small synthetic substrates or denatured proteins. It was shown that ISP cleaves denatured carboxymethylated RNase A with the rate comparable with that of subtilisin BPN', whereas the rate of cleavage of the native RNase with ISP and subtilisin BPN' differs approximately 10-fold. The comparable results were obtained for other pairs of native-denatured proteins.

Hence, the dimerization might be considered as a specific structural feature of the intracellular enzyme designed to prevent the occasional proteolysis in vivo of the native intracellular proteins, but to permit the selective proteolytic elimination of the denatured and, possibly, aberrant proteins. In accord with ISP of bacilli, a number of known intracellular E.coli proteases (e.g., proteases I, II and L) (10-12) showing negligible activity against native proteins wère reported to exist as dimers in the active state. A comparable mechanism of restricting for proteolytic action was recently reported by Levine & Walsh (13) for acrosin-like intracellular protease from sea urchin sperm - an enzyme, composed of two non-identical subunits, the catalytic and the regulatory one. In conclusion, it might be reasonable to assume that mechanisms, presumably, non-proteolytic ones (14), controlling at least partial denaturation of the proteins might govern the rate and selectivity of the intracellular protein turnover.

References

1. Izotova, L.S., Gorodetsky, D.I., Yanonis, V.V., Baratova, L.A., Belyanova, L.P., Timokhina, E.A., Strongin, A.Ya., Stepanov, V.M.: Bioorg.Khimiya 4, 397-403 (1978).

2. Strongin, A.Ya., Izotova, L.S., Abramov, Z.T., Gorodetsky, D.I., Ermakova, L.M., Baratova, L.A., Belyanova, L.P., Stepanov, V.M.: J.Bact. 133, 1401-1411 (1978).

3. Strongin, A.Ya., Gorodetsky, D.I., Kuznetsova, I.A., Yanonis, V.V., Abramov, Z.T., Belyanova, L.P., Baratova, L.A., Stepanov, V.M.: Biochem.J.: 179, 333-339 (1979).

4. Strongin, A.Ya., Abramov, Z.T., Yaroslavtseva, N.G., Baratova, L.A., Shaginyan, K.A., Belyanova, L.P., Stepanov, V.M.: J.Bact. 137, 1017-1019 (1979).

5. Stepanov V.M.: In "Enzyme regulation and mechanism of action" (Eds. P.Mildner, B.Ries), Pergamon Press, Oxford-New York, v.60, 331-337, 1980

6. Markaryan, A.N., Ostoslavskaya, V.I., Svyadas, V.K., Yakusheva, L.D., Lyublinskaya, L.A., Strongin, A.Ya., Stepanov, V.M.: Int.J.Biochem. 13, 201-206 (1981).

7. Strongin, A.Ya., Izotova, L.S., Abramov, Z.T., Gorodetsky, D.I., Stepanov, V.M.: Mol.Gen.Genet. 159, 337-339 (1978).

8. Stepanov, V.M., Strongin, A.Ya., Izotova, L.S., Abramov, Z.T., Ermakova, L.M., Baratova, L.A., Belyanova, L.P., Lyublinskaya, L.A.: Biochem.Biophys.Res.Commun. 77, 298-305 (1977).

9. Strongin, A.Ya., Izotova, L.S., Abramov, Z.T., Ermakova, L.M., Gorodetsky, D.I., Stepanov, V.M.: Arch.Microbiol. 119, 287-293 (1978).

10. Pacaud, M., Richaud, C.: J.Biol.Chem. 250, 7771-7779 (1975).

11. Shaginyan, K.A., Izotova, L.S., Strongin, A.Ya, Stepanov, V.M.: Biokhimiya 45, 695-703 (1980).

12. Strongin, A.Ya., Gorodetsky, D.I., Stepanov, V.M.: J.Gen.Microbiol. 110, 443-451 (1979).

13. Levine, A.E., Walsh, K.A.: J.Biol.Chem. 255, 4814-4820 (1980).

14. Ciechanover, A., Heller, H., Elias, S., Hass, A.L., Hershko, A.: Proc.Natl.Acad.Sci. U.S.A. 77, 1365-1368 (1980).

STRUCTURE AND BIOLOGICAL FUNCTION OF PROTEINASE INHIBITORS FROM YEAST

Konrad Maier and Helmut Holzer

Gesellschaft für Strahlen- und Umweltforschung, Abteilung für Enzymchemie, D-8042 Neuherberg, West-Germany

Renate Barth, Peter Bünning and Eiki Kominami

Biochemisches Institut der Universität Freiburg, D-7800 Freiburg, West-Germany

Introduction

Endogenous inhibitors for two endopeptidases and a carboxypeptidase have been identified in yeast. The specific inhibitors for proteinase A, I^A2 and I^A3, and proteinase B, I^B1, I^B2 and I^B3 are heat-stable and acid-resistant polypeptides with molecular weights between 7,600 and 11,500 (1-3). In contrast, the carboxypeptidase Y inhibitor I^C is inactivated by heat or acid treatment and exhibits a higher molecular weight of 23,800 (4). The proteinase inhibitors are located in the cytosol of the yeast cell (3,5), whereas the respective proteinases occur in the vacuoles (6). The properties of the yeast proteinase inhibitors are summarized in Table I.

Results

Structure of the Yeast Proteinase Inhibitors

The amino acid compositions of five of the yeast proteinase inhibitors are shown in Table II (7-12). The proteinase A inhibitor I^A3 and the proteinase B inhibitors I^B1 and I^B2

Chemistry of Peptides and Proteins, Vol. 1

Table I. Macromolecular Proteinase Inhibitors from Yeast.

Inhibitor of	IP	MW	Hydrolyzed by	Stability to heat and acid
Proteinase A				
I^A2	5.7	7,600	Proteinase B	Resistant
I^A3	6.3			
Proteinase B				
I^B1	8.0	8,500		
I^B2	7.0	8,500	Proteinase A	Resistant
I^B3	4.6	11,500		
Carboxypeptidase Y				
I^C	6.6	23,800	Proteinases A and B	Labile

contain neither cystine/cysteine, tryptophan nor arginine. Furthermore, proline is absent from I^A3 and methionine from I^B1 and I^B2. In contrast, from I^B3 only tryptophan is absent and I^C does not lack any of these amino acids.

Table II. Amino Acid Compositions of I^A3 (I. Nunez), I^B1/I^B2 (K. Maier), I^B3 (H. Matern) and I^C (R. Barth).

Amino Acid	I^A3	I^B1/I^B2	I^B3	I^C
Asp	10	12	15	22
Thr	3	5	5	15
Ser	6	2	13	23
Glu	13	6/7	12	21
Pro	0	2	3	16
Gly	5	3	7	13
Ala	5	2	6	17
Cys	0	0	1	8
Val	2	9/8	7	10
Met	2	0	1	5
Ile	1	6	7	5
Leu	2	5/6	9	12
Tyr	2	1	5	6
Phe	2	3	2	13
Lys	13	12/11	8	20
His	2	6	2	5
Arg	0	0	2	3
Trp	0	0	0	3
Total	68	74	105	217

The determination of the amino acid sequence has revealed that I^B1 and I^B2 differ in two positions only. Val-32 and Lys-34 in I^B1 are replaced by Leu-32 and Glu-34 in I^B2, respectively (Table III; 8,9). The almost identical primary structures of these two proteinase B inhibitors indicate that they are isoinhibitors. In contrast, the amino acid composition of I^B3 differs characteristically from those of the two other proteinase B inhibitors. Furthermore, an antibody prepared against I^B2 does not crossreact with I^B3 (4). These results exclude the possibility that I^B3 is a precursor of I^B1 and I^B2. Recent experiments on the structure of the proteinase A inhibitors have indicated that I^A2 and I^A3 have similar amino acid compositions and, therefore, may also be called isoinhibitors (10,11,13).

In comparison with other proteinase inhibitors, such as the well investigated trypsin inhibitors, the proteinase A and B inhibitors from yeast are unique because they contain no disulfide bridges. According to the prediction rules of Chou and Fasman (14) the absence of disulfide bridges should favour a

TABLE III. AMINO ACID SEQUENCES OF YEAST PROTEINASE B INHIBITORS 1 AND 2.

	5 10 15 20
I^B1	H_3^+N-THR-LYS-ASN-PHE-ILE-VAL-THR-LEU-LYS-LYS-ASN-THR-PRO-ASP-VAL-GLU-ALA-LYS-LYS-PHE-
I^B2	H_3^+N-THR-LYS-ASN-PHE-ILE-VAL-THR-LEU-LYS-LYS-ASN-THR-PRO-ASP-VAL-GLU-ALA-LYS-LYS-PHE-
	25 30 35 40
I^B1	LEU-ASP-SER-VAL-HIS-HIS-ALA-GLY-GLY-SER-ILE-VAL-HIS-LYS-PHE-ASP-ILE-ILE-LYS-GLY-
I^B2	LEU-ASP-SER-VAL-HIS-HIS-ALA-GLY-GLY-SER-ILE-LEU-HIS-GLU-PHE-ASP-ILE-ILE-LYS-GLY-
	45 50 55 60
I^B1	TYR-THR-ILE-LYS-VAL-PRO-ASP-VAL-LEU-HIS-LEU-ASN-LYS-LEU-LYS-GLU-LYS-HIS-ASN-ASP-
I^B2	TYR-THR-ILE-LYS-VAL-PRO-ASP-VAL-LEU-HIS-LEU-ASN-LYS-LEU-LYS-GLU-LYS-HIS-ASN-ASP-
	65 70 74
I^B1	VAL-ILE-GLU-ASN-VAL-GLU GLU-ASP-LYS-GLU-VAL-HIS-THR-ASN-COO^-
I^B2	VAL-ILE-GLU-ASN-VAL-GLU-GLU-ASP-LYS-GLU-VAL-HIS-THR-ASN-COO^-

↓ CHYMOTRYPSIN

⇀ AUTOMATED EDMAN DEGRADATION

← CARBOXYPEPTIDASE A (BOVINE PANCREAS), pH 8.5

← CARBOXYPEPTIDASE A (BOCINE PANCREAS), pH 4.2

← CARBOXYPEPTIDASE Y (YEAST), pH 4.2

random-coil-like conformation of these inhibitors which was confirmed for I^B2 by NMR-studies.
Disulfide bridges are also regarded to be essential for inhibitor activity of the soybean trypsin inhibitor and related inhibitors from other sources. Ozawa and Laskowski have proposed that a specific peptide bond in the active center of the native soybean trypsin inhibitor is cleaved by the respective proteinase (15). The two fragments of the inhibitor are hold together by disulfide bridges. The absence of disulfide bridges in the yeast inhibitors points to quite a different mechanism of inhibition.
The degradation of I^B2 by carboxypeptidase A has shown that the COOH-terminal region is essential for inhibition of

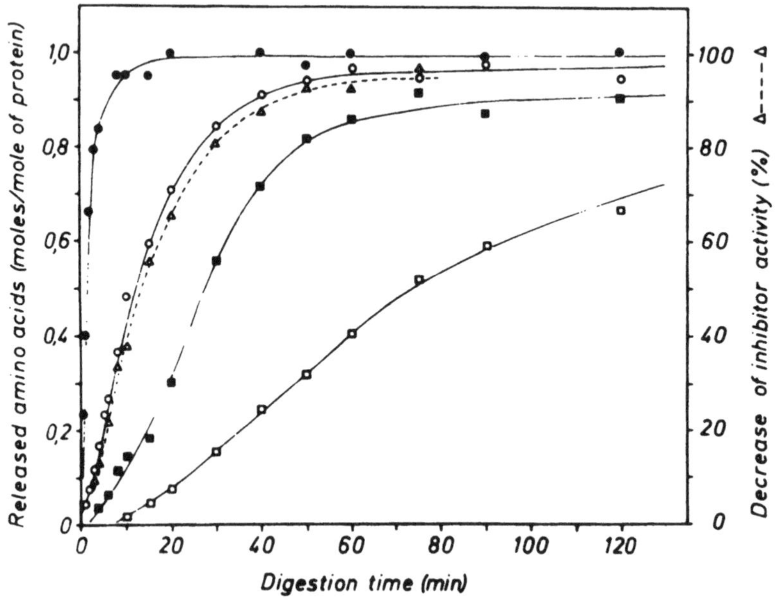

Figure 1. Degradation and inactivation of I^B2 by bovine carboxypeptidase A. The reaction was performed in 0.2 M N-ethylmorpholine acetate, pH 8.5, at 25°C, using an enzyme-substrate ratio of 1 : 390 (mol/mol). The following amino acids were released: ●——● Asn, ○——○ Thr, ■——■ His, □——□ Val. Proteinase B inhibitor activity: △---△.

proteinase B. When threonine, the second amino acid from the COOH-terminus is removed the inhibitor looses its activity completely (Figure 1). On the other hand, a specific proteolytic cleavage at position 40 also inactivates the inhibitor suggesting that the NH_2-terminal half of the molecule is involved in the formation of the proteinase inhibitor complex. A variety of chemical modification reactions were carried out with the proteinase B inhibitor I^B2 to obtain more detailed information on the mechanism of inhibition. The inhibitor contains a single tyrosyl residue in position 41 (Table III).

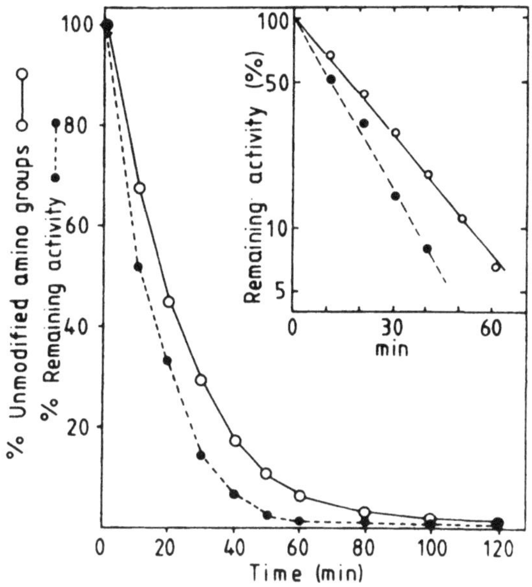

Figure 2. Time course of the I^B2 modification with trinitrobenzenesulfonic acid and loss of proteinase B inhibitor activity. The inhibitor was incubated with a 130-fold molar excess of reagent at pH 8.5 and 37°C. At the indicated incubation times the reaction was stopped. The number of modified amino groups was determined spectrophotometrically. From an aliquot of the samples proteinase B inhibitor activity was assayed. o—o unmodified amino groups, ●- -● proteinase B inhibitor activity. Inset: Semilogarithmic plots of modified residues and loss of proteinase B inhibitor activity versus time.

Specific modification of this residue was achieved by nitration with tetranitromethane. As a result of the modification the association constant of the proteinase inhibitor complex decreased by an order of magnitude (16). The inhibitor contains eleven lysyl residues which were modified with trinitrobenzenesulfonic acid and O-methylisourea. When I^B2 was incubated with a 130-fold excess of trinitrobenzenesulfonic acid the rate of modification paralleled the diminution of inhibitor activity with time (Figure 2). The fully modified inhibitor is totally inactive. Semilogarithmic plots of the number of modified residues and of the loss of inhibitor acitivity versus reaction time yield linear relations. These results show that all lysyl residues are modified with similar reaction rates which indicates that there is no particular lysyl residue with a prominent role for the inhibition process. The loss of inhibitor activity by modification with trinitrobenzenesulfonic acid may be caused either by replacement of positive charges, by negative ones or by insertion of bulky groups which lead to steric hindrance. Chemical modification of lysyl residues with O-methylisourea converts lysine to homoarginine, a reaction which does not change the positive charge of the residue. In contrast to the modification with trinitrobenzenesulfonic acid, quantitative guanidation with O-methylisourea only leads to 50 % loss of inhibitor activity. These results indicate that ionic interactions between proteinase and inhibitor may be essential for the inhibition mechanism.

Biological Function of Yeast Proteinase Inhibitors

The three types of inhibitors exhibit absolute specificity toward their respective proteinases. Furthermore, the inhibitors differ characteristically in their susceptibility to hydrolytic inactivation by proteinases A and B (Table I). Incubation of I^B2 with proteinase A at pH 7.0 leads to limited proteolysis and inactivation of the inhibitor (Figure 3). When proteinase B is added to the degradation products they

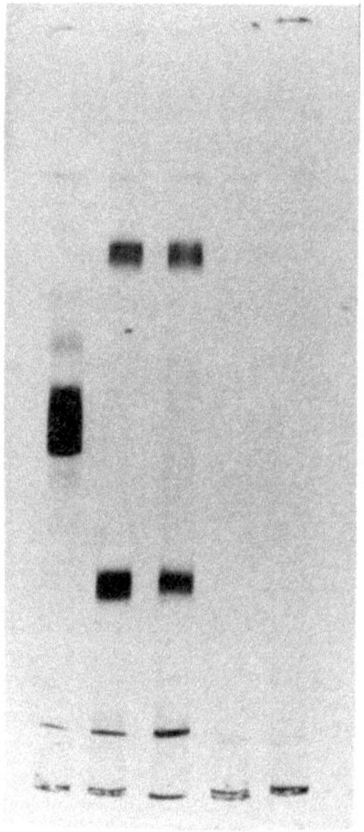

a b c d e

Figure 3. Action of yeast proteinases A and B on I^B2. (a) I^B2 without proteinase, (b,c) incubation of I^B2 with proteinase A for 30 and 60 min, respectively. (d,e) subsequent incubation of (c) with proteinase B for 30 and 60 min, respectively.

are rapidly hydrolyzed to small peptides, whereas at pH 7.0 proteinase B does not attack native I^B2. At pH 4.0, however, proteinase B inactivates native I^B2 (Figure 4). This hydrolysis of I^B2 by proteinase B may be caused by a lower association constant of the inhibitor complex at this pH value. The characteristic susceptibility to proteolysis on the one hand and the absolute specificity of the inhibitors on the other provide the yeast cell with an effective mechanism for controlling proteolytic activities. Direct evidence for such a regulation mechanism was obtained from experiments with a proteinase A inhibitor minus mutant in which the turnover rate of the proteinase B inhibitor is much higher than the respective wild type strain (17). Hence it may be possible that the

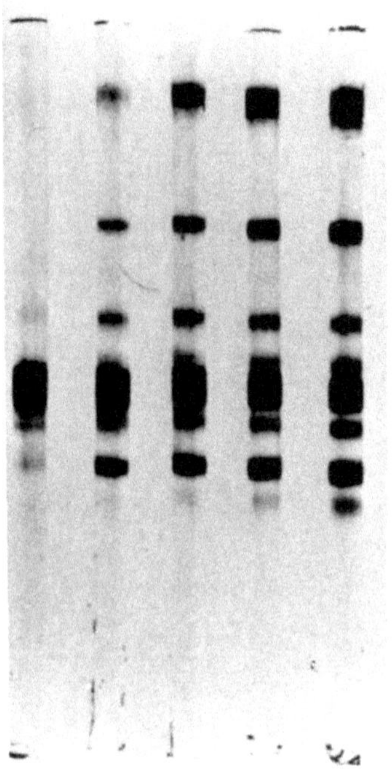

Figure 4. Time course of the proteolysis of proteinase B inhibitor 2 by proteinase B. $I^{B}2$ was incubated with 0.4 mol% of proteinase B in sodium acetate buffer, pH 4.0, at 25°C. At the times indicated, the reaction was stopped. Aliquots were analyzed by electrophoresis on 7.5 % polyacrylamide gels at pH 8.9. Further aliquots were assayed for remaining proteinase B inhibitor activity.

proteinase inhibitors are involved in the control of both, protein turnover and specific, limited proteolysis.

References

1. Saheki, T., Matsuda, Y., Holzer, H.: Eur. J. Biochem. 47, 325-331 (1974).

2. Betz, H., Hinze, H., Holzer, H.: J. Biol. Chem. 249, 4515-4521 (1974).

3. Matern, H., Weiser, U., Holzer, H.: Eur. J. Biochem. 101, 325-332 (1979).

4. Matern, H., Hoffmann, M., Holzer, H.: Proc. Natl. Acad. Sci. USA 71, 4874-4878 (1974).

5. Matern, H., Betz, H., Holzer, H.: Biochem. Biophys. Res. Commun. 60, 1051-1057 (1974).

6. Matile, P., Wiemken, A.: Arch. Microbiol. 56, 148-155 (1967).

7. Maier, K., Müller, H., Holzer, H.: J. Biol. Chem. 254, 8491-8497 (1979).

8. Maier, K., Müller, H., Tesch, R., Trolp, R., Witt, I., Holzer, H.: J. Biol. Chem. 254, 12555-12561 (1979).

9. Maier, K., Müller, H., Tesch, R., Witt, I., Holzer, H.: Biochem. Biophys. Res. Commun. 91, 1390-1398 (1979).

10. Biedermann, K., Montali, U., Martin, B., Svendsen, I., Ottesen, M.: Carlsberg Res. Commun. 45, 225-235 (1980).

11. Nunez de Castro, I., Holzer, H.: Hoppe-Seyler's Z. Physiol. Chem. 357, 727-734 (1976).

12. Matern, H., Barth, R., Holzer, H.: Biochim. Biophys. Acta 567, 503-510 (1979).

13. Meussdoerffer, F.: Biochem. Biophys. Res. Commun. 97, 423-429 (1980).

14. Chou, P. Y., Fasman, G. D.: Ann. Rev. Biochem. 47, 251-276 (1978).

15. Ozawa, K., Laskowski, M., Jr.: J. Biol. Chem. 241, 3955-3961 (1966).

16. Bünning, P., Holzer, H.: J. Biol. Chem. 252, 5316-5323 (1977).

17. Beck, I., Fink, G. R., Wolf, D. H.: J. Biol. Chem. 255, 4821-4828 (1980).

INHIBITORS OF HUMAN NEUTRAL GRANULOCYTIC PROTEINASES FROM THE LEECH: BIOCHEMICAL CHARACTERIZATION AND PATHOBIOCHEMICAL ASPECTS

Ursula Seemüller

Abteilung für Klinische Chemie und Klinische Biochemie in der Chirurgischen Klinik der Universität München, D-8000 München, Germany

Manfred Eulitz

Institut für Hämatologie der Gesellschaft für Strahlen- und Umweltforschung, Abteilung Immunologie, D-8000 München, Germany

Introduction

From the leech Hirudo medicinalis a group of proteinase inhibitor proteins, the eglins, was isolated and characterized in detail (1). The eglins exert strong inhibitor activity towards chymotrypsin, subtilisin, and the human granulocytic proteinases elastase and cathepsin G. The dissociation constants of the enzyme-inhibitor complexes are in the range of 10^{-10} mol/L. The eglins consist of a single polypeptide chain with a molecular weight of 8 100 as calculated from the amino acid composition (Table 1). Striking features of these inhibitors are the lack of cysteine or cystine and their strong resistance against denaturation by heat and acid

Primary structure

Our recent studies have been concerned with the primary structure of eglin c. The sequence was elucidated by automatic Edman degradation of the native protein up to position 40, the tryptic peptides (designed as T1 - T7), and a chymotryptic pep-

Chemistry of Peptides and Proteins, Vol. 1

tide derived from limited proteolysis (Ch1) (Figure 1). Determination of the C-terminal sequence was performed by degradation with carboxypeptidase Y in the presence of 0.5 % sodium dodecylsulfate(SDS). Interestingly, without SDS the eglin was resistant to enzymatic attack.

Table 1. Amino acid composition of eglin c.

	Eglin c		Residues/Mol
	20 h	72 h	
Asp	6.76	6.70	7
Thr	4.52	4.08	5
Ser	2.93	2.31	3
Glu	7.16	7.11	7
Pro	5.9	6.22	6
Gly	5.18	5.27	5
Ala	1.16	1.17	1
Cys[a)]			
Val	10.1	10.84	11
Met[a)]			
Ile			
Leu	4.83	4.83	5
Tyr	5.12	4.61	5-6
Phe	4.93	4.93	5
His	2.99	3.26	3
Lys	2.18	2.23	2
Arg	2.60	3.84	4
Trp[b)]	0.72		0-1
N-Terminus			Thr
Total			70
Molecular weight			8 099
Isoelectric Point			6.45 ± 0.3

a) after performic acid oxidation

b) photometrically determined

Figure 1. Primary structure of eglin c.

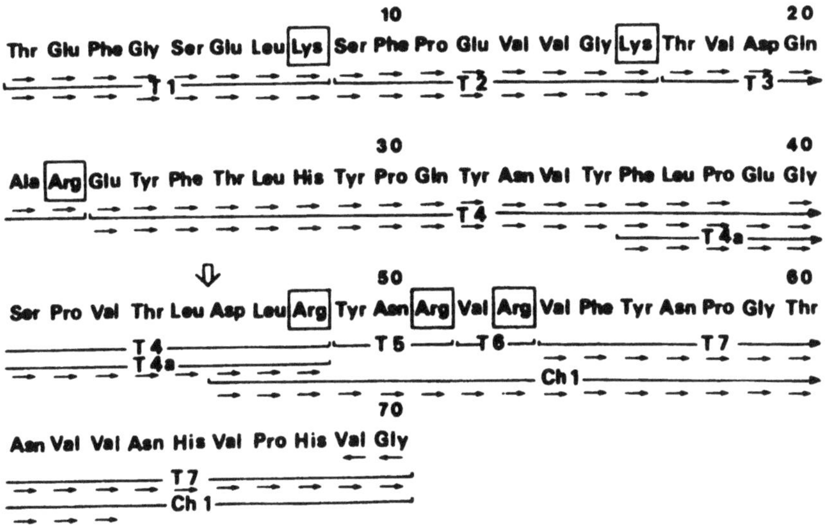

T1 - T7: tryptic peptides; Ch: chymotryptic peptide;
⟶ residue identified by automatic Edman degradation;
⟵ residue identified after cleavage with carboxypeptidase Y.
Tryptic cleavage sites are indicated by the boxed amino acids (except Tyr^{35}); the arrow denotes the chymotrypsin-reactive bond.

The chymotryptic peptide Ch1 was obtained by incubation of eglin c with 50 mol % chymotrypsin and fractionation of the incubation mixture by gel chromatography in acidic medium.

Inhibition mechanism

In preceding experiments it could be demonstrated that incubation of eglin with varying amounts of chymotrypsin and separation of the mixtures by SDS gel electrophoresis yields a low molecular weight peptide.

Figure 2a shows the incubation of eglin c with 5 mol % chymo-

Figure 2. Incubation of eglin c with chymotrypsin and separation of the mixture by SDS gel chromatography (in the absence of reducing agents).

a) Incubation with 5 mol % chymotrypsin

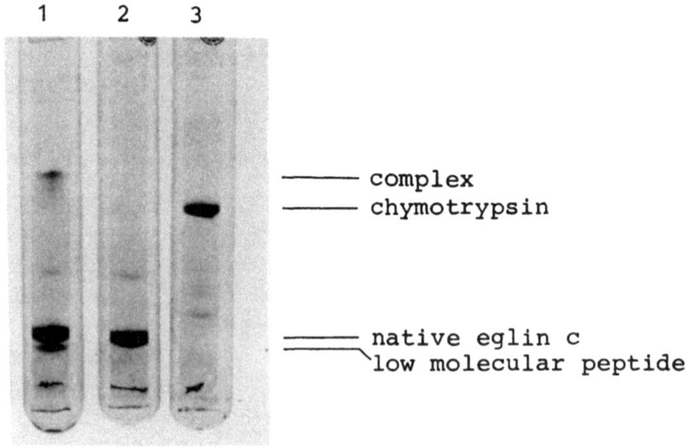

b) Incubation with 50 mol % chymotrypsin

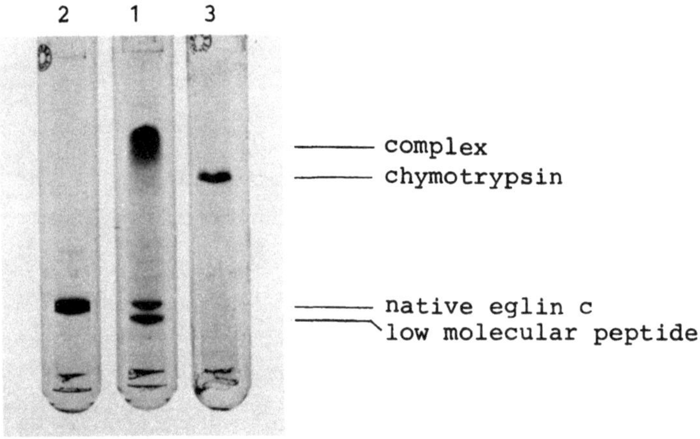

1 incubation mixture of eglin c with chymotrypsin
2 native eglin c
3 chymotrypsin

trypsin. Three bands appeared: the upper band with a molecular weight of about 30 000 corresponds to the eglin-chymotrypsin complex; the second band with a molecular weight of 8 000 corresponds to native eglin c; and the last band corresponds to the low molecular weight peptide. The same band pattern is obtained from the incubation mixture of eglin c with 50 mol % enzyme (Figure 2b). The only difference is that the intensity of the eglin band is decreased whereas the intensity of the complex and the low molecular weight peptide band is increased. From sequence data we could identify the low molecular weight peptide as the C-terminal fragment of eglin c comprising positions 46 - 70 (Ch1).

From these results it seems likely that during complex formation of eglin c with chymotrypsin limited proteolysis occurs resulting in the release of a C-terminal peptide. The N-terminal part of the inhibitor molecule remains firmly bound to the enzyme in the complex. This complex might be covalently linked as it does not dissociate in SDS gel electrophoresis or during gel chromatography in acidic medium.

Generally, proteinase inhibitor proteins do not form covalent complexes with proteinases. They react according to the following mechanism:

$$E + I \underset{k_{-1}}{\overset{k_1}{\rightleftharpoons}} L \underset{k_{-2}}{\overset{k_2}{\rightleftharpoons}} C \underset{k_{-3}}{\overset{k_3}{\rightleftharpoons}} A \underset{k_{-4}}{\overset{k_4}{\rightleftharpoons}} L^* \underset{k_{-5}}{\overset{k_5}{\rightleftharpoons}} E + I^*$$

E - enzyme; I - native inhibitor; I^* - modified inhibitor (reactive site peptide bond hydrolysed); L and L^* loose complexes consisting of enzyme and native inhibitor or modified inhibitor, respectively; C - stable enzyme-inhibitor complex; A - acyl-enzyme-complex.

Inhibitor and enzyme form via a loose intermediate (L) the noncovalent, catalytically inactive complex C (2). This complex is very stable as expressed by a low dissociation constant. Limited proteolysis, i.e. hydrolysis of the reactive site peptide bond of the inhibitor may occur resulting in the formation of the modified inhibitor. The modified inhibitor is also inhibitory active, although the velocity of complex formation is normally reduced. It may be transformed again to native inhibitor by thermodynamically controlled resynthesis of the cleaved peptide bond.

All inhibitors shown so far to obey the above given standard mechanism are characterized by a "reactive peptide bond which is encompassed in at least one disulfide loop. This disulfide loop ensures that during conversion of virgin to modified inhibitor the two peptide chains cannot dissociate" (3). However, the eglins do not contain a disulfide bridge, i.e. there are characteristic differences between the eglins and the "standard" proteinase inhibitor proteins:

(i) The enzyme-inhibitor complex formed by the eglins seems to be a covalent complex, the one formed by the standard inhibitors is a noncovalent complex.
(ii) Complex formation is reversible for standard inhibitors but obviously not for the eglins.

(iii) The degree of limited proteolysis of the eglin reactive site peptide bond correlates with the applied enzyme concentration but depends on a true thermodynamic equilibrium between native and modified inhibitor for the standard inhibitors.

Thus, the inhibition mechanism of the eglins resemble that of plasma inhibitor proteins, e.g. α-1-proteinase inhibitor (4) and antithrombin III (5).

Pathobiochemical aspects

Biochemical and clinical investigatione confirm the assumptions that the neutral granulocytic proteinases elastase and cathepsin G are involved in inflammatory processes, e.g. in lung emphysema, glomerulonephretis, and septic shock. These pathological events are distinguished by tissue destruction and degradation of plasma components, respectively. In a recent study it was observed that in the course of experimental endotoxaemia consumption of plasma factors proceeds not only via the classical routes, i.e. by activation of the clotting, fibrinolytic and complement cascades by system-specific proteinases such as thrombokinase or the plasminogen activator. It is due also to an appreciable degree to unspecific proteolysis by lysosomal enzymes, especially elastase and cathepsin G (6,7). In this model medication with an inhibitor from soybeans, which strongly inhibits granulocytic elastase and cathepsin G, reduces degradation of these plasma factors significantly.

Hence, eglin is a promising tool for the therapeutical application in inflammatory processes and the elucidation of the biochemical basis of such diseases.

References

1. Seemüller, U., Meier, M., Ohlsson, K., Müller, H.-P., Fritz, H.: Hoppe-Seyler's Z. Physiol. Chem. 358, 1105-1117 (1977).

2. Laskowski, jr., M., Sealock, R. W.: The Enzymes Vol. III 3rd ed. (ed. Boyer, P. D.). Academic Press, New York, 375-473, 1971

3. Laskowski, jr., M., Kato, I.: Ann. Rev. Biochem. 49, 593-627 (1980).

4. Morii, M., Odani, S., Ikenaka, T.: J. Biochem. 86, 915-921 (1979).

5. Longas, M. O., Finlay, T. H.: Biochem. J. 189, 481-489 (1980).

6. Witte, J.: Habilitationsschrift, Medical Faculty, University Munich, 1979

7. Fritz, H.: Protein Degradation in Health and Disease, Ciba Foundation 75 (New Series), Excerpta Medica, Elsevier/North Holland, 351-379, 1980

ISOLATION OF A SPECIFIC PROTEIN INHIBITOR OF FUNGAL PROTEINASE AND YEAST PROTEINASE B FROM THE KIDNEY BEAN SEEDS

V.V.Mosolov, E.L.Malova, A.N.Tcheban, V.M.Lakhtin
A.N.Bakh Institute of Biochemistry, Moscow

Introduction

Protein - proteinase inhibitors are widely distributed in plants. Leguminosae seeds are especially rich sources of proteinase inhibitors (I). Not much is known about the physiological role of these proteins in plants. There are a few reports of specific inhibitors which are active against proteinases of micro-organisms (2,3,4). The present communication describes the isolation of a new specific protein which inhibits Aspergillus oryzae proteinase (EC 3.4.2I.I4). The purified protein was also effective inhibitor of the proteinase B from baker's yeast (EC 3.4.22.9).

Results

Finely ground kidney bean seeds (Phaseolus vulgaris L., cv. Tweed wonder) were extracted with distilled water in the ratio of I:IO (w/v) at room temperature for I h. After centrifugation the supernatant was brought to pH 5.0 with I M HCl and left standing overnight at 4°C. The precipitate was removed by centrifugation and supernatant was freeze-dried. The preparation was dissolved in 0.05 M Tris-HCl buffer, pH 8.0, containing 0.5 M NaCl. The solution was applied to the column of trypsin-Sepharose 4B equilibrated with the same buffer. The trypsin inhibitors were removed quantitatively

Chemistry of Peptides and Proteins, Vol. 1

by this procedure. The Aspergillus - proteinase inhibitor activity was not bound by trypsin-Sepharose and was further purified by ion-exchange chromatography on DEAE-cellulose, and gel filtration on Sephadex G-75. Some properties of the isolated proteins are summarized in table I.

Table I. Characteristics of proteinase inhibitors.

	Trypsin inhibitors*	Aspergillus- proteinase inhibitor
Molecular weight		
gel filtration	-	I0 700
SDS-gel electrophoresis	I0 500	-
amino acid analysis	I0 200	II 000
pI (isoelectric focusing)	4.3-5.I	4.7

* Four isoinhibitors had isoelectric points of 4.3, 4.5, 4.7 and 5.I.

Both inhibitors have high heat and acid stability. No activity losses were observed during a I5 minute period of incubation at 95°C. The amino acid analysis of the inhibitory proteins is presented in table 2. The fungal proteinase inhibitor is composed of I0I residues. Its amino acid composition is markedly different from that of the trypsin inhibitors. Particularly the Aspergillus- proteinase inhibitor contains only six half-cystine residues per molecule.

Purified proteinase from Asp.oryzae was obtained using affinity chromatography on immobilized duck ovomucoid. Chromatography of the purified enzyme on a calibrated Sephadex G-75 column indicated an apparent molecular weight of 25 000. The proteolytic activity was assayed by Kunitz procedure using casein as substrate (5). Inhibition of the Aspergillus proteinase by increasing levels of inhibitor was linear up to approximately 85 % and extrapolation of the data indicates that one mole of the inhibitor reacts with one mole of prote-

Table 2. Amino acid composition of the inhibitors.

Amino acid	Trypsin inhibitor (pI 4.3)	Aspergillus -proteinase inhibitor
ASX	I6	I4
THR	7	8
SER	I6	I3
GLX	5	I3
PRO	6	4
GLY	2	5
ALA	5	6
VAL	-	4
MET	-	-
ILE	6	4
LEU	3	4
TYR	I	2
PHE	2	2
HIS	5	5
LYS	6	8
ARG	2	3
CYS	I4	6
Total	96	IOI

inase. The stoichiometry of the inhibition reaction was also established by gel filtration. The apparent molecular weight of approximately 34 000 is that expected for the formation of I : I molar inhibitor-proteinase complex.

The effects of Aspergillus- proteinase inhibitor upon other proteolytic enzymes were studied (table 3). The inhibition spectra showed that the Aspergillus -proteinase inhibitor was highly specific and had no activity toward other proteinases tested. The only exception was the proteinase B from baker's

Table 3. Inhibition of various proteinases by the two inhibitors

Enzyme	Trypsin inhibitor	Aspergillus-proteinase inhibitor
Trypsin	+	-
Chymotrypsin	+	-
Elastase	-	-
Subtilisin	-	-
Asp.oryzae proteinase	-	+
Proteinase B from yeast	-	+

yeast (partially purified preparation, QAE-Sephadex step according (6) was used). The proteolytic activity was assayed with Azocoll (7).

Prof. Holzer and his coworkers described several proteinase B inhibitors from yeasts (8, 9, I0). One of them (I^B3) has approximately the same isoelectric point and molecular weight as the Aspergillus -proteinase inhibitor described here (I0). It is interesting to note that the proteinase resembling yeast proteinase B was found recently in plant tissue (II). The presence of specific inhibitors in seeds which are acting upon proteinases of micro-organisms suggests the possible role of the proteinase inhibitors in the defence of plants against parasitic micro-organisms.

References

I. Richardson, M.: Phytochemistry I6, I59-I69 (I977).
2. Mikola, J., Suolinna, E.-M.: Arch.Biochem.Biophys. I44,

566-575 (I97I).

3. Yoshikawa, M., Iwasaki, T., Fujii, M., Oogaki, M.: J.Biochem. 79, 765-773 (I976).

4. Seidl, D.S., Abreu, H., Jaffe, W.G. : FEBS Letters 92, 245-250 (I978).

5. Kunitz, M. : J.Gen.Physiol. 30, 29I-3IO (I947).

6. Sanada, Y., Fujishiro, K., Tanaka, H., Katunuma, N. : Biochem.Biophys.Res.Commun. 86, 8I5-82I (I979).

7. Saheki, T., Holzer, H.: Eur.J.Biochem. 42, 62I-626(I974).

8. Betz, H., Hinze, H., Holzer, H.: J.Biol.Chem. 249, 45I5-452I (I974).

9. Meier, K., Müller, H., Holzer, H.: J.Biol.Chem. 254, 849I -8497 (I979).

IO. Matern, H., Weiser, U., Holzer, H.: Eur.J.Biochem. IOI, 325-332 (I979).

II. Wallace, W.: Biochim.Biophys.Acta 524, 4I8-427 (I978).

ISOLATION OF MEMBRANE GLYCOPROTEINS FROM PARAMYXOVIRUS SV5 FOR ANALYSIS OF THEIR CARBOHYDRATE STRUCTURE

Peter Prehm§
The Rockefeller University, New York, USA

Introduction

Paramyxoviruses contain two glycoproteins (HN and F) in their viral membrane. The HN protein has hemagglutinating and neuraminidase activity and is responsible for virus attachment to the host cell. The F protein mediates viral penetration, cell fusion and hemolysis. These properties are activated by the cleavage of a precursor F_o into two disulfide-linked glycopeptides F_1 and F_2 by a host cell protease (1). For the structural analysis of the carbohydrates these proteins had to be purified. Since the original method of purification (2) did not yield pure preparations of the HN and F proteins (3), a modified procedure was developed which provided both proteins in high quantity and apparent homogeneity (4).

Results

The glycoproteins HN and F were extracted from purified virus preparations by the nonionic detergent Emulphogen PC-720 and chromatographed on fetuin coupled Sepharose at a low ionic strength and 4°C (Fig. 1). The unadsorbed fraction represented pure F protein. Elution of enzymatically active HN protein was

§present address: Max Planck Institut für Biochemie
D-8033 Martinsried, FRG

Chemistry of Peptides and Proteins, Vol. 1

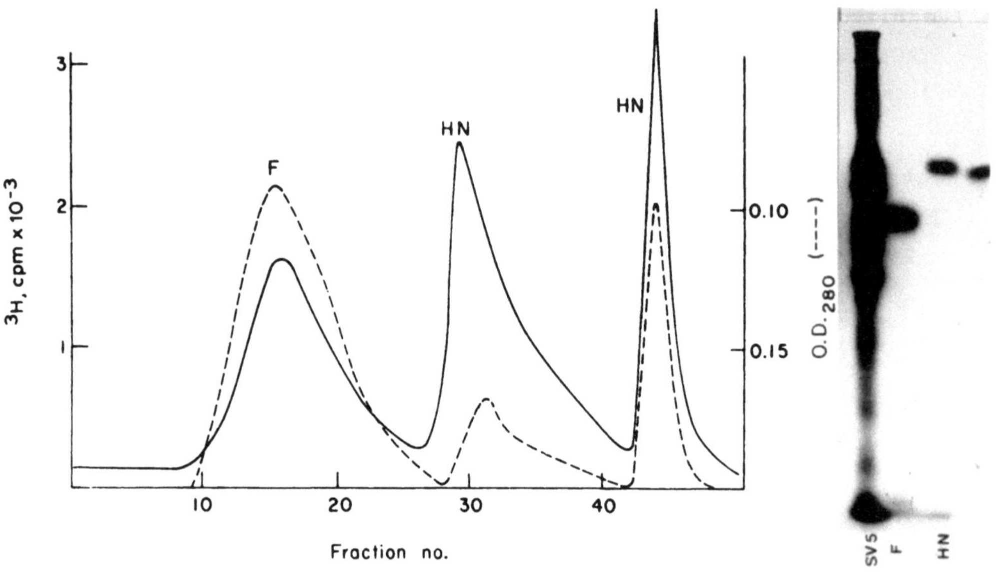

Fig. 1. Separation of SV5 glycoproteins on fetuin Sepharose and SDS polyacrylamide gel electrophoresis of isolated SV5 glycoproteins.

achieved by raising the ionic strength to 150 mM sodium chloride. A second fraction of the proteins was obtained by elution with borate buffer at pH 8.5. The purity of the prepartions was estimated by polyacrylamide gel electrophoresis (Fig. 1).

Further separtion of F_1 and F_2 polypeptides was impeded by extensive aggregation at protein concentration above 0.1 mg/ml. A similar aggregation had been observed with the HA glycoprotein of influenza virus (5). It seems to be a characteristic feature of membrane proteins containing a hydrophobic polypeptide region (6).

Aggregation of the F protein could be reduced by the following chemical modification (Fig. 2). The F protein was delipidated by precipitation with butanol, solubilized in 8 M urea and carbamidomethylated.

After reduction with dithiothreitol and addition of SDS a subsequent polyacrylamide gel electrophoresis revealed that

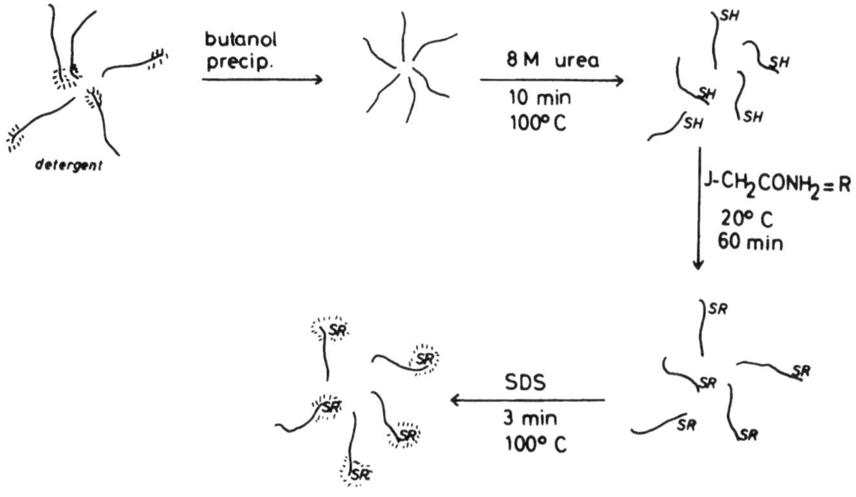

Fig. 2. Chemical modification of the F protein.

the chemical modification, indeed, reduced aggregation (Fig. 3). The two polypeptides could then be separated by hydroxylapatite chromatography (Fig. 4).

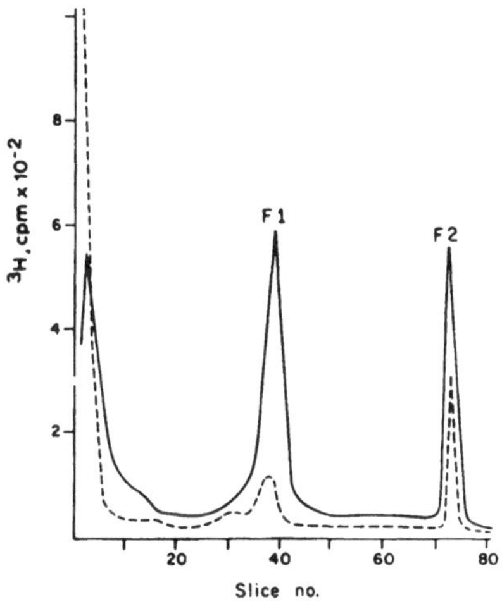

Fig. 3. Polyacrylamide gel electrophoresis of F protein before (——) and after (···) chemical modification.

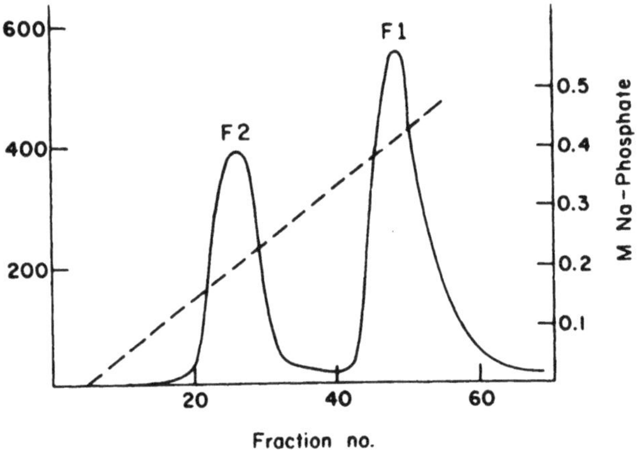

Fig. 4. Separation of F_1 and F_2 glycopeptides by hydroxylapatite chromatography.

The purified proteins were desalted by dialysis at low concentration and then concentrated by vacuum evaporation after addition of 1/3 volume i-butanol at 40°C. This prevents adsorption of the proteins to column supports of dialysis tubing and assures a better recovery.

The purified glycoproteins HN, F_1 and F_2 were subjected to carbohydrate analysis. Because only limited amounts were available and the conventional methods for quantitative analysis required 0.2 - 1 mg of material, a sensitive method was developed which allows the analysis of 5 - 10 µg glycoprotein (7) (Fig. 5).

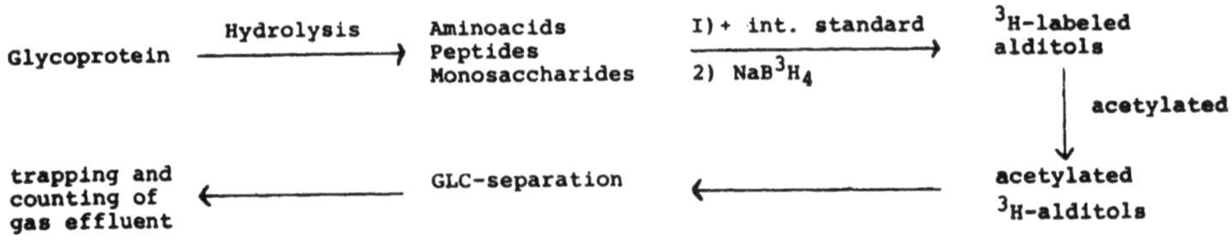

Fig. 5. Sensitive method for the analysis of carbohydrates.

Table 1. Carbohydrate composition of SV5 glycoproteins.

Compostion of glycopeptides	
Fucose	1
Mannose	3
Galactose	4
Glucosamine	2
Asparagine	1
Threonine	1
Alanine	<1
Glutamic acid	<1
Proline	<1
Sulfate	0.05

Carbohydrate chains in glycoproteins	number	%
HN	3	9.2
F_1	3	13.5
F_2	1	14.0

The glycoproteins were hydrolyzed and the monosaccharides were labeled by reduction with ^3H-$NaBH_4$. The radioactive alditols were acetylated and subjected to radio-gas chromatography. The quantities of the separated alditol acetates were determined after preparative fractionation of the gas effluent. This method was applied to the analysis of SV5 glycoproteins. The analytical values are listed in Table 1.

The SV5 glycopeptides were further analyzed for their carbohydrate structure by hydrazinolysis, deamination, methylation analysis and CrO_3-oxidation.

The three glycopeptides HN, F_1 and F_2 contain identical carbohydrate units as shown in Fig. 6. The structure is compared to known complex carbohydrate structures from other glycoproteins. The SV5 carbohydrate structure is unique in the sulfate esters substituted at the glucosamine residues and in the peripheral portion of two galactoses which are terminated by acetaldehyde in acetal linkages. The glycopeptide proved to be so labile in alkali that it could only be methylated by methyl iodide in DMSO/sodium after acidic desulfurization and deacetalization. A mild procedure is now available which allows the methylation of such labile carbohydrates (8). The unusual carbohydrate structure was also found on cell surface glycoproteins of uninfected host cells. Thus, the host cells seem to determine the carbohydrate structure of the

Fig. 6. Structure of the SV5 glycopeptide.

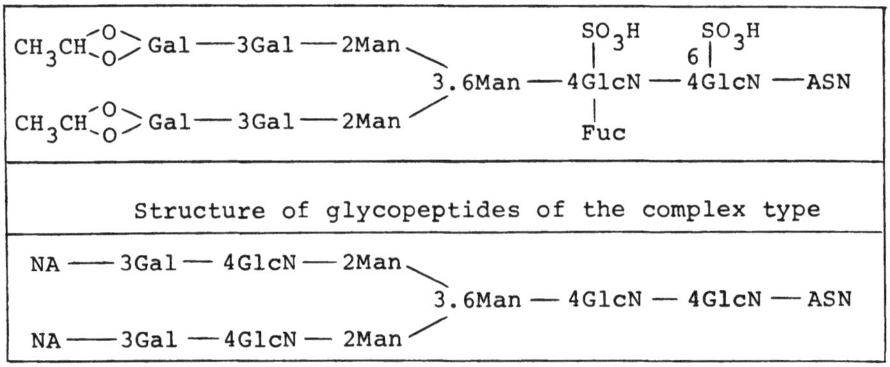

virus.

Paramyxoviruses are efficiently propagated only in certain epidermal cells, but not in fibroblast cells such as BHK-21 which apparently are unable to terminate the carbohydrate structure by acetalization. This terminal portion seems to be necessary for maturation of those viruses which contain a neuraminidase activity. The viral neuraminidase has been shown to function intracellularly (9) and may interfere with the release of the virus from those host cells which terminate the carbohydrate structure only by neuraminic acid and not by acetaldehyde. A scheme of paramyxovirus maturation is shown in Fig. 7.

The viral polypeptides are synthesized in the endoplasmatic reticulum, glycosylated in the Golgi and terminated by acetaldehyde on their way to the cell membrane. The viruses are liberated by budding from the plasma membrane. The infectious cycle is again initiated by adsorption through the hemagglutinating activity of the HN protein to uninfected host cells presenting neuraminic acid residues on their surface.

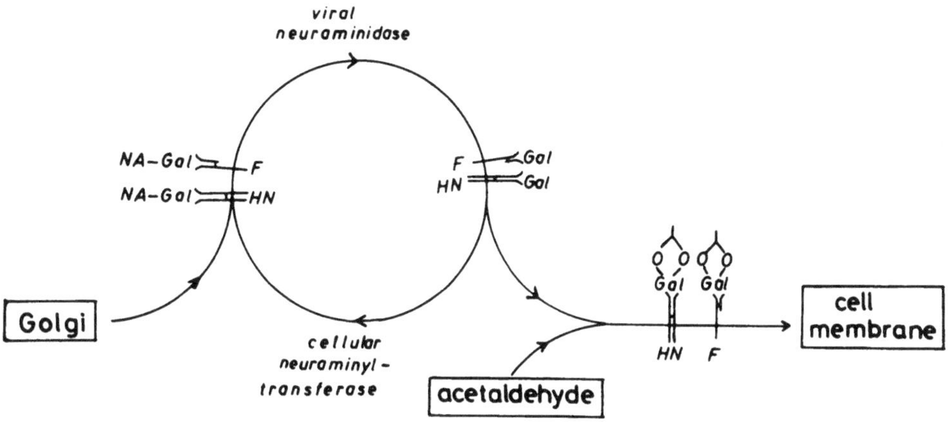

Fig.7. Maturation of Paramyxovirus SV5.

References

1. Choppin, P.W., Compans, R.W.: Comprehensive Virology (Fraenkel-Conrat,K., Wagner, R.R., eds.) Vol 4, 95-178 (1975).
2. Scheid,A., Choppin, P.W.: Virology 62,125-133 (1974).
3. Fukami, Y. Hosaka, Y. Yamamoto, K.: Febs. Lett. 114, 342-346 (1980).
4. Prehm, P., Scheid, A., Choppin, P.W.: J. Biol. Chem.254, 9669-9677 (1979).
5. Laver, W.G.: Virology 45, 275-288 (1971).
6. Klingenberg, M.: Nature 290, 449-454 (1981).
7. Prehm, P. Scheid,A.: J. Chromatogr. 166, 461-467 (1978).
8. Prehm, P.: Carbohydr. Res. 78, 372-374 (1980).
9. Merz, D.C., Prehm, P., Scheid, A., Choppin, P.W.: Virology 112, 296-305 (1981).

HAEMOGLOBIN POLYMORPHISM IN CHIRONOMUS (DIPTERA)

Primary Structure, Allergenic Determinants and Monomer-Dimer Equilibrium

Heinz Aschauer, Traute Kleinschmidt, Wolfgang Steer, Gerhard Braunitzer

Max-Planck-Institut für Biochemie, D-8033 Martinsried b.München

Introduction

The family Chironomidae (midges) belongs to the order Diptera. There are some 75 000 species spread over the whole world and in Europe alone over 1 500 species have been described (1). Most of them have extracellular haemoglobin in the larval stage. The haemoglobin concentration varies between 2-6 % depending upon stage of development and environmental factors. These haemoglobins have low molecular weights, in contrast to most invertebrate haemoglobins which have molecular weights of approximately 10^6 daltons. Chironomus haemoglobins form monomers and dimers with molecular weights of 16 000 and 32 000 daltons (2,3). The haemoglobin of one species can be separated in disc gel electrophoresis into many components (Fig. 1). In the case of Chironomus thummi thummi 12 different haemoglobins have been characterized (4). This polymorphism is found in all species depending upon developmental stage and environmental and population conditions. This high degree of polymorphism is interesting from a structural, fuctional,genetic and evolutionary standpoint.

Chemistry of Peptides and Proteins, Vol. 1

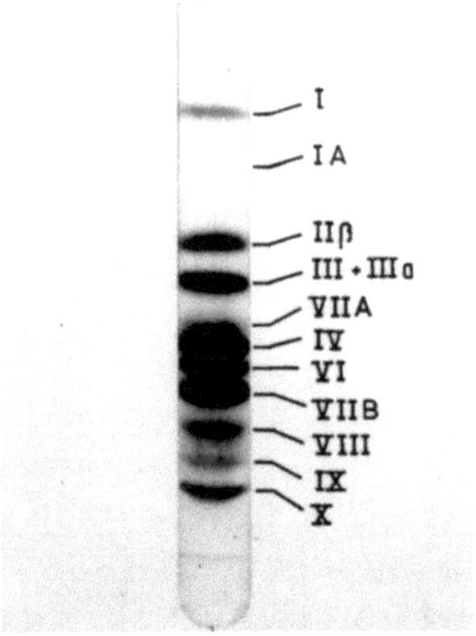

Fig. 1
Disc gel electrophoresis of the haemolymph of Chironomus thummi thummi. The 12 polymorphic haemoglobins are indicated.

The presence of giant chromosomes in the salivary glands of Chironomus enabled, for the first time, the localization of a structural gene (5,6). Comparison of the disc gel electrophoresis pattern of haemoglobins with the chromosome structure of parent and hybridized generations, showed that the haemoglobin gene was localized on one chromosome (chromosome III) out of a total of four. Single haemoglobin components were even localized to particular bands or groups of bands on that chromosome. The existance of the haemoglobin locus on one single chromosome indicates that the polymorphism of haemoglobin probalby evolved through gene duplication.

Results

The sequences of all 12 haemoglobins of Chironomus thummi thummi have been determined. The primary structures of CTT I and CTT IA are identical and here we suspect structural isomersim. All other 11 haemoglobins are different. The 12 haemoglobins represent the determination of the sequence of 1738 amino acids. Significant partial sequences are shown in figures 2,3 and 4.

Monomers:

CTT I	Gly-Pro-Ser-Gly-Asp-Gln-Ile-Ala-Ala-Ala-Lys-Ala-Ser-Trp-
CTT IA	Gly-Pro-Ser-Gly-Asp-Gln-Ile-Ala-Ala-Ala-Lys-Ala-Ser-Trp-
CTT IIIA	Val-Ala-Thr-Pro-Ala-Met-Pro-Ser-Met-Thr-Asp-Ala-Gln-Val-Ala-Ala-Val-Lys-Gly-Asp-Trp-
CTT III	Leu-Ser-Ala-Asp-Gln-Ile-Ser-Thr-Val-Gln-Ala-Ser-Phe-
CTT IV	Leu-Thr-Ala-Asp-Gln-Ile-Ser-Thr-Val-Gln-Ser-Ser-Phe-

Dimers:

CTT IIβ	Ala-Pro-Leu-Ser-Ala-Asp-Glu-Ala-Ser-Leu-Val-Arg-Gly-Ser-Trp-
CTT VI	Ala-Val-Leu-Thr-Thr-Glu-Gln-Ala-Asp-Leu-Val-Lys-Lys-Thr-Trp-
CTT VIIA	Ala-Pro-Leu-Ser-Ala-Asp-Gln-Ala-Ser-Leu-Val-Lys-Ser-Thr-Trp-
CTT VIIB	Ser-Pro-Leu-Thr-Ala-Asp-Glu-Ala-Ser-Leu-Val-Gln-Ser-Ser-Trp-
CTT VIII	Ala-Val-Thr-Pro-Met-Ser-Ala-Asp-Gln-Leu-Ala-Leu-Phe-Lys-Ser-Ser-Trp-
CTT IX	Asp-Pro-Val-Ser-Ser-Asp-Glu-Ala-Asn-Ala-Ile-Arg-Ala-Ser-Trp-
CTT X	Asp-Pro-Glu-Trp-His-Thr-Leu-Asp-Ala-His-Glu-Val-Glu-Gln-Val-Gln-Ala-Thr-Trp-

Fig.2. N-terminal sequence of Chironomus thummi thummi hemoglobin.

Monomers:

CTT I	-Trp-Ala-Pro-Val-Leu-Asp-Phe-Val-Phe-Ser-Thr-Leu-Lys-Asn-Glu-Leu (143)
CTT IA	-Trp-Ala-Pro-Val-Leu-Asp-Phe-Val-Phe-Ser-Thr-Leu-Lys-Asn-Glu-Leu (143)
CTT IIIA	-Trp-Asn-Ala-Thr-Val-Asp-Leu-Met-Phe-His-Val-Ile-Phe-Asn-Ala-Leu-Asp-Gly-Thr-Pro-Val (151)
CTT III	-Trp-Gly-Ala-Thr-Leu-Asp-Thr-Phe-Phe-Gly-Met-Ile-Phe-Ser-Lys-Met (136)
CTT IV	-Trp-Gly-Ala-Thr-Leu-Asp-Ala-Phe-Phe-Gly-Met-Val-Phe-Ala-Lys-Met (136)

Dimers:

CTT IIβ	-Trp-Thr-Gln-Gly-Leu-Asp-Asn-Val-Phe-Asn-Met-Met-Phe-Ser-Tyr-Leu (145)
CTT VI	-Trp-Thr-Leu-Ala-Leu-Asp-Thr-Thr-Tyr-Ala-Met-Leu-Phe-Ser-Ala-Met-Asp-Ser (147)
CTT VIIA	-Trp-Thr-Ala-Gly-Leu-Asp-Asn-Ile-Phe-Gly-Leu-Leu-Phe-Ala-Ala-Leu (145)
CTT VIIB	-Trp-Ser-Lys-Ala-Leu-Asp-Asn-Thr-Tyr/Phe-Ala-Ile-Val-Val-Pro-Arg-Leu (145)
CTT VIII	-Trp-Asp-Leu-Ala-Leu-Asn-Asn-Met-Phe-Phe-Tyr-Ile-Leu-His-Ala-Leu-Glu-Val-Ala-Gln (151)
CTT IX	-Trp-Thr-His-Gly-Leu-Asp-Asn-Ile-Phe-Gly-Met-Ile-Phe-Ala-His-Leu (145)
CTT X	-Trp-Asp-Asp-Ala-Phe-Asp-Lys-Ala-Phe-Ser-Val-Ile-Phe-Glu-Val-Leu-Glu-Ser (151)

Fig.3. C-terminal sequence of Chironomus thummi thummi hemoglobin.

	E7	E11
Monomers:		
α	-Ala-Gln-Val-Lys-Gly-His-Gly-Lys-Lys-Val-Ala-Asp-Ala-Leu-Thr-Asn-	
CTT I	-Pro-Asp-Phe-Ser-Lys-His-Ala-Gly-Arg-Val-Val-Gly-Leu-Phe-Ser-Glu-	
CTT IIIA	-Ala-Glu-Phe-Lys-Asp-Gln-Ala-Asp-Lys-Ile-Ile-Ala-Phe-Leu-Gln-Gly-	
CTT III	-Ala-Pro-Phe-Glu-Thr/Ile-His-Ala-Asn-Arg-Ile-Val-Gly-Phe-Phe-Ser-Lys-	
CTT IV	-Ala-Asp-Phe-Ser-Ala-His-Ala-Asn-Lys-Ile-Val-Gly-Phe-Phe-Ser-Lys-	
Dimers:		
CTT IIβ	-Gly-Gln-Phe-Ala-Thr-His-Ala-Gly-Arg-Ile-Val-Gly-Phe-Val-Ser-Glu-	
CTT VI	-Ala-Ala-Phe-Ala-Thr-His-Ala-Thr-Arg-Ile-Val-Ser-Phe-Leu-Ser-Glu	
CTT VIIA	-Gly-Ala-Phe-Ala-Thr-His-Ala-Thr-Arg-Ile-Val-Ser-Phe-Leu-Ser-Glu	
CTT VIIB	-Gly-Ala-Phe-Ala-Thr-His-Ala-Thr-Arg-Ile-Val-Ser-Phe-Leu-Ser-Glu-	
CTT VIII	-Ala-Asp-Phe-Ala-Val-His-Ser-Gly-Arg-Ile-Val-Gly-Phe-Phe-Ser-Glu-	
CTT IX	-Gly-Ala-Phe-Ala-Thr-His-Ala-Gly-Arg-Ile-Val-Gly-Phe-Ile-Ser-Glu-	
CTT X	-Ala-Asp-Phe-Ala-Val-His-Ala-Ser-Arg-Ile-Ile-Gly-Phe-Phe-Gly-Glu-	

Fig.4. Sequence of the E7 region (distal histidine); α = Human α chains.

A comparison of the primary structures of the globins shows that the differences are relatively large. The chain length varies between 136 - 151 residues. There are N-terminal extensions of up to 6 amino acids and deletions of up to 2 amino acids (Fig. 2). There are also C-terminal extensions as well as insertions and deletions within the chains. There are only 20 positions in which the amino acid is the same in all 12 components. If any two chains are compared there is about a 50% idenitiy. Despite this high variability there is a region, in the CD segment, which is pratically constant:

CD D
-Phe-Pro-Glu-Phe-Ala-Gly-Lys-Asp-Leu-

The reason for this constance is not clear. The X-ray structure of the monomer component CTT III, in various liganded forms, is known at a 1,4 Å resoltuion (7). The tertiary structure is very similar to that of myoglobin. However, there are significant differences in the haem complex as compared to the well known haem complexes of the vertebrate haemoglobins and myoglobins. The haem group is turned 180° around the α,γ meso axis. However NMR measurements on monomeric CTT III and CTT IV haemoglobins show a signal which indicates that up to 10% of the haem has its normal orientation (8,9). The haem is bound to the protein through the proximal histidine F8. The distal histidine E7 on the other hand, although present in all CTT haemoglobins, does not occupy the sixth coordination position of the iron in the deoxy form but forms a salt bridge with a propionic acid group of the haem. On the distal side is residue E11, isoleucine which, because of its bulky side chain, pushes the histidine out of the haem pocket.

The Chironomus haemoglobin is a potent antigen in man (10). Allergic reactions occur in people who have had contact with insect larvae, either by the preparation and use of fish food or in research laboratories. Using a Radio-Allergo-Sorbent-Test (RAST) we could detect specific Ig E antibodies against Chironomus thummi thummi haemoglobins (CTT IA was not investigated). It was also shown that the haemoglobin fractions were active when a larvae extract was separated using gel chromatography. The allergic action of the separated haemoglobin components was also confirmed by a skin test. In one component, the dimer CTT VI, a region was identified in which at least one antigenic determinant is present. This region contains residues 63 - 98 (E11 - FG2). Cleavage of this region at the C-terminal of Lys 89 (F2) caused complete loss of antigenic activity. One could therefore expect the antigenic determinant to be in the region of the cleavage point and have the sequence:

F1 90
Leu-Ala-Lys-Glu-Leu-Ala- .

The haemoglobins of Chironomus thummi thummi form monomers and dimers. The dimers can dissociate, depending upon the pH value, haemoglobin concentration and ligand (Fig. 5) (11,12).

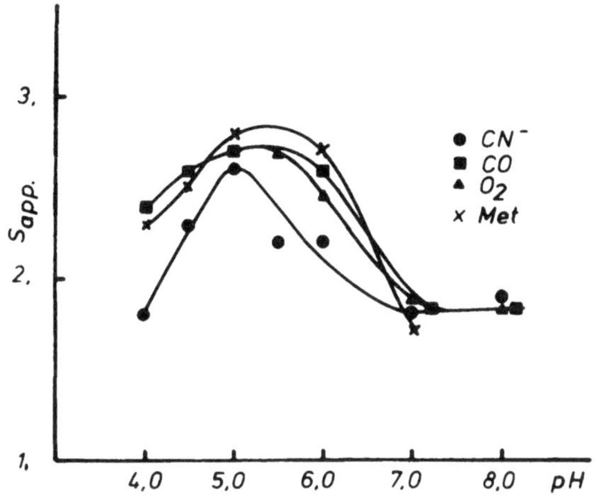

Fig. 5. Monomer-dimer equilibrium of the haemoglobin component X of Chironomus thummi thummi and its dependence upon pH and ligand.

Concerning the pH-dependence of the equilibrium, it is different for each haemoglobin component.
The CTT IIß component exists as a dimer in the cyanomet form until pH 8-9, while the CTT X component is a dimer up to pH 7.0. The pH dependence of the monomer-dimer equilibrium gives an important indication as to the nature of the binding sites. We assume that the ionic interactions play an important role. To further investigate the salt bridge the side chains of histidine (13), lysine (14) and arginine (15) were modified. After selective reaction of histidine residues with diethylpyrocarbonate, we found that only CTT IIß remained as a dimer while all other compoenents were present as monomers. In position G19, CTT IIß has a lysine whereas all other dimers have histidine in this position. Conversely, only CTT IIß existed as a monomer after reaction of lysine residues with maleic anhydride. We believe that position G19 (His, Lys) is a residue that forms a salt bridge and thereby has an important effect on the dimerisation reaction. This would explain the stability of the CTT II/3 dimer at high pH values. As the dimers of CTT haemoglobins are formed from identical, even numbered subunits, they should be formed by isologous association as described by Monod (16). By comparing the primary structure, we believe that glutamic acid G7 is the isologous partner in the salt bridge. This suspicion is strengthened by the observation that some of the monomer components also have histidine in position G19 but no glutamic acid at G7.

Discussion

In our model, one can see, in the quarternary structure, similarities to the ß-ß or α-α subunits of mammalian haemoglobins, but not in the binding regions. However they differ considerably, through tetramer formation, in the inclination of the subunits in the direction of the G helix (17,18). Assuming that salt bridges in the CTT dimers in this region are formed, an inclination of the subunits and therefore tetramer formation is impossible

Goodman, Moore and Matsuda deduced from phylogenetic data that the heterotetramer evolved via a preceding homotetramer (19). They further thought that the homotetramer was similar to the ß chains of mammalian haemoglobin as only ß chains and not α chains can form homotetramers. Have we discovered, in the homodimers of Chironomus haemoglobins, a further ancestor of the homotetramers and a still earlier developmental form of haemoglobin?

The phylogenetic tree of haemoglobin shows a very early separation of the insect haemoglobins from others. This separation must have taken place approximately 650 million years ago, that is, before the accepted formation of hematetramers that took place between the separation of myoglobins about 500 million years ago and the separation into α and ß chains about 400 million years ago. In our work we could find no relationship between the binding sites. Therefore we assume that the dimers of Chironomus thummi thummi haemoglobins are not ancestors of the mammalian heterotetramer haemoglobins and that the binding sites have no phylogenetic relationship. Further the dimer binding sites have different characteristics and were therefore not able to evolve into tetrameric haemoglobins.

References

1. Thienemann, A., in: Die Binnengewässer, Vol. XX: Chironomus, Schweizerbart'sche Verlagsbuchhandlung, Stuttgart, 1954
2. Svedberg, T., Erikson-Quensel, I.B: J. Am. Chem. Soc. 56 1700 (1934).
3. Braunitzer, G., Braun, V.: Hoppe-Seyler's Z. Physiol. Chem. 340, 80 (1965).
4. Braun, V., Chrichton, R.R., Braunitzer, G.: Hoppe-Seyler's Z. Physiol. Chem. 349, 197 (1968).
5. Tichy, H.: Chromosoma (Berlin) 29, 131 (1970).
6. Tichy, H.: J. Mol. Evol. 6, 39 (1975).
7. Steigemann, W., Weber, E.: J. Mol. Biol. 127, 309 (1979).

8. Ribbing, W., Krümpelmann, D., Rüterjans, H.: FEBS Letters 92, 105 (1978).
9. La Mar, G.N., Smith, K.M., Gersonde, K., Sick, H., Overkamp, M.: J. Biol. Chem. 255, 66 (1979).
10. Baur, X., Ziegler, D., Reichenbach-Klinke, H.H., Aschauer, H., Braunitzer, G.: Naturwiss. 67, 365 (1980).
11. Behlke, J., Scheler, W.: Eur. J. Biochem. 3, 152 (1967).
12. Behlke, J., Pfeil, W., Blanck, J., Ristan, O., Rein, H., Müller, K., Mohr, P., Scheler, W.: Acta Biol. Med. Ger. 34, 365 (1975).
13. Melchior, W.B., Jr., Fahrney, D.: Biochemistry 9, 251 (1970).
14. Kowalski, D., Laskowski, M., Jr.: Biochemistry 11, 3451 (1972).
15. Dietl, T., Tschesche, H.: Hoppe-Seyler's Z. Physiol. Chem. 357, 657 (1976).
16. Monod, J., Wyman, J., Changeux, J.P.: J. Mol. Biol. 12, 88 (1965).
17. Perutz, M.F.: Nature 228, 726 (1970).
18. Perutz, M.F., Rossmann, M.G., Gullis, A.F., Muirhead, H., Will, G., Worth, A.C.T.: Nature 185, 416 (1960).
19. Goodman, M., Moore, G.W. Matsuda, G.: Nature 253, 603 (1975).

HIGH ALTITUDE RESPIRATION OF THE BAR-HEADED GOOSE (ANSER INDICUS) AND THE DIFFERENT EVOLUTION OF THE α- AND β-CHAINS IN AVIAN HEMOGLOBINS

Investigations on hemoglobins of the Grey-Lag Goose (Anser anser) (1), Bar-headed Goose (Anser indicus), Canada Goose (Branta canadensis) and Ostrich (Struthio camelus)

Gerhard Braunitzer and Walter Oberthür

Max-Planck-Insitut für Biochemie
D-8033 Martinsried bei München

Introduction

We have investigated the high altitude respiration of the Bar-headed Goose (Anser indicus) and have described the primary structure of the main components of the hemoglobin of the Grey-Lag Goose (Anser anser), Bar-headed Goose (Anser indicus), Canada Goose (Branta canadensis) and Ostrich (Struthio camelus).

To determine the primary structure of the hemoglobins, the α- and β-chains were fragmented and the products isolated. The complete amino acid sequences of the α- and β-chains were de-

Abbreviations:

Quadrol = N,N,N',N'-Tetrakis(2- hydroxypropyl)ethylendiamine,
Reagent I = sodium 1-(isothiocyanato)benzenesulphonate,
Reagent IV = trisodium 7-(isothiocyanato)naphthalene-1,3,5-trisulphonate,
IPP = myoinositol-1,3,4,5,6-pentaphosphate,
DPG = 2,3-diphosphoglycerate.

Chemistry of Peptides and Proteins, Vol. 1

termined by automatic sequencing of the intact chains and isolated fragments. The different evolution of the α- and β-chains and the problem of high altitude breathing of the Bar-headed Goose are discussed (2,3).

Methods

Crude hemoglobin was prepared by lysing erythrocytes and removing cell membranes and nuclei. After cleaving off the heme-group, the α- and β-chains of the main component were isolated using ion exchange chromatography on CM-cellulose((Servacel CM-52 (2)).

Fragments for sequence analysis were prepared and isolated using the following methods as described earlier (2,3): cyanogen bromide cleavage (4); tryptic digestion and limited tryptic digestion after blocking the ε-amino-groups of lysine with maleic anhydride (5); acid hydrolysis of Asp-Pro-bonds (6); digestion with Staphylococcus aureus protease V8 (7).

Amino acid analysis: analyses were carried out using a Beckman Model 121 C amino acid analyser (Palo Alto, Calif., USA).

Amino acid sequence analysis: amino acid sequences were determined automatically in a Beckman Sequenator, Model 890 C (8). Lysin peptides (0,5 μmol) were coupled with 8 μmol of reagent I (9) or reagent IV (10) and sequenced using a Quadrol programme. Arginine peptides were sequenced using a 1 M diethylaminopropine buffer (11,12). Amino acid derivatives from the Sequenator were converted using 3 M trifluoracetic acid at 80° C for 14 minutes and the resulting phenylthiohydantoin amino acids identified using either thin layer chromatography or gradient elution HPLC (13).

Results and Discussion

Biology of the Geese and of the Ostrich

The Grey-Lag Goose (Anser anser) and the Bar-headed Goose (Anser indicus) belong the the genus Anser (real geese), the Canada Goose (Branta canadensis) to the genus Branta (see geese in the family Anatidae (ducks). According to Johnson (14) the important phylogenetic steps in the evolution of the existing geese took place at the Tertiary period.
The Grey-Lag Goose (1) lives in the milder areas of the Boreal Zone of the Palaearctic. Today the breeding area of the Grey-Lag Goose covers a wide zone stretching from Iceland through northern Eurasia to Salachin. In Europe they breed mostly in low-lying areas and on the cost. In contrast, the Bar-headed Goose breeds on highland lakes in central Asia (4000-6000 m) and crosses the Himalaya to winter in India (15-17). The Canada Goose is the most common goose in America and gives rise to most of the subspecies. They inhabit, like the Grey-Lag Goose, low-lying areas.
The African Ostrich (Struthio camelus) is the only living member of the suborder Ostriches (Struthiones, family Struthionidae) whose relatives have inhabited large areas of Asia, Europe and Africa since the Eocene period. Today it is the largest living bird but its position in the family tree of birds is still disputed.

Physiology of the Avian Hemoglobins.

In all birds (18,19) IPP is the major allosteric effector of hemoglobins (ca. 5 µmol) together with low levels of ATP, ADP, DPG and other organic phosphates. Investigations on the affinity of oxygen for hemoglobins in the Grey-Lag Goose, Bar-headed Goose and Canada Goose showed, in the absence of phosphate, only small differences but on the addition of phosphate the differences became highly significant (20). The difference between the $P_{50}O_2$ for Canada and Bar-headed Goose hemoglobins, without phosphate, was ca. 2 torr but in the presence of phosphate 12 torr. The $P_{50}O_2$ for the blood of the Bar-headed Goose, a highland bird, is 29.7 torr which indicates a very high oxygen affinity. In contrast, the blood of the lowland Canada and Grey-Lag Geese have $P_{50}O_2$ values of 39.5 and 42.0 torr respectively, indicating a much lower oxygen affinity (20). These results correlate with investigations on hemoglobins from the Llama and Camel, foetal and adult humans (21).

The Primary Structures

The amino acid sequence of the α- and β-chains of the major component of hemoglobins from Anser anser, Anser indicus, Branta canadensis and Struthio camelus are given in figures 1 to 4.

Comparing the homology of the hemoglobin chains of various birds, including chicken (22,23) the following differences can be seen:

Table 1:

	A Amino acids			B Nucleotides		
	α	β	α/β	α	β	α/β
GG/BG	3	1	3	3	1	3
CG/BG	5	3	1.67	6	3	2
CG/GG	2	3	0.67	3	3	1
OS/GG	15	4	3.75	16	4	4
OS/BG	16	5	3.2	17	5	3.4
OS/CG	16	6	2.67	17	6	2.84
CH/GG	30	6	5	39	6	6.5
CH/BG	32	7	4.71	41	7	5.86
CH/CG	32	5	6.4	41	5	8.2
CH/OS	29	7	4.15	39	7	5.57

CH = Chicken, GG = Grey-Lag Goose, BG = Bar-headed Goose, CG = Canada Goose, OS = Ostrich

In A the number of amino acid exchanges are shown, in B the corresponding number of nucleotide exchanges, assuming a minimum of base exchanges. For both amino acid and nucleotide exchanges, the relationship between the differences in the α- and β-chains is calculated as a factor α/β.

In the α-chains, the majority of the mutations are in the N-terminal region, half of the chain whereas the C-terminal is relatively constant. None of the α-chains contain tryptophan. In the β-chains, the majority of the mutations also lie in the N-terminal region. There are no substitutions in the D to G helices. Three mutations are present in the interhelical regions and the beginning of the H helix. Only very few of the total mutations are found in the β-chains.

Phosphate Binding Sites and Unequal Evolution of the α- and β-Chains

The oxygen affinity of hemoglobins is controlled by phosphate (18), in birds IPP. The binding sites were determined by comparing the X-ray analysis of human hemoglobin (24) with the amino acid sequence of the β-chain of chicken hemoglobin. In all birds that have been investigated (except in one case where a conservative exchange was reported) no subtstitutions have been found in these positions.
Until now it was thought that the mutation rate in the β-chain of hemoglobin was somewhat higher than that of the α-chain. However, by comparing the primary structure of the α- and β-chains of five birds, we have found a significantly lower mutation rate in the β-chains (see table).
How can one explain this surprising result? Family trees are constructed using palaeontological and primary structural data. The availability of sequence data enables quantitative comparisons to be made. Thus, for hemoglobin, a mutation rate of one every 6-10 million years has been calculated (25). As fossil finds of birds are very rare, paleontologists can say very little about the family tree. The common ancestor of, for example, the chicken and the goose, is thought to be 80-120 million years old. By a mutation rate of 6 million years, this

would mean a normal mutation rate for the α-chain (2 x 15 exchanges ≙ 90 million years) but for the β-chain, however, a significantly reduced, and therefore an abnormal mutation rate. A possible explanation for this could be the increased phosphate binding site in avian hemoglobins (12 molecules instead of 6 as in mammals). We believe that these different rates of evolution are caused by the requirement for IPP which fixes the structure of the β-chain through its 12 binding sites (23). The rate of evolution of the β-chain will be reduced in comparison to the α-chain which has no binding sites and will therefore normally evolve.

High Altitude Respiration of the Bar-headed Goose

The higher oxygen affinity of the hemoglobin of the Bar-headed Goose compared to those of the Grey-Lag Goose and Canada Goose can not be caused by different phosphate binding sites, as it is the case, for example, in the Llama/Camel and HbA/HbF, as here no differences were found. Of all the mutations the exchange α63 (E12) Val in the α-chain of the Bar-headed Goose causes the most drastic alteration to the structure of hemoglobin. This surprising finding is a result of a written discussion with M. F. Perutz. "The position α63 E12 is occupied by Ala in all mammals (except the rabbit where Ser was found) and in all other birds. The β-methyl group of this alanine touches the C=O carbon of alanine α21 B2 so that in the normal structure there is no room for a longer side chain in this position. The exchange of Ala for Val therefore causes the helices E and B to be pushed at least 1.5 Å apart which represents a large alteration in the tertiary structure."
It should be noted that the E-helix in E7 carries the distal histidine of the heme complex. Additional mutations at the α and β binding sites, together with the substitution described above, probably cause a different conformational change in hemoglobin when IPP is bound and thus allows a molecular interpretation of the high altitude respiration of the Bar-headed Goose.

```
H                 Pro                 10     Ala-Ala-Trp-Gly    Val    Ala  20    Gly                Ala     30
GGα  Val-   -Leu-Ser-Ala-Ala-Asp-Lys-Thr-Asn-Val-Lys-Gly-Val-Phe-Ser-Lys-Ile-Gly-Gly-His-Ala-Glu-Glu-Tyr-Gly-Ala-Glu-Thr-Leu-Glu-
GGβ  Val-His-Trp-Ser-Ala-Glu-Glu-Lys-Gln-Leu-Ile-Thr-Gly-Leu-Trp-Gly-Lys-Val-Asn-     -Val-Ala-Asp-Cys-Gly-Ala-Glu-Ala-Leu-Ala-
H           Leu-Thr-Pro          Ser  10 Val     Ala                                 20 Asp-Glu-Val     Gly           Gly

               Leu-Ser-Phe     Thr     40                           Ser  50                               Val
     -Arg-Met-Phe-Thr-Ala-Tyr-Pro-Gln-Thr-Lys-Thr-Tyr-Phe-Pro-His-Phe-   -Asp-Leu-Gln-His-               -Gly-Ser-Ala-Gln-Ile-
     -Arg-Leu-Leu-Ile-Val-Tyr-Pro-Trp-Thr-Gln-Arg-Phe-Phe-Ser-Ser-Phe-Gly-Asn-Leu-Ser-Ser-Pro-Thr-Ala-Ile-Leu-Gly-Asn-Pro-Met-Val-
      30       Val                  40        Glu          Asp          Thr  51 Asp     Val-Met             Lys  60

         Gly          60          Asp         Thr-Asn     70 Ala     Val        Met-Pro-Asn      80    Ala
     -Lys-Ala-His-Gly-Lys-Lys-Val-Ala-Ala-Ala-Leu-Val-Glu-Ala-Val-Asn-His-Ile-Asp-Asp-Ile-Ala-Gly-Ala-Leu-Ser-Lys-Leu-Ser-Asp-Leu-
     -Arg-Ala-His-Gly-Lys-Lys-Val-Leu-Thr-Ser-Phe-Gly-Asp-Ala-Val-Lys-Asn-Leu-Asp-Asn-Ile-Lys-Asn-Thr-Phe-Ala-Gln-Leu-Ser-Glu-Leu-
      Lys                     Gly-Ala  71 Ser    Gly-Leu-Ala-His          80 Leu     Gly          Thr         90

           His  90                              Leu 101 Ser        Leu        Thr-Leu 110 Ala     Leu     Ala-Glu-Phe
     -His-Ala-Gln-Lys-Leu-Arg-Val-Asp-Pro-Val-Asn-Phe-Lys-Phe-Leu-Gly-His-Cys-Phe-Leu-Val-Val-Val-Ala-Ile-His-His-Pro-Ser-Ala-Leu-
     -His-Cys-Asp-Lys-Leu-His-Val-Asp-Pro-Glu-Asn-Phe-Arg-Leu-Leu-Gly-Asp-Ile-Leu-Ile-Ile-Val-Leu-Ala-Ala-His-Phe-Ala-Lys-Glu-Phe-
                              100                         Asn-Val 110 Val-Cys          His        Gly 120

           Ala 121                      129 Ala-Ser    Ser               Ser     140
     -Thr-Pro-Glu-Val-His-Ala-Ser-Leu-Asp-Lys-Phe-Leu-Cys-Ala-Val-Gly-Thr-Val-Leu-Thr-Ala-Lys-Tyr-Arg
     -Thr-Pro-Glu-Cys-Gln-Ala-Ala-Trp-Gln-Lys-Leu-Val-Arg-Val-Val-Ala-His-Ala-Leu-Ala-Arg-Lys-Tyr-His
           Pro-Val             Tyr 131     Val     Ala-Gly         Asn 140        His
```

Fig. 1: Comparison of the α- and β-chains of the Grey-Lag Goose and Human hemoglobins.

```
GG                                    10                        Gly    20                         30
BGα   Val-   -Leu-Ser-Ala-Ala-Asp-Lys-Thr-Asn-Val-Lys-Gly-Val-Phe-Ser-Lys-Ile-Ser-Gly-His-Ala-Glu-Glu-Tyr-Gly-Ala-Glu-Thr-Leu-Glu-
BGβ   Val-His-Trp-Ser-Ala-Glu-Glu-Lys-Gln-Leu-Ile-Thr-Gly-Leu-Trp-Gly-Lys-Val-Asn-      -Val-Ala-Asp-Cys-Gly-Ala-Glu-Ala-Leu-Ala-
GG                                  10                                            20

                                    40                                     50
     -Arg-Met-Phe-Thr-Ala-Tyr-Pro-Gln-Thr-Lys-Thr-Tyr-Phe-Pro-His-Phe-   -Asp-Leu-Gln-His-                -Gly-Ser-Ala-Gln-Ile-
     -Arg-Leu-Leu-Ile-Val-Tyr-Pro-Trp-Thr-Gln-Arg-Phe-Phe-Ser-Ser-Phe-Gly-Asn-Leu-Ser-Ser-Pro-Thr-Ala-Ile-Leu-Gly-Asn-Pro-Met-Val-
      30                                  40                                     50                                    60

                  60         Ala                     70                                    80
     -Lys-Ala-His-Gly-Lys-Lys-Val-Val-Ala-Ala-Leu-Val-Glu-Ala-Val-Asn-His-Ile-Asp-Asp-Ile-Ala-Gly-Ala-Leu-Ser-Lys-Leu-Ser-Asp-Leu-
     -Arg-Ala-His-Gly-Lys-Lys-Val-Leu-Thr-Ser-Phe-Gly-Asp-Ala-Val-Lys-Asn-Leu-Asp-Asn-Ile-Lys-Asn-Thr-Phe-Ala-Gln-Leu-Ser-Glu-Leu-
                                    70                                     80                                    90

              90                                  100                                    110
     -His-Ala-Gln-Lys-Leu-Arg-Val-Asp-Pro-Val-Asn-Phe-Lys-Phe-Leu-Gly-His-Cys-Phe-Leu-Val-Val-Val-Ala-Ile-His-His-Pro-Ser-Ala-Leu-
     -His-Cys-Asp-Lys-Leu-His-Val-Asp-Pro-Glu-Asn-Phe-Arg-Leu-Leu-Gly-Asp-Ile-Leu-Ile-Ile-Val-Leu-Ala-Ala-His-Phe-Ala-Lys-Glu-Phe-
                               100                                    110                                    120

        Pro 120                               130                                    140
     -Thr-Ala-Glu-Val-His-Ala-Ser-Leu-Asp-Lys-Phe-Leu-Cys-Ala-Val-Gly-Thr-Val-Leu-Thr-Ala-Lys-Tyr-Arg
     -Thr-Pro-Asp-Cys-Gln-Ala-Ala-Trp-Gln-Lys-Leu-Val-Arg-Val-Val-Ala-His-Ala-Leu-Ala-Arg-Lys-Tyr-His
            Glu                  130                                    140
```

Fig. 2: Comparison of the α- and β-chains of the Bar-headed Goose and Grey-Lag Goose hemoglobins.

```
GG             Ala-Ala              10        Val            Gly-Gly  20                           30
OSα   Val-   -Leu-Ser-Gly-Thr-Asp-Lys-Thr-Asn-Val-Lys-Gly-Ile-Phe-Ser-Lys-Ile-Ser-Ser-His-Ala-Glu-Glu-Tyr-Gly-Ala-Glu-Thr-Leu-Glu-
OSβ   Val-Gln-Trp-Ser-Ala-Glu-Glu-Lys-Gln-Leu-Ile-Ser-Gly-Leu-Trp-Gly-Lys-Val-Asn-       -Val-Ala-Asp-Cys-Gly-Ala-Glu-Ala-Leu-Ala-
GG        His                     10     Thr                              20

           Thr-Ala            40                               Gln  50
      -Arg-Met-Phe-Ile-Thr-Tyr-Pro-Gln-Thr-Lys-Thr-Tyr-Phe-Pro-His-Phe-   -Asp-Leu-His-His-                  -Gly-Ser-Ala-Gln-Ile-
      -Arg-Leu-Leu-Ile-Val-Tyr-Pro-Trp-Thr-Gln-Arg-Phe-Phe-Ala-Ser-Phe-Gly-Asn-Leu-Ser-Ser-Pro-Thr-Ala-Ile-Leu-Gly-Asn-Pro-Met-Val-
        30                              40        Ser                       50                                        60

                 60           Ala         Val         70                        Ala         80
      -Lys-Ala-His-Gly-Lys-Lys-Val-Ala-Asn-Ala-Leu-Ile-Glu-Ala-Val-Asn-His-Ile-Asp-Asp-Ile-Ser-Gly-Ala-Leu-Ser-Lys-Leu-Ser-Asp-Leu-
      -Arg-Ala-His-Gly-Lys-Lys-Val-Leu-Thr-Ser-Phe-Gly-Asp-Ala-Val-Lys-Asn-Leu-Asp-Asn-Ile-Lys-Asn-Thr-Phe-Ala-Gln-Leu-Ser-Glu-Leu-
                                      70                                   80                                      90

             90                               Phe 101    His                         110
      -His-Ala-Gln-Lys-Leu-Arg-Val-Asp-Pro-Val-Asn-Phe-Lys-Leu-Leu-Gly-Gln-Cys-Phe-Leu-Val-Val-Val-Ala-Ile-His-His-Pro-Ser-Ala-Leu-
      -His-Cys-Asp-Lys-Leu-His-Val-Asp-Pro-Glu-Asn-Phe-Arg-Leu-Leu-Gly-Asp-Ile-Leu-Ile-Ile-Val-Leu-Ala-Ala-His-Phe-Thr-Lys-Glu-Phe-
                               100                                  110                                  Ala 120

          120                            130        Gly-Thr                     140
      -Thr-Pro-Glu-Val-His-Ala-Ser-Leu-Asp-Lys-Phe-Leu-Cys-Ala-Val-Ser-Ala-Val-Leu-Thr-Ala-Lys-Tyr-Arg
      -Thr-Pro-Glu-Cys-Gln-Ala-Ala-Trp-Gln-Lys-Leu-Val-Arg-Val-Val-Ala-His-Ala-Leu-Ala-Arg-Lys-Tyr-His
                              130                                140
```

Fig. 3: Comparison of the α- and β-chains of the Ostrich and Grey-Lag Goose hemoglobins.

```
GG                                    10                               20     Glu                              30
CG α  Val-   -Leu-Ser-Ala-Ala-Asp-Lys-Thr-Asn-Val-Lys-Gly-Val-Phe-Ser-Lys-Ile-Gly-Gly-His-Ala-Asp-Glu-Tyr-Gly-Ala-Glu-Thr-Leu-Glu-
CG β  Val-His-Trp-Thr-Ala-Glu-Glu-Lys-Gln-Leu-Ile-Thr-Gly-Leu-Trp-Gly-Lys-Val-Asn-       -Val-Ala-Asp-Cys-Gly-Ala-Glu-Ala-Leu-Ala-
GG              Ser                10                                    20

                Thr                40                                50
     -Arg-Met-Phe-Val-Ala-Tyr-Pro-Gln-Thr-Lys-Thr-Tyr-Phe-Pro-His-Phe-   -Asp-Leu-Gln-His-                -Gly-Ser-Ala-Gln-Ile-
     -Arg-Leu-Leu-Ile-Val-Tyr-Pro-Trp-Thr-Gln-Arg-Phe-Phe-Ser-Ser-Phe-Gly-Asn-Leu-Ser-Ser-Pro-Thr-Ala-Ile-Leu-Gly-Asn-Pro-Met-Val-
      30                                40                               50                               60

                    60                              70                              80
     -Lys-Ala-His-Gly-Lys-Lys-Val-Ala-Ala-Ala-Leu-Val-Glu-Ala-Val-Asn-His-Ile-Asp-Asp-Ile-Ala-Gly-Ala-Leu-Ser-Lys-Leu-Ser-Asp-Leu-
     -Arg-Ala-His-Gly-Lys-Lys-Val-Leu-Thr-Ser-Phe-Gly-Asp-Ala-Val-Lys-Asn-Leu-Asp-Asn-Ile-Lys-Asn-Thr-Phe-Ala-Gln-Leu-Ser-Glu-Leu-
                                   70                              80                              90

                 90                            100                              110
     -His-Ala-Gln-Lys-Leu-Arg-Val-Asp-Pro-Val-Asn-Phe-Lys-Phe-Leu-Gly-His-Cys-Phe-Leu-Val-Val-Val-Ala-Ile-His-His-Pro-Ser-Ala-Leu-
     -His-Cys-Asp-Lys-Leu-His-Val-Asp-Pro-Glu-Asn-Phe-Arg-Leu-Leu-Gly-Asp-Ile-Leu-Ile-Ile-Val-Leu-Ala-Ala-His-Phe-Ala-Lys-Asp-Phe-
                               100                              110                              120 Glu

             120                          130                              140
     -Thr-Pro-Glu-Val-His-Ala-Ser-Leu-Asp-Lys-Phe-Leu-Cys-Ala-Val-Gly-Thr-Val-Leu-Thr-Ala-Lys-Tyr-Arg
     -Thr-Pro-Asp-Cys-Gln-Ala-Ala-Trp-Gln-Lys-Leu-Val-Arg-Val-Val-Ala-His-Ala-Leu-Ala-Arg-Lys-Tyr-His
             Glu             130                             140
```

Fig. 4: Comparison of the α- and β-chains of the Canada Goose and Grey-Lag Goose.

Figures 1-4

The amino acid sequences of the α- and β-chains of the hemoglobin main components from the Grey-Lag Goose, Bar-headed Goose, Canada Goose and Ostrich are compared. The sequences are aligned in pairs to show homology and only substituted positions are shown in the second sequence. Numbers above the main sequence indicate positions in the α-chain and below in the β-chain.

H = Human, GG = Grey-Lag Goose, BG = Bar-headed Goose, CG = Canada Goose, OS = Ostrich.

The work described here is the basis for the complete explanation of the molecular mechansim of the high altitude respiration of the Bar-haeded Goose.

We thank Prof. Irene Würdinger, Hildesheim, for valuable discussions.

References

1. Lorenz, K., Kalas, S., Kalas, K.: Das Jahr der Graugans, Piper & Co. Verlag, München (1978).

2. Braunitzer, G., Oberthür, W.: Hoppe-Seyler's Z. Physiol. Chem. 360, 679 (1979).

3. Oberthür, W., Voelter, W., Braunitzer, G.: Hoppe-Seyler's Z. Physiol. Chem. 361, 969-975 (1980).

4. Gross, E., Wittkop, B.: J. Am. Chem. Soc. 83, 1510-1511 (1961).

5. Fritz, H., Fink, E., Gebhard, M., Hochstrasser, K.,Werl, E.: Hoppe-Seyler's Z. Physiol. Chem. 350, 933 (1969).

6. Jauregui-Adell, J., Marti, J.: Anal. Biochem. 69, 468-473 (1975).

7. Houmard, J., Drapeau, G.R.: Proc. Natl. Acad. Sci. 69, 3506 (1972).

8. Edman, P., Begg, G.: Eur. J. Biocnem. 1, 80-91 (1967).

9. Braunitzer, G., Schrank, B., Ruhfus, A, A: Hoppe-Seyler's Z. Physiol. Chem. 351, 1589-1590 (1970).

10. Braunitzer, G., Schrank, B., Petersen, S., Petersen, U.: Hoppe-Seyler's Z. Physiol. Chem. 352, 1730-1732 (1971).

11. Braunitzer, G., Schrank, B.: Hoppe-Seyler's Z. Physiol. Chem. 351, 417-418 (1970).

12. Braunitzer, G., Schrank, B., Stangl, A., Scheithauer, U.: Hoppe-Seyler's Z. Physiol. Chem. 359, 137-146 (1978).

13. Zimmermann, C.L., Pisano, J.: Meth. Enzymol. 57, 45 (1977).

14. Johansen, H.: Acta Arctica 8, 1-2 (1956).

15. Würdinger, I.: Int. Zoo Yearbook 13, 116 (1973).

16. Swan, L.W.: Nat. Hist. 70, 68 (1970).

17. Kydryaliew, A.: Ornitologija 8, 245 (1967).

18. Rapoport, S., Guest, G.M.: J. Biol. Chem. 138, 269 (1941).

19. Johanson, L.F., Tate, M.E.: Can. J. Chem. 47, 63 (1969).

20. Petschow, D., Würdinger, I., Baumann, R., Duhm, J., Braunitzer, G., Bauer, C.: J. Appl. Physiol. 42, 139 (1977).

21. Braunitzer, G., Schrank, B., Stangl, A., Bauer, C.: Hoppe-Seyler's Z. Physiol. Chem. 359, 547 (1978).

22. Matsuda, G., Takei, H., Wu, K.C., Shiozawa, T.: Int. J. Protein Res. 3, 173 (1971).

23. Matsuda, G., Maita, T., Mizuno, K., Ota, H.: Nature (London) New Biol. 244, 244 (1973).

24. Arnone, A., Perutz, M.F.: Nature (London) 249, 34-35 (1974).

25. Zuckerkandl, E., Pauling, L.: Horizons in Biochemistry pp. 189, Acad. Press, New York 1962

THE STRUCTURE OF THE DICYCLOHEXYLCARBODIIMIDE-BINDING SUBUNIT

E.Wachter, T.Graf, G.Wild, W.Sebald

Institut für Physiologische Chemie, Physikalische Biochemie und Zellbiologie der Universität München, 8000 München, Germany

ATP synthetases are located in the membranes of bacteria, chloroplasts and mitochondria. The enzyme is composed of several subunits, the number of which depends on the organism under investigation. A few subunits are coded for by the mitochondrial genome, wereas the others are synthesized on cytoplasmic ribosomes in the case of the mitochondrial ATPase complex. Apart from the synthesis of ATP this rather complex enzyme acts as a proton pump. Thus the ATP synthetase may be considered also as an H^+ ATPase. The enzyme can be separated into two defined parts: Firstly, the hydrophilic F_1 part (F_1 ATPase) which has been identified to comprise the binding site for ADP, P_i and the catalytic site, and secondly, the hydrophobic part (F_o) which is integrated within the lipid bilayer of the membrane and is part of the proton translocating machinery. Proton translocation and in consequence ATP synthesis is effectively inhibited by the hydrophobic reagent dicyclohexylcarbodiimide. This inhibitor reacts specifically with one acidic residue, glutamic or aspartic acid, within the polypeptide chain of a membrane integrated ATPase subunit. This subunit is called proteolipid because of its unusual solubility properties. The polarity of proteolipids is extremely low. The oligomycin sensitivity is attributed to a specific amino acid sequence close to the N-terminal part of certain proteolipids.

The stoichiometry of the proteolipid subunit of the ATPase

Chemistry of Peptides and Proteins, Vol. 1

complex seems to vary from organism to organism and is still under discussion. However, there is accumulating evidence that the DCCD-binding polypeptide is present in an oligomeric, most probably hexameric arrangement.
One approach to get a more detailed view of the molecular mechanism of the proton translocation across the membrane is the elucidation of the primary structure of the proteins involved. Thus the DCCD-binding subunit of the ATPase complex from eight different organisms has been analyzed. The strictly conserved residues are discussed with respect to a possible mechanism of proton translocation.
A further tool may be the determination of the amino acid replacements in mutant strains of bacteria such as E.coli which are deficient or depressed in ATP synthesis or in mutant strains for example of Neurospora crassa and yeast which are resistant to the antibiotic oligomycin.
The topography of the F_o part of the ATPase complex has been investigated by labelling with reagents some of which are capable to penetrate the membrane. These studies were done with whole mitochondria, submitochondrial particles and artificial vesicles.

STRUCTURAL STUDIES OF THE ACTIVE ION TRANSPORT SYSTEMS

Nikolai Modyanov, Alexei Babakov, Natalia Arzamazova, Karine Dzhandzhugazyan, Svetlana Kocherginskaya

Shemyakin Institute of Bioorganic Chemistry, USSR Academy of Sciences, Moscow, USSR

Introduction

Among the systems of active ion transport across biological membranes ATPases - molecular pumps - occupy the key position. The detailed information is now available on functional properties of Na^+-K^+, and H^+-ATPases. These enzymes are membrane proteins which catalyze vectorial ion transfer, using the same energy source - ATP. Still the ATPases differ considerably in structural organization and kinetic properties.

To find out the structural basis of the functions of ATPases of both types we have started the studies on the structure and spatial organization of the systems of active proton, sodium, and potassium transport on examples of mitochondrial H^+-ATPases from bovine heart and glycolysing bacteria *Streptococcus faecalis* and Na^+-K^+-ATPase from pig kidney.

Results

Immunochemically homogenous Na^+-K^+-ATPase was isolated from outer medulla of pig kidney using modified Jorgensen's procedure (1). The enzyme was shown to consist of two subunits with M.W. resp. 96 kD (α) and 47 kD (β), the last being glycoprotein (2). The fluorimetric determination of the amount of dansyl derivative of N-terminal amino acid residue yielded a M.W. 40 kD for the protein moiety of β-subunit. The hydrocarbon

Chemistry of Peptides and Proteins, Vol. 1

component has M.W. 7 kD. It should be mentioned that the usage of SDS-electrophoresis in polyacrylamide gel for the determination of M.W. of β-subunit leads (as well as in the case of many other glycoproteins)to overestimated value of M.W. (58 kD). The preparative separation of subunits of the Na^+-K^+-ATPase was carried out using the recently developed method involving chromatography on hydroxylapatite in the presence of SDS followed by gel-filtration (Bio-Gel A-1,5 m). The homogeneity of the isolated samples was proved by electrophoresis in polyacrylamide gel in the presence of SDS and by determination of N-terminal amino acid residues.

We have determined the amino acid and hydrocarbon composition of the subunits: α (~880 residues) Asx (80), Thr (52), Ser (60), Glx (91), Pro (41), Gly (77), Ala (72), Val (59), Met (21), Ile (55), Leu (84), Tyr (23), Phe (41), His (18), Lys (46), Arg (41), Cys (19), Trp (3); β (~365 residues) Asx (31), Thr (23), Ser (25), Glx (42), Pro (24), Gly (35), Ala (24), Val (24), Met (6), Ile (16), Leu (30), Tyr (16) Phe (16), His (6), Lys (22), Arg (15), Cys (9), Trp (2), Gal (7), Man (6), Fuc (1), GlcNAc (10), NeuNAc (4).

The presence of N-acetylglucosamine (not N-acetylgalactosamine) in the hydrocarbon component testifies N-glycosidic bonds between hydrocarbons and polypeptide chain of β-subunit.

SDS-dansyl-Edman degradation was used to determine the N-terminal sequence of α-subunit NH_2-Gly-Arg-Asx-Lys-Tyr- . All the attemps to determine the sequence on the sequenator failed even when detergent was used . It is interesting that the sequence fully coincides with that of α-subunit of Na^+-K^+-ATPase from duck salt gland (3).

The sequence of β-subunit was determined with the use of a sequenator (NH_2-Ala-Asn(CHO)-Gly-Thr-Ala). No PTH-derivative of any amino acid could have been traced during the second degradation cycle. The dansyl-Edman method enabled the localization of Asx in this position. The Asn-X-Thr sequence is characteristic of the binding sites of hydrocarbons in the case of N-glycosidic bonds,thus we assumed that the second position is occupied by

glycosylated asparagine. Indeed, hydrocarbon analysis of the water-soluble thiazolinone, extracted at this stage, pointed to sugars, and amino acid analysis - to Asp. Hence, one of the hydrocarbon chains of β-subunit binds to Asn-2 residue.
Scanning of the electrophoregram of the enzyme in polyacrylamide gel in the presence of SDS with the isolated subunits as standards yielded the equimolar ratio of the subunits in Na^+-K^+-ATPase.
Two independent approaches were used to estimate M.W. of the Na^+-K^+-ATPase. Freeze-fracture analysis showed that the enzyme formed globulas ca 100 Å in diameter, buried in membrane matrix. The calculations based on the estimation of lipid:protein ratio in the sample and membrane surface on electron microphotographs gave the value of ca 550kD for M.W. of the observed intramembrane globular structure.
Na^+-K^+-ATPase modification by alkylating ATP analog (ATP-γ-4-(N-2-chloroethyl-N-methylamino)benzylamidate) resulted in complete inhibition of ATPase activity by binding 1 mol of the reagent with the protein. These data permitted to make the assumption that the enzyme molecule consists of 4 α- and 4β-subunits.
Some information on the structural organization was derived from the following results. Localization of glycosylated asparagine of N-terminal part of β-subunit showed that this protein domain is located on the outer surface of the cell membrane. Labelling of the membrane-bound enzyme and "inside-out" proteoliposomes pointed out that liposomal β-subunit was intact by modification. So β-subunit (or at least its NH_2 groups) are not exposed at the inner membrane surface.
It was proved immunochemically (by haemagglutination and complement-mediated lysis) that antibodies against α-subunit interact with the outer surface of sheep erythrocytes (their Na^+-K^+-ATPase is antigenically similar with the isolated enzyme). As α-subunit contributes to the formation of the catalytic site of the ATPase it means that subunit spans the plasmatic membrane and may contribute to the formation of the

ionic channel across the membrane.

Proton ATPases differ from Na^+-K^+-ATPases in more complex subunit composition. H^+-ATPases of various eukaryotic and prokaryotic cells have quite a similar structural organization. We have isolated H^+-ATPase complex from *S. faecalis* consisting of 7 subunits, five of which (α - 55 kD, β - 51 kD, γ - 35 kD, δ - 20 kD, ε - 13 kD) formed the catalytic moiety (factor F_1) and two other (h_1-16 kD, h_2-7,3 kD) participated in the hydrophobic component, forming the proton transferring channel *via* the membrane (factor F_o). Stoichiometry α_3:β_3:γ_1:δ_1:ε_1 was proposed for F_1, F_o stoichiometry is still unclear.

H^+-ATPase of glycolysing bacteria *in vivo* are known only to perform proton transport, however *S. faecalis* ATPase reconstituted in proteoliposomes can effectively generate membrane potential and synthesize ATP as well (5).

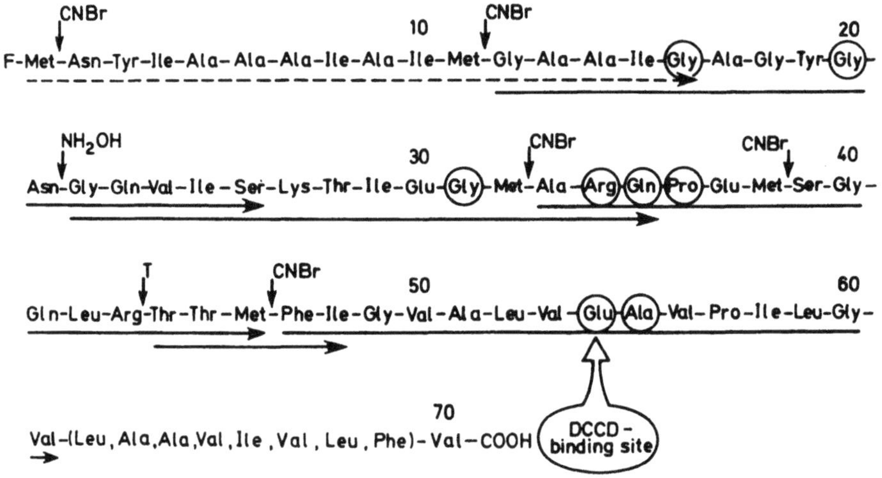

Fig. 1. Primary structure determination of DCCD-binding subunit of H^+-ATPase from *Streptococcus faecalis*.
o - conservative residues, (T) - trypsin. The arrows indicate residues identified by automated solid-phase Edman degradation (⟶) or manual procedure with DABITC (⟶).

Topographic study of the enzymes was carried out with the aid of antibodies against α, β, γ and ε-subunits. Antibodies against α-subunit were proved (contrary to those against β) not to inhibit the ATPase activity. Antibodies against γ did not practically interact with the enzyme and antibodies against ε, revealed the same inhibiting capacity of ATP hydrolysis by the enzyme and F_1, however, F_1 was precipitated to a larger extent. Thus, it was assumed that ε was near F_1-F_o contact, α and β were exposed to the molecule surface while γ was buried inside F_1 (6).

Dicyclohexylcarbodiimide (DCCD)-binding protein (subunit h_2) is the main component of F_o, taken now as responsible for proton translocation. This subunit of the membrane part of the ATPase is integral membrane protein. Its amino acid composition has a pronounced hydrophobic character and the content of hydrophylic residues does not exceed 23%. The primary structure of DCCD-binding protein (Fig. 1) was determined mainly with the aid of chemical methods of splitting using cyanogen bromide and hydroxylamine. The amino acid sequence was determined mainly by Edman's degradation method with DABITS (4-N,N-dimethylaminoazobenzene-4-isothiocynate) (7). Thin-layer chromatography or HPLC (in the case of Ile and Leu residues) were used for identification of amino acid residues.

N-terminal sequence was established with the aid of solid-phase sequenator after deformylation of the protein. The overlapping sequence between CNBr peptides 33-46 and 47-71 was obtained from analysis of tryptic hydrolysis products. C-terminal CNBr fragment was subjected to additional hydrolysis with pepsin. It was shown that inactivation of ATPase by DCCD was due to binding of the latter to Glu-54. It should be emphasized that there exists the pronounced structural homology of various protein regions with the structures of DCCD-binding subunits of other ATPases determined earlier by Wachter and Sebald (8). Thus, the residues marked by circles are identical or isofunctional in all the proteins.

Amino acid residues of DCCD-binding protein are arranged along

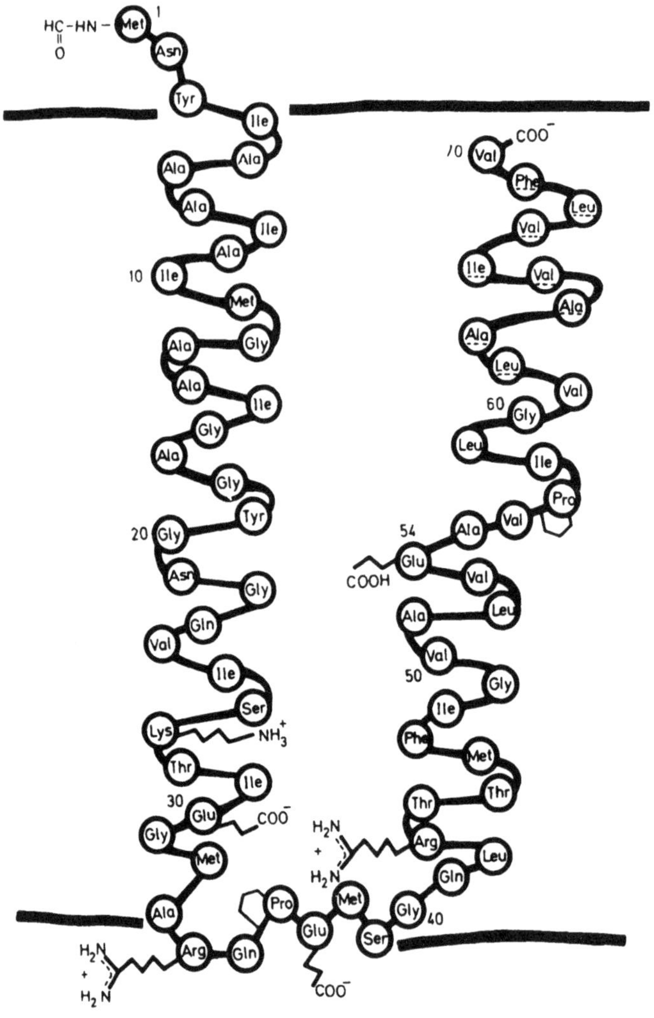

Fig. 2. Possible arrangement of DCCD-binding subunit of H^+-ATPase in the membrane. Amino acid sequence of C-terminal region (62-69) is not yet determined.

the polypeptide chain in two hydrophobic clusters divided by a hydrophobic region. The important Glu residue is located in the middle of C-terminal rod. Such a strict clusterisation permits quite a definite arrangement of the polypeptide chain in the membrane (Fig. 2). Thus hydrophobic segments span a lipid bilayer, whereas the polar segment is either exposed to the aqueous phase or involved into specific contacts with other subunits of the ATPase complex.

The enzyme molecule contains a number of DCCD-binding protein copies. This oligomer seems to form the transmembrane proton channel. Proton transfer might occur in this channel due to the system of hydrogen bonds in immobilized water. This mechanism of proton transfer is likely realized in the membrane channel formed by gramicidin A, which is also permeable to monovalent metal cations (9). However, we have shown that F_o reconstituting in liposome membrane provided proton permeable channel non-permeable to K^+ ions. Proton transfer by conjugated hydrogen bonds formed by polar amino acid groups could hardly be assumed due to the lack of these groups. To engage the supposed packing of DCCD-binding protein in the membrane with the high proton selectivity, the mixed mechanism of proton permeability might be assumed involving the chain of hydrogen bonds formed by immobilized water and polar amino acid residues as well.

It was shown by the study of the effect of the limited proteolysis on the interaction between membrane and catalytic moiety of ATPase complex that after tryptic treating of subbacterial particles (carrying no F_1 factor) the membrane sector loses its affinity to F_1. It is not proved that DCCD-binding protein is split by the treatment; it may be surmised that h_1 is located on F_o surface, thus, protecting DCCD-binding protein and contributing in the formation of the binding site for the catalytic moiety of the ATPase.

We hope that the used approach - parallel studies on several active ion transporting systems - will permit the definition of the basic principles of the structural organization of ionic channels and will facilitate the more complete understanding of the molecular mechanism of function of these enzymes.

References

1. Jørgensen, P.L.: Biochim. Biophys. Acta 356, 36-52 (1974).
2. Dzhandzhugazyan, K.N., Modyanov, N.N., Ovchinnikov, Yu.A.: Bioorg. Khim. 7, 847-857 (1981).
3. Hopkins, B.E., Wagner, H.Jr., Smith, T.W.: J. Biol. Chem. 251, 4365-4371 (1976).
4. Babakov, A.V., Vasilov, R.G.: Bioorg. Khim. 5, 119-125 (1979).
5. Vasilov, R.G., Babakov, A.V., Gevondyan, N.M., Terekhov, O.P.: the Soviet-Swedish Symposium on "Chemistry and Physics of Biological Membranes", Stockholm-Lidingö (1979).
6. Babakov, A.V., Gavrilova, E.G., Gevondyan, N.M., Polishchuk, G.I., Terekhov, O.P., Usmanova, R.G.: Bioorg. Khim. 5, 611-620 (1979).
7. Chang, I.Y.: Biochim. Biophys. Acta 578, 175-187 (1979).
8. Sebald, W., Hoppe, J., Wachter, E.: "Function and Molecular Aspects in Biomembrane Transport" (E.Quagliariello et al., eds) Elsevier, North-Holland, pp. 63-74 (1979).
9. Hladky, S.B., Haydon, D.A.: Biochim. Biophys. Acta 274, 294-312 (1972).

ISOLATION OF THE TTX-SENSITIVE PROTEIN(S) OF EXCITABLE TISSUES

V.K.Lishko, M.K.Malysheva, A.V.Stefanov, A.M.Chagovetz, A.A.Bogomoletz
Institute of Physiology, Academy of Sciences, Ukr.SSR, Kiev

Introduction

Investigation of the structure-function relationship of ionophores of the excitable membranes is one of the central problems of neuroscience. A necessary step of these investigations lies in isolation of purified ionophores. Recently it was shown that vesicular preparations of different excitable membranes preserve sensitivity to TTX and veratrine, the modificators of "fast" sodium channels (1,2). Similar results were obtained for liposomes treated with a detergent-soluble membrane fraction (3). Meanwhile, experiments with cytosol of internodal cells of Nitella demonstrated restoration of the excitable membrane (4). We have also demonstrated the effects of TTX and veratrine on the ^{22}Na influx in liposomes after incorporation of the soluble fraction of bovine brain and heart muscle (5). The present paper describes partial purification of the TTX-sensitive protein(s) from the bovine brain and heart muscle extracts.

Results

The methods of preparation of the soluble fraction of bovine brain and heart muscle were described earlier (5). Crude lipids of bovine brain were obtained by the method of Bligh and

Chemistry of Peptides and Proteins, Vol. 1

Dyer (6). Liposomes were prepared by sonication of the lipid suspension (22 kHz, 5 min) in the buffer of the following composition: 100 mM NaCl, 40 mM Tris-Cl, pH 7.4. For reconstruction of the soluble fraction of bovine brain (heart muscle) it was incubated with liposomes overnight. The method for determining ^{22}Na influx was as in previous investigations (5). Tetrodotoxin (TTX) and veratrine were applied in concentrations of 200 nM and 200 µM respectively. The protein content was estimated by the method of Lowry et al. (7).

Fig.1 presents a typical recording of DEAE Servacel 23 SH chromatography of the brain extract. Testing of different fractions for the presence of TTX-sensitive protein(s)(Fig.2) demonstrates that this component was eluted mainly in the fractions corresponding to the concentration interval of 250-300 mM NaCl. The absence of the TTX-sensitive protein(s) in the "free volume" fractions shows its acidic character. The total protein content of the "active" fraction was about 1 % of the initial one. Minor effects of veratrine and TTX were obtained in other fractions.

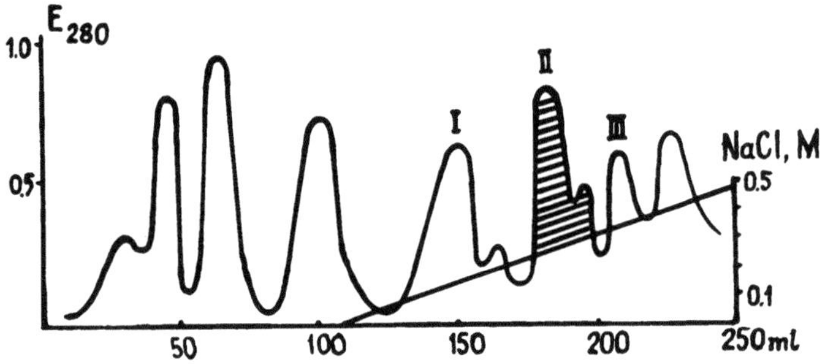

Fig.1. DEAE Servacel 23 SH chromatography of the bovine brain extract. Column: 200 x 25 mm, flow rate: 10 ml/h; starting buffer: 40 mM Tris-Cl, pH 7.4; elution by linear gradient of NaCl (0 - 500 mM).

Fractionating of the bovine (heart muscle) extracts with $(NH_4)SO_4$ was performed by the addition of certain amounts

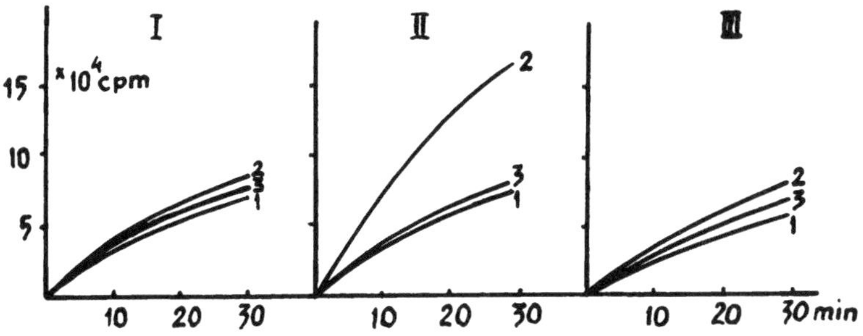

Fig.2. Testing of fractions I, II, III (Fig.1) by TTX and veratrine. 1: control influx of ^{22}Na; 2: influx in liposomes treated with veratrine; 3: influx in liposomes treated with veratrine and TTX.

of crystalline salt at permanent stirring. The pellets were solibilized and gel-filtrated on a Sephadex G-100 column (600 x 30 mm) equilibrated with 100 mM NaCl, 40 mM Tris-Cl, pH 7.4. Results of testing different fractions (Fig. 3) show the presence of TTX-sensitive protein(s) in the fraction, corresponding to the concentration interval of 0 - 30 % $(NH_4)_2SO_4$ saturation.

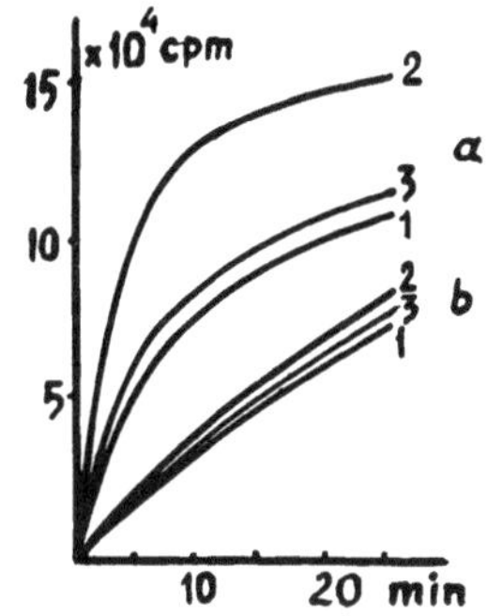

Fig.3. Testing of fractions after $(NH_4)_2SO_4$ treatment. 1: control influx of ^{22}Na; 2: influx in liposomes, treated with veratrine and TTX; 3: influx in liposomes treated with veratrine.
a) fraction 0 - 30 % of $(NH_4)_2SO_4$ saturation
b) fraction 30 - 50 % of $(NH_4)_2SO_4$ saturation.

DEAE Servacel 23 SH chromatography of the "active" fraction after $(NH_4)_2SO_4$ treatment (Fig. 4) demonstrated that the TTX-sensitive component is eluted at 300 mM NaCl.

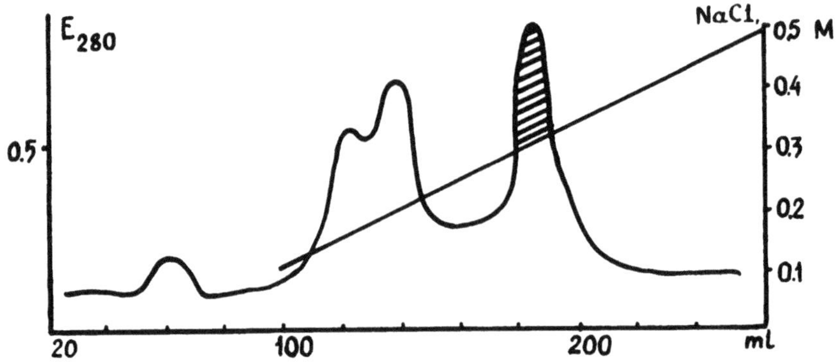

Fig.4. DEAE Servacel chromatography of the "active fraction after $(NH_4)_2SO_4$ treatment. Column: 200 x 15 mm; flow rate: 8 ml/h: starting buffer: 100 mM NaCl, 40 mM Tris-Cl, pH 7.4. Elution by linear gradient of NaCl (100 - 500 mM). "Active" fraction is marked.

Experiments with the extracts of non-excitable tissues (liver, bovine serum) exhibited none of the effects of TTX and veratrine with these proteoliposomes. Lack of Na^+, K^+-ATPase activity in the extracts excludes the possibility of the presence of membrane fragments in the soluble fractions. Data reported suggest two possibilities of interpretation: either isolated TTX-sensitive protein(s) are the soluble precursors of the "fast" sodium channels of excitable membranes or they are washed out of membranes during homogenization.

References

1. Barnola, F.V. and Villegas, R.: Gen.Physiol.67, 81-90 (1976).
2. Lishko, V.K., Malysheva, M.K., Stefanov, A.V.: Neuroscience 2, 1043-1048 (1977).
3. Lishko, V.K. Malysheva, M.K., Chagovetz, A.M.: Biochim. Biophys.Acta 602, 70-77 (1980).
4. Inoue, I., Ueda, T., Kobatake, Y.: Biochim.Biophys.Acta 298, 653 (1973).
5. Lishko, V.K., Malysheva, M.K., Stefanov, A.V.: Dokl.Academ. Nauk SSSR 248, 1257-1259 (1979).

6. Bligh, E.Q., Dyer, W.J., Can, J.: Biochem.Physiol.37, 911-917 (1959).

7. Lowry, O.H., Rosenbrough, N.J., Far, A.L., Randa, U.R.J.: J.Biol.Chem.193, 265-275 (1951).

FRAGMENTS FORMED FROM NEUROPEPTIDES UPON ACTION OF HYPOTHALAMIC ENDOPEPTIDASES

Tatos Akopyan, Ara Arutunyan, Alla Oganisyan
Institute ofBiochemistry, Sevag str. 5/I, Yerevan, 375044, USSR

Introduction

The metabolism of peptide hormones plays an important role in the control of numerous physiological mechanisms. In recent years, many studies showed that possibly most peptide hormones derive from larger molecular weight protein precursors. In turn, the enzymatic hydrolysis of some peptide hormones results in the formation of new physiologically active fragments. In this aspect, the degradation of luliberin, somatostatin, substance P and ß-LPH by soluble protein fraction and partially purified enzymes of brain has been studied (1-5). In the present work, we report our data concerning the action of hypothalamic neutral and acid proteinases on luliberin, somatostatin, substance P, ACTH and ß-LPH.

Results

For the purification of bovine hypothalamic neutral proteinase luliberin was used as substrate (6). The enzyme was purified about 100-fold. The molecular mass of enzyme was approximately 40 000; the pH optimum was about 7; K_m for luliberin $-1.3\ 10^{-5}$M. The effects of heavy metal ions, p-CMB and dithiothreitol on the hydrolysis of luliberin indicate that the enzyme is thiol-dependent.

Incubation of luliberin with hypothalamic neutral proteinase led to the appearance of glycine and leucine as N-terminal

Chemistry of Peptides and Proteins, Vol. 1

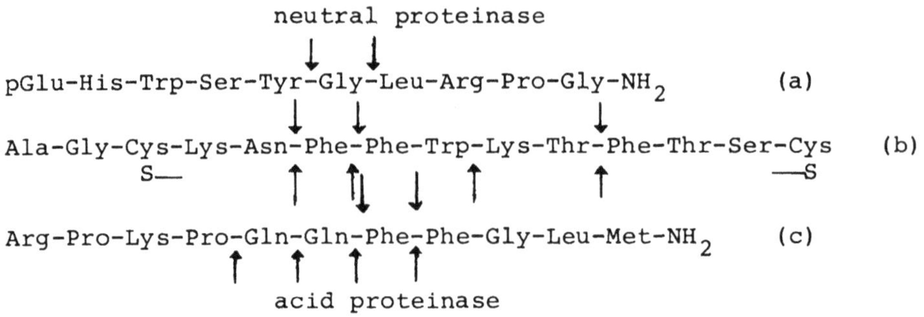

Fig.1. The action of hypothalamic neutral and acid proteinases upon luliberin (a), somatostatin (b) and substance P (c).

amino acids, determined by dansyl technique. Thin-layer chromatography of the products of enzymatic hydrolysis showed the absence of glycinamide. These facts indicate that in luliberin the enzyme splits Tyr-Gly and Gly-Leu peptide bonds (Fig.1). In the case of somatostatin, appearance of N-terminal phenylalanine indicates the possible cleavage of Asn-Phe, Phe-Phe and Thr-Phe peptide bonds. The enzyme also cleaves substance P. The detection of N-terminal phenylalanine after enzyme action indicates the cleavage of Gln-Phe or Phe-Phe bonds. It is more likely that enzyme acts on both peptide bonds as the chromatography of enzymatic digest shows the presence of four fluorescamine positive spots.

The acid proteinase from bovine hypothalamus was purified about 1000-fold. The molecular weight of the enzyme was near 50 000. Maximal activity against hemoglobin was obtained at pH 3.2 - 3.5. Hypothalamic acid proteinase was partially inhibited by ß-phenylpyruvate and was completely inhibited by low concentration of pepstatin ($K_i = 4.10^{-9}$M). This enzyme was identified as cathepsin D (7). Using sensitive and convenient method of endopeptidase assay (a method based on fluorescent properties of globin modified by pyridoxal-5-phosphate) it was established that cathepsin D exists in bovine hypothalamus in the form of six isoenzymes (8). Our experiments show

10 ↓
Glu-Leu-Ala-Gly-Ala-Pro-Pro-Glu-Pro-Ala-Arg-Asp-Pro-Glu-↓Ala

20 ↓ ↓30
Pro-Ala-Glu-Gly-Ala-Ala-Ala-Arg-Ala-Glu-↓Leu-Glu-Tyr-Gly-↓Leu-

↓ 40
Val-↓Ala-Glu-Ala-Gln-Ala-Ala-Glu-Lys-Lys-Asp-Glu-Glu-Pro-Tyr-

↓50 60
Lys-Met-Glu-His-↓Phe-Arg-Trp-Gly-Ser-Pro-Pro-Lys-Asp-Lys-Arg-

↓ 70
Tyr-Gly-Gly-↓Phe-Met-Thr-Ser-Glu-Lys-Ser-Gln-Thr-Pro-Leu-Val-

↓ 80 90
Thr-Leu-↓Phe-Lys-Asn-Ala-Ile-Val-Lys-Asn-Ala-His-Lys-Lys-Gly-

Gln (ß-LPH)

↓ 10
Ser-Thr-Ser-Met-Glu-His-↓Phe-Arg-Trp-Gly-Lys-Pro-Val-Gly-Lys-

20 30 ↓
Lys-Arg-Arg-Pro-Val-Lys-Val-Tyr-Pro-Asn-Gly-Ala-Glu-Asp-Glu-↓

↓ ↓
Leu-↓Ala-Gln-Ala-↓Phe-Pro-Leu-Glu-Phe (ACTH)

Fig.2. The action of hypothalamic acid proteinase upon pig ß-LPH and ACTH.

that cathepsin D cleaves somatostatin and substance P but not luliberin. The probable sites of enzyme action on these peptides were determined (Fig.1) (7).

Graf et al. studied the degradation of ß-LPH by hypophyseal cathepsin D (9). They established that enzyme can split 32-33 and 77-78 peptide bonds. It was established that hypothalamus contains ACTH and ß-LPH activities as well.

Our investigations showed that both hormones can be degraded by hypothalamic cethepsin D (10). In the experimental conditions used this enzyme splits 14-15, 25-26, 29-30, 31-32, 49-50, 63-64 and 77-78 peptide bonds in ß-LPH and 6-7, 30-31, 31-32 and 34-35 peptide bonds in ACTH (Fig.2). Both substrates contain locuses which are cleaved more intensively by cathepsin D: For ß-LPH this locus corresponds to the sequence 25-32 and for ACTH to 30-35.

The obtained data indicate that both hypothalamic endopeptides exhibit preferable specificity against peptide bonds which are formed by hydrophobic amino acid residues.

Acknowledgments
We express our gratitude to Dr.L.Graf (Institute of Pharmaceutical Chemistry, Budapest), who kindly gave us ß-LPH.

References

1. Marks, N., Stern, F.: Biochem.Biophys.Res.Commun. 61, 1458-1463 (1974).
2. Marks, N., Stern, F.: FEBS Lett. 66, 220-224 (1975).
3. Koch, Y., Baram, T., Chobsieng, P., Fridkin, M.: Biochem. Biophys.Res.Commun. 61, 95-103 (1974).
4. Martins, A., Caldo, H., Coelho, H., Moreira, A., Antunes-Rodriges, J., Greene, L., Martins de Camargo, A.: J.Neurochem. 34, 100-107 (1980).
5. Benuk, M., Marks, N.: Biochem.Biophys.Res.Commun.65,153-160 (1975).
6. Akopyan, T., Arutunyan, A., Oganisyan, A., Lajtha, A., Galoyan, A.: J.Neurochem. 32, 629-631 (1979).
7. Akopyan, T., Arutunyan, A., Lajtha, A., Galoyan, A.: Neurochem.Res.251, 4528-4536 (1977).
8. Akopyan, T., Barchudaryan, N., Karabashyan, L., Arutunyan, A., Lajtha, A., Galoyan, A.: J.Neuroscience Res. 4, 365-370 (1979).
9. Graf, L., Kenessey, A., Patthy, A., Grynbaum,A., Marks,N., Lajtha, A.: Arch.Biochem.Biophys. 193, 101-109 (1979).
10. Arutunyan, A., Akopyan, T., Oganisyan, A., Galfayan, V., Galoyan, A.: Voprosi Biochem.Mozga 14, 51-57 (1980).

NEW DATA ABOUT THE SPECIFIC PROTEINS OF HYPOTHALAMUS - CARRIERS OF CARDIOACTIVE COMPOUNDS

Armen Galoyan, Rimma Srapionian
Institute of Experimental Biology of Acad. Sci. ArmSSR
Yerevan, 375044 USSR

Introduction

It is already more than 20 years we postulated that hypothalamus produces not only releasing hormones, but also organotropic cardioactive neurohormones lettered "K", "C"and "G" (1). The existence of new specific proteins - carriers of the above mentioned substances in the hypothalamus was shown (2). These proteins possess not only a transporting role, but are precursors of the cardioactive neurohormones also (3). It was established (4,5) that the new specific proteins differ by their physicochemical and biological properties from known specific proteins of the brain (6,7,8). Our aim in the present investigation is to isolate preparative quantities of these proteins for studying some physicochemical properties of the neurohormone "C"-like cardioactive fragment.

Methods

From cattle hypothalamus an extract of water-soluble proteins, which was fractionated by ammonium sulfate, was isolated by the method described earlier (4). The extraction of coronary active proteins from the acetone powder of hypothalamic tissue was made by phosphate buffer, pH 7,38 (9). As a source of coronary active proteins a fraction with salt

Chemistry of Peptides and Proteins, Vol. 1

saturation of about 75-100% was used. Further purification of the proteins was made by the scheme described in our previous investigations (3,5). Amino acid analysis was made on an automatic analyser D-500 (Durrum, with ion-exchange resin DC-4). The protein hormone complex was dissociated by dialysis (48 h) against 0.1 N CH_3COOH. The protein was denaturated in 6 M urea and treated with trypsine in ammonium bicarbonate, pH 8,3, at 37°C for 22 h.

Mass spectrometric analysis of the cardioactive fragments was made on quadrupole gas chromatograph mass spectrometer "Finnigan-4000". The temperature rate was 150°-290°C (6° per minute) and the gas-carrier helium. Mass spectra were obtained by electron bombardment at an ionization energy of 70 electron volt in a range of 41-420 mass numbers.

Subcellular components were isolated by differential centrifugation in a saccharose gradient by a method representing a combination of Whittaker (10), Babich (11) and Hayos (12).

Antisera to proteins are obtained by rabbit immunization with corresponding antigens (colloidal mixture with Freund's adjuvant). The interaction of immune sera and antibodies was analysed immunoelectrophoretically by the Scheidegger method (13).

The bioassay of active substances was made by measuring the blood flow from coronary vessels per unit of time (14) and by determination of the 3',5'-cAMP phosphodiesterase activity (15).

Results and discussion

These methods allowed us to isolate 4 coronary active proteins, lettered according to their binding property with the corresponding coronary active neurohormones as PCC - protein carrier of the neurohormone "C", PCK - protein carrier of the neurohormone "K", PCG - protein carrier of the neurohormone "G" and PCX - protein carrier of the not yet identified com-

pound.
The immunochemical investigation of the mentioned proteins shows once more the existence of cardioactive proteins in the hypothalamus, and the negative results of cross-reactions testify to their strong specifity.
From 1 kg of fresh hypothalami 12 mg of homogenous coronary active proteins were isolated. The amount of PCK prevails, being almost 35% of the total sum of proteins. Proteins differ by their electrophoretic mobility, though, they have a narrow pH interval of 5,7-6,3, belonging to a group of weakly acidic proteins.
According to investigations by disk electrophoresis in 10% polyacrylamide gel the molecular mass is 20,000 for PCC, 24,000 for PCG and 39,000 for PCK.
When comparing the obtained data, testifying the definite differences between proteins in amino acid composition, we must note the similiarity between the proteins. First of all they have high contents of serine and glycine (42-44%) and up to 38% of alanine, glutamic acid, asparagine and histidine.
In order to show the presence of neurohormone "C"-like substances in the primary structure of PCC, we made a trypsinolysis of the denaturated (6-8 M urea) protein.
The studies of the biochemical, physicochemical and biological properties of one of the compounds, the cardioactive fragment of PCC, lettered "C_4", revealed its similiarity to the neurohormone "C" dissociated from the protein carrier. These data show the above mentioned compounds being glycopeptides.
In table 1 mass spectroscopic data are given of two isoforms of the neurohormone "C", dissociated from the protein-carriers (C_7, C_{10}) and of the cardioactive fragment C_4.
Similarity of the mass spectra and identical retention time of these substances make the presence of structural isomeres of the neurohormone "C" in the protein carrier probable.
The investigation of the subcellular distribution of these

Table 1. MASS SPECTROSCOPIC DATA OF C_7, C_{10} and C_4

Fraction	Retention time in min	Area value of nominal units by m/e			Area value, %	
		S_{331}	S_{169}	S_{211}	S_{331}/S_{169}	S_{211}/S_{169}
C_7	34,79	9410	64000	50000	14,7	78,1
C_{10}	34,34	7152	40000	21500	17,9	53,7
C_4	34,15	9568	63000	37000	15,2	58,6

proteins points to their predominant localization in neurosecretory granules and synaptosomes. Though the function of cardioactive proteins in synaptosomes is not clear at present, their prevailing localization in that subcellular fraction might be indicative of their participation in synaptic transmission.

References

1. Galoyan, A. A.: Dokl. Akad. Nauk ArmSSR 34, 109-111 (1962).
2. Galoyan, A. A.: Dokl. Akad. Nauk ArmSSR 38, 305-307 (1964).
3. Srapionian, R. M., Sahakian, S. A., Sahakian, F. M., Galoyan, A. A.: Problems of Brain Biochemistry, ed. A.Galoyan, Yerevan, 13, 67-78 (1978).
4. Galoyan, A. A., Srapionian, R. M.: Dokl. Akad. Nauk ArmSSR 42, 210-213 (1966).
5. Srapionian, R. M., Sahakian, S. A., Galoyan, A. A.: Problems of Brain Biochemistry, ed. H.Buniatian, Yerevan, 11, 97-104 (1976).
6. Moore, B. W., McGregor, D.: J. Biol. Chem. 240, 1647-1652 (1965).
7. Perez, V., Moore, B. W.: J. Neurochem. 15, 971-976 (1968).
8. Cicero, T., Suntzeff, V., Moore, B. W.: J. Neurochem. 19, 2119-2123 (1972).
9. Srapionian, R. M., Sahakian, S. A., Medvedev, F. A.,

Galoyan, A. A.: Dokl. Akad. Nauk ArmSSR 70, 182-186 (1980).

10. Whittaker, V., Michelson, Y., Kirkland, R.: Biochemical Journal 90, 293-303 (1964).

11. Babich, J.: Brain Research 49, 135-150 (1973).

12. Hayos, F.: Brain Research 93, 485-489 (1975).

13. Scheidegger, J.: Intern. Arch. Allergy 1, 103-107 (1955).

14. Morawitz, P., Zahn, A.: Deutsch. Arch. Klin. Med. 116, 364-367 (1914).

15. Galoyan, A. A., Gurvits, B. Y., Poggossian, M. A.: Problems of Brain Biochemistry, ed. H.Buniatian, Yerevan, 11, 197-207 (1976).

ISOLATION AND STRUCTURAL STUDY OF NEUROTOXIN FROM THE VENOM OF SPIDER LATRODECTUS TREDECIMGUTTATUS

Shavkat Salikhov, Mukhrodj Tashmukhamedov, Marat Adylbekov, Janna Abdurakhmanova, Alexander Korneyev, Abid Sadykov
Institute of Bioorganic Chemistry, Tashkent, USSR

At present at least 8 Latrodectus species are known and experimental data on phisiological effects of the whole venom are abundant (1-11). Isolation of homogeneous protein toxin and study of its mechanism of action were reported in several studies (12-17). Thus, a protein neurotoxin with a m.w. 130,000, designated α- latrotoxin, has been isolated from Latrodectus tredecimguttatus venom (12,14). It induces massive release of ACh, noradrenaline and γ- aminobutyric acid from various nerve terminals. A protein presynaptically acting on cockroach neuromuscular junction (m.w. 125,000) has been also isolated (13). But we have no data so far for the discussion of structural and functional differences of toxins isolated by different investigators.

The present paper is devoted to the determination of some physical and chemical characteristics, the study of the quarternary structure of the protein neurotoxin from the venom of Latrodectus spiders from Central Asia. The following reagents and sorbents have been used for chromatography: Sephadex (Pharmacia,Sweden), Bio-Gel (Bio-Rad,USA), DEAE-cellulose (Whatmann,England), polyacrylamide gel plates PAAG-4 (Pharmacia,Sweden), reagents for disc-electrophoresis in the presence of sodium dodecyl sulphate (Bio-Rad,USA), dansyl chloride (Serva,FRG).

The pH of the buffer solutions has been determined by a pH meter PHM-63 (Radiometer,Denmark), the protein concentration has been measured according to Lowry and by optical density

Chemistry of Peptides and Proteins, Vol. 1

at 280 nm using the spectrophotometer SF-16 (USSR) and Uvicord-II (LKB, Sweden).
The venom has been purified from adult females by application of helicera and their subsequent homogenization at 4°C and centrifugation at 20,000 for 30min. The supernatant was separated and subjected to lyophilization. The amount of dried venom obtained from one spider was approximately 0.2 mg depending upon the season of collection and the spider size. The supernatant dried by lyophilization was kept at 0°C for several months without noticeable loss of activity.
The venom showed toxic effects at pH values between 5.0-12.0. The venom is rather thermostable and does not lose its activity even being heated up to 60°C, nevertheless the procedures of isolation and fractionation have been carried out at 4°C. When studying the whole venom electrophoretically by PAAG in the presence of sodium dodecyl sulphate 15 coloured bands with m.w. from 10,000 to 20,000 have been observed.
First the whole venom fraction was treated with ammonium sulphate. 20-60% ammonium sulphate solutions precipitated toxic protein fractions, which were further separated on Sephadex G-100 (fine) in 0.05 M Tris-HCl buffer of pH 8.2. Three fractions, the first being toxic (fig.1), were received. We succeeded in fractionating the whole venom into 9 components using Bio-Gel P-60, however it was estimated electrophoretically that under these conditions the isolation of only low molecular weight components has been improved. The toxic fraction was further subjected to ion-exchange chromatography on DEAE-cellulose DE-52 in Tris-HCl buffer of pH 8.2 using linear NaCl concentration gradient from 0.05 to 0.5 M (fig.2).
10 fractions were received, the fraction eluting at 0.28 M NaCl being the most toxic. After rechromatography one symmetrical peak was received. Electrophoresis in PAAG (pH 8.9) in the presence of sodium dodecyl sulphate yielded only one band. By N-terminal analysis no N-terminal amino acid was detected. Considerable amounts of Dns-Lys and Dns-O-Tyr, detected on

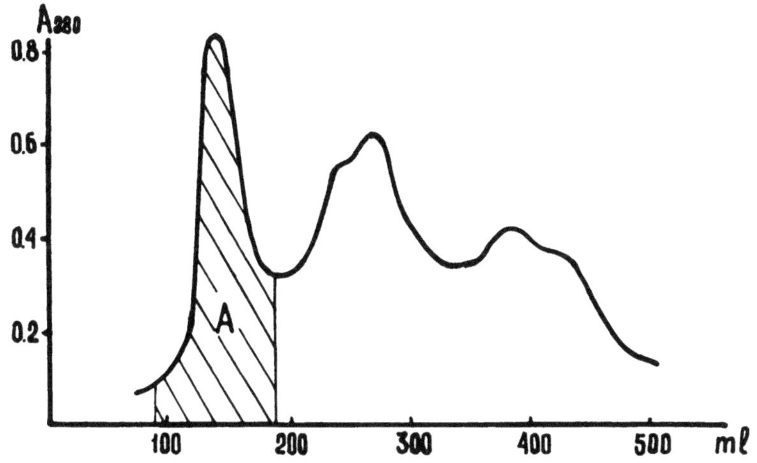

Fig.1. Gel filtration of Latrodectus tredecimguttatus venom following fractional precipitation by ammonium sulphate on a Sephadex G-50 (fine) column (2,5x90cm) in 0.05M Tris-HCl buffer, pH 8.2. Elution rate 20ml/h. A-Fraction with toxic activity.

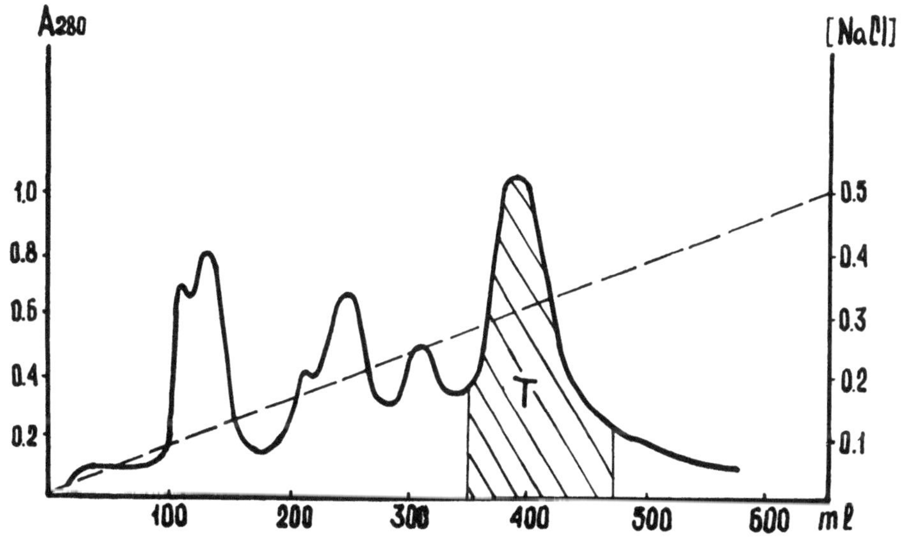

Fig.2. Chromatography of fraction A on a column (2.5x20cm) with cellulose in Tris-sodium chloride buffer of pH 8.2.

the plate by thin-layer chromatography during the identification of Dns amino acids, suggests that the N-terminal amino group of the toxin is blocked. The antiserum of the Latrodectus tredecimguttatus venom formed only one precipitation line which demonstrates homogeneity of the toxin. Quarternary structure of toxin molecule has been proved by the data obtained during the isolation and purification procedures. From the R_f value of gel chromatography on Sephadex G-100 a m.w. of approximately 200,000 can be calculated. From electrophoretic data in 10% PAAG in the presence of sodium dodecyl sulphate a m.w. of about 75,000 is estimated. When toxin was subjected to chromatography on a Sephadex G-200 column in 0.1 M ammonium acetate buffer at various pH values two protein peaks of different intensity have been observed (fig.3).

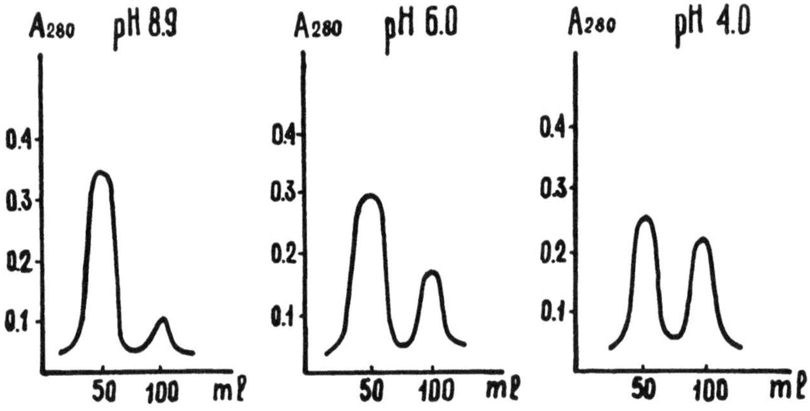

Fig.3. Fraction T gel filtration on a 1.0x100cm Sephadex G-200 column; eluent: 0.1 M ammonium acetate buffer at various pH values.

At increasing pH the biological activity was recovered very slowly. It was proven electrophoretically that the toxin exists as a monomer with a m.w. of about 75,000, a dimer with a m.w. of 150,000 and a tetramer with a m.w. of 300,000. Individual states of these forms have been received by preparative electrophoresis in PAAG. Each form turned into an initial mixture of these components under usual electrophoretic conditions but under desaggregation conditions only one band with the same mobility was observed. All three toxic forms had similar amino acid composition. The authors suggest that the toxin molecule takes the form of a biologically active dimer and tetramer at neutral and weak alkaline pH but is capable to dissociate into a non-active monomer at low pH in the presence of detergents. Table 1 shows amino acid composition based on the results after a 24-,48- and 72-hour hydrolysis.

Table 1. Amino acid composition of Latrodectus tredecimguttatus toxin.

Amino acid	Number of residues	By Grasso (14)	Amino acid	Number of residues	By Grasso
Asx	80	126	Ile	38	162
Thr	29	45	Leu	59	125
Ser	33	117	Tyr	25	4
Glx	67	127	Phe	34	30
Pro	34	47	His	23	12
Gly	38	172	Lys	36	43
Ala	40	90	Arg	37	41
Val	49	71	Trp	4	not detect.
Met	12	7	All in all:	638	1219

Data in reference (14) are given for comparison. It is obvious that neurotoxins considerably differ from each other in amino acid composition and molecular weight. The molecular weight, estimated from amino acid composition, is in good agreement with the value based on disc electrophoretic data in the presence of sodium dodecyl sulphate.

Full hydrolysis of the carboxymethylated toxin preparation by trypsin has been undertaken as the first step for establishing the complete primary structure of Latrodectus tredecimguttatus toxin. It has been carried out under usual conditions: 37°C, 12h, pH 8.4, enzyme-substrate ratio 1:50. Acidic treatment of the hydrolysate produced unsoluble products. The resultant mixture of soluble peptides was first fractionated by gel filtration on Bio-Gel P-10. Three combined fractions have been obtained: PT_1, PT_2 and PT_3.

For further separation of fractions PT_2 and PT_3 we used ion--exchange chromatography (Aminex-6, concentration gradient, pH of pyridine acetate). As a result PT_2 yielded 27 combined fractions and PT_3 - 8.

The composition of the fractions was established by chromatography on a thin layer of cellulose. 11 fractions were shown to be homogenous (II-5, II-7, II-8, II-14, II-16, II-17, II-18, II-21, II-22, II-25, III-8) and do not need further purification. The remaining fractions were mixtures of a relatively small number of peptides. For further separation of combined fractions as well as for purification of peptides we used one of the most effective methods of peptide separation - chromatography on paper, and for fractions II-1, II-15 and II-20 we used also the method of peptide maps.

The N-terminal amino acid sequence of the peptides was established by the Edman method, the amino acids were determined as Dns or/and Pth derivatives. The C-terminal sequence was determined using carboxypeptidases A and B. The structure of soluble peptides after tryptic hydrolysis of fractions T-II and T-III determined by the dansyl modification of the Edman method is given in Tables 2 and 3.

Table 2. Tryptic hydrolysate of Latrodectus tredecimguttatus toxin (fraction PT-II, soluble peptide amino acid sequence).

Peptide	Amino acid sequence
II-1-1	Ala-Ser-Gly-Glu-Asn-Ile-Pro-Thr-Asp-Gln-Ile-Asp-Glu -Arg
II-1-2	Ala-Thr-Glx-Thr-Ile-Gly-Asx-Glu-Leu-(Val,Ser,Lys)
II-1-3	Phe-Gln-Asp
II-2-1	Leu-Lys-Pro-Tyr-Asp-Asp-Asp-Glu-Leu-Gly-Phe-Tyr
II-2-2	Lys-Gly-Glu-Leu-Pro-Gly-Gly-Ser-Gln-Asp
II-2-3	His-Gly-Lys
II-3-1	Gly-Val-Phe-Val-Pro-Pro-Leu-Glu-Glu-Ile-Phe-Pro
II-3-2	Tyr-Asp-Gly-Pro-Asp-Val-Leu-Leu-Asp-Leu-Ile-Asp
II-3-3	Asp-Leu-Val-Ala-Gly-Leu-Phe-Asp-Gly-Gly
II-4-1	Glu-Gly-Gly-Glu-Leu-Thr-Leu-Gly-Glx-Asx-Ala
II-4-2	Leu-Leu-Gln-Ala-Arg
II-5	Asp-Leu-Gln-Glu-Ala-Glu-Lys
II-6-1	Ile-Ile-Val-Ala-Thr-Asp-Ser-Met-Axs-Glx-Glx-Lys
II-6-2	Phe-Val-Asp-Leu-Cys-Glu-Met-Ala-Arg
II-7	Asn-Gln-Asp-Leu-Asn-Leu-Gln-Cys-Glx-Lys
II-8	Cys-Val-Asx-Val-Glx-Asx-Met-Glx-His-Arg-Lys
II-9-1	Gly-Asp-Asp-Lys
II-9-2	Glu-Ala-Thr-Lys
II-9-3	Lys-Leu-Asp-Asp-Leu-Pro-Asp-Cys-Ser-Val-Gln-Asp-Leu - Glu-Arg
II-9-4	Leu-Ile-Asx-Val-Glx-Ser-Arg
II-10-1	Glu-Asp-Arg
II-10-2	Ser-Ile-Glu-Asp-Ala-Asp-Lys
II-10-3	Leu-Ala-Val-Asp-Leu-Asp-Glx-Ser-Gly-Arg
II-11-1	Glu-Ser-Lys
II-11-2	Tyr-Ala-Glx-Thr-Arg
II-11-3	Cys-Ala-Arg
II-11-4	Leu-Ile-Glx-Ser-Pro-Glx-Ile-Asx-Ile-Asx- -(Ser,Gly,Val,Arg)
II-12-1	Gly-Glu-Gly-Ala-Asp-Arg

Table 2. Continuation.

II-12-2	Leu-Asn-Gln-Glu-Arg
II-13-1	Tyr-Asp-Cys-Gln-Arg
II-13-2	Tyr-Asp-Glu-Leu-Gly-Asn-Glu-Ile-Lys
II-13-3	Phe-Ile-Glx-Asx
II-14	Ala-Glu-Arg
II-15-1	Cys-Arg-Val-Asx-Glx-Asx-Arg
II-15-2	Ser-Ser-Gly-Thr-Ser-Leu
II-15-3	Ser-Leu-Cys-Lys
II-15-4	Leu-Asx-Ile-Glx-Gly-Ala-Asx-Ile-Gly-Ser-Asx-Glx-Leu--Gly-Gly-Lys
II-16-1	Leu-Asp-Ser-Arg
II-17	Glu-Lys
II-18	Glu-Ile-Trp-Lys
II-19-1	Ser-Thr-Glu-Leu-Asn-Gln-Pro-Asp-Lys
II-19-2	Ala-Val-Pro-Ala-Ala-Asx
II-19-3	Glu-Leu-Leu-Glu-Asn-Lys
II-19-4	Phe-Val-Pro-Ala-Asx-Ile-Leu-Asx-Lys
II-20-1	Gly-Gly-Glu-Arg
II-20-2	Leu-Phe-Glu-Gln-Pro-Leu-Glu-Gly-Asp-His-Val-Lys
II-20-3	Ile-Val-Val-Ser-Pro-Gly-Asx-Asx-Asx-Gly-Lys
II-20-4	Val-Asx-Asx-Ser-Gly-Asx-Arg
II-21	Ser-Lys
II-22	Asp-Asn-Leu-Tyr-Pro-Asp-Arg
II-23-1	Ala-Arg
II-23-2	Lys-Asp-Cys-Arg
II-23-3	Glu-Arg-Ala-Lys
II-23-4	Thr-Ile-Val-Ala-Gln-Ser-His-Ser-Asp-Glu-Phe-Leu-Gly--Pro-Met-Glu-Lys
II-24-1	Leu-Lys
II-24-2	Ile-Ser-Glu-Ile-Leu-Val-Arg
II-25	Leu-Asp-Pro-Glu-His-Gln-Arg
II-26-1	Ile-Trp-Glu-Glu-Leu-Ile-Gly-Lys
II-26-2	Glu-Leu-Leu-His-Gly-Gln-Asp-Leu-Arg

Table 2. Continuation.

II-26-3 Ser-Glu-Phe-His-Ser-Gln-Gly-Asp-Phe-Gly-Glu-Leu-Leu--(Val,Pro,Cys,Lys)

Table 3. Tryptic hydrolysate of Latrodectus tredecimguttatus toxin (fraction PT-III, soluble peptide amino acid sequence)

Peptide	Amino acid sequence
III-1-1	Gly-Glu-Gln-Lys
III-1-2	Leu-Gln-Phe-Lys
III-2-1	Ser-Arg
III-2-2	Leu-Ile-Lys
III-2-3	Pro-Leu-Ile-Gly-Ala-(His,Lys,Arg,Ser,Asx,Glx,Val)
III-3-1	Asp-Arg
III-3-2	Ala-Ile-Lys
III-3-3	Thr-Ala-Lys
III-3-4	Ser-Thr-Val-Arg
III-4-1	Ala-Asx-Thr-Val-(Ser,Glx,Gly,Tyr,Ala,Lys,Met)
III-4-2	Val-Leu-Arg-Lys
III-5-1	Ile-Ile-Leu-Arg
III-5-2	Asp-Lys-Arg
III-6-1	Ala-Lys
III-7-1	Gly-Ser-Lys
III-7-2	Gly-Pro-Lys
III-8	Gly-Pro-Arg

As pertains the tables 2 and 3, Pth derivatives of dicarboxylic acids and their amides were identified by the reaction with ninhydrin in the solution alcohol:collidine (95:5). The arrows show degradation steps according to Edman: Pth identification (—•), Dns (—►), Pth and Dns (⇉) derivatives, (←—) amino acids determined by carboxypeptidases A and B.

The application of only this method was sufficient for complete sequencing of the peptides II-11-3, II-15-2, II-15-3, II-19-2, II-21, II-23, III-2-2, III-3-2, III-3-3, III-3-4, III-4-2, III-5-1, III-6-2, III-7-1, III-7-2, III-8. The structure of the peptides II-1-3, II-9-2, II-10-1, II-14,

II-16, II-17 and II-20-1 has been proved by Edman degradation and by the results of amino acid analysis following hydrolysis of the peptides by aminopeptidase M. Enzymatic methods of cleavage were also applied for structural studies of some smaller and larger peptides.
It should be noted that the protein under investigation has some peculiarities one of which may be explained by free Arg and Lys residues in the proteolysis products, namely the sequences Arg-Arg-, Arg-Lys or Lys-Arg-, Lys-Lys.
Generally speaking the tryptic hydrolysis of the Latrodectus tredecimguttatus toxin has been carried out under certain conditions and it showed that the majority of the peptides have lysine and arginine as C-terminal residues. Isolated soluble tryptic peptides contained altogether 519 amino acid residues which constitute about 80% of the whole protein molecule. Hydrolysis of the toxin by BrCN and staphylococcus protease has been carried out to obtain overlapping areas.

References

1. Clark A.W., Hurlbut W.P. and Mauro A.: J. Cell. Biol. 52, 1-14 (1972).
2. Clark A.W., Mauro A., Longenecker H.E., Jr., Hurlbut W.P.: Nature (Lond.) 225, 703-705 (1970).
3. Cull-Candy S.G., Neal H. and Usherwood: Nature (Lond.) 241, 353-354 (1973).
4. Frontali N.: Brain Res. 37, 146-148 (1972).
5. Frontali N., Granata F., Parisi P.: Biochem. Pharmacol. 21, 969-974 (1972).
6. Frontali N. Grasso A.: Arch. Biochem. Biophys. 106, 213-218 (1964).
7. Longenecker H.E., Jr., Hurlbut W.P., Mauro A., Clark A.W.: Nature (Lond.) 225, 701-703 (1970).

8. Majori G.C., Bettini S., Casaglia O.: J. Insect Phisiol. 18, 913-927 (1972).
9. Cecarelli B., Grohobaz F., Hurlbut W.P., Iezzi N.: J. Cell. Biol. 81, 163-177 (1979).
10. Cecarelli B., Grohobaz F., Hurlbut W.P., Iezzi N.: J. Cell. Biol. 81, 177-192 (1979).
11. Gorio A., Hurlbut W.P., Cecarelli B.: J. Cell. Biol. 78, 716-733 (1978).
12. Frontali N., Cecarelli B., Gorio A., Mauro A., Siekevitz P., Mu-Chin Tzeng., Hurlbut W.P.: J. Cell. Biol. 68, 462-479 (1976).
13. Ornberg R.L., Smyth T., Jr., Benton Q.W.: Toxicon 14, 329-333 (1976).
14. Grasso A.: Biochem. Biophys. Acta. 439, 406-412 (1976).
15. Fritz L.C., Mu-Chin Tzeng., Mauro A.: Nature 283, 486--487 (1980).
16. Finkelstein A., Rubin L., Mu-Chin Tzeng.: Science 193, 1009-1011 (1976).
17. Grasso A., Ruffini S., Senni M.I.: FEBS Letters 85, 241--244 (1978).

SEA ANEMONE TOXINS; A MINIREVIEW

Lászlo Beress
Dept.of Toxicology,University of Kiel, D-2300 Kiel,Germany

Introduction

Since the beginning of this century several important scientific discoveries have been made using marine animals as research objects; for example, Landsteiner discovered the antigen-antibody precipitation reaction in the early 1900's using then blue haemocyanin of the lobster and an antihaemocyanin. Richet discovered in 1902 by repeated injection of sea anemone (Anemonia sulcata) extracts into dogs the phenomenon of anaphylaxy. Early pancreas studies were performed by Best on skates. The use of the giant axon of the squid allowed Hodgkin and Huxley (1952) to perform their fundamental studies on description of membrane current and its application to conduction and excitation in the nerve. Some marine toxins - for example tetrodotoxin, saxitoxin (for review see 20) and sea anemone toxins - exert their toxic effect by modifying the properties of voltage-sensitive sodium channels involved in action potential generation in nerve, heart and skeletal muscle.

Results

Since the early work of Richet (19,19) on the extraction of two toxic substances thalassin and congestin from tentacles of the sea anemone Anemonia sulcata (Fig.1) there has been a steady increase of work on the isolation of pharmacologically active compounds from coelenterates (11).

Chemistry of Peptides and Proteins, Vol. 1

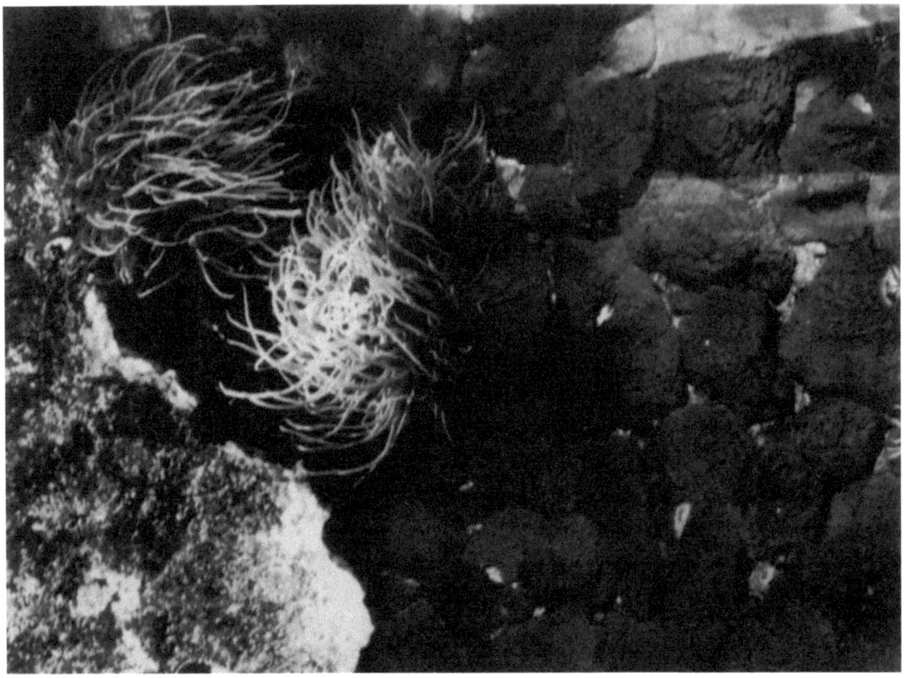

Figure 1. Anemonia sulcata (photo Edith Mempel, Kiel, Meeresaquarium).

A systematic study devoted to the isolation and characterization of sea anemone toxins, especially those from Anemonia sulcata was initiated in 1966/67 at the Zoological Station in Neaples. This work led to the discovery and partial purification of two of its toxins, Toxin I (ATX I) and Toxin II (ATX II) (5). Both are polypeptides and have potent paralysing effect on crustaceans and mammals. It was later shown that a third toxin, Toxin III (ATX III) was also present in Anemonia sulcata. Improved methods (6, 7) resulted in complete purification of the three toxins on a preparative scale. With slight modification on these techniques four toxins from the sea anemones Condylactis aurantiaca (CTX I; II; III and IV) and two toxins from Bolocera tuediae were also prepared (9, 8). Three further sea anemone toxins (APA, APB, and APC) were recently purified and characterized by Norton et al. (13, 22). All investigated sea anemone toxins are polypeptides, their molecular weights ranging from

2933 Dalton (ATX III) to 5630 Dalton (CTX III). The sequences of some of the sea anemone toxins have been determined (table 1).

Table 1. Amino acid sequences of some sea anemone toxins.

	1				5					10					15
ATX I:	Gly	Ala	Ala / Pro	Cys	Leu	Cys	Lys	Ser	Asp	Gly	Pro	Asn	Thr	Arg	Gly
ATX II:	Gly	Ile / Val	Pro	Cys	Leu	Cys	Asp	Ser	Asp	Gly	Pro	Ser	Val	Arg	Gly
AP - C:	Gly	Val	Pro	Cys	Leu	Cys	Asp	Ser	Asp	Gly	Pro	Ser	Val	Arg	Gly
AP - A:	Gly	Val	Ser	Cys	Leu	Cys	Asp	Ser	Asp	Gly	Pro	Ser	Val	Arg	Gly
ATX III:	Arg	Ser	Cys	Cys	Pro	Cys	Tyr	Trp	Gly	Gly	Cys	Pro	Trp	Gly	Gln

					20					(25)				(29)
ATX I:	Asn	Ser	Met	Ser	Gly	Thr	Ile	Trp	Val			Phe	Gly	Cys
ATX II:	Asn	Thr	Leu	Ser	Gly	Ile	Ile	Trp	Leu			Ala	Gly	Cys
AP - C:	Asn	Thr	Leu	Ser	Gly	Ile	Leu	Trp	Leu			Ala	Gly	Cys
AP - A:	Asn	Thr	Leu	Ser	Gly	Thr	Leu	Trp	Leu	Tyr	Pro	Ser	Gly	Cys
ATX III:	Asn	Cys	Tyr	Pro	Glu	Gly	Cys	Ser	Gly			Pro	Lys	Val

	(30)					(35)					(40)					(45)				(49)
ATX I:	Pro	Ser	Gly	Trp	Asn	Asn	Cys	Glu	Gly	Arg	Ala		Ile	Ile	Gly	Tyr	Cys	Cys	Lys	Gln
ATX II:	Pro	Ser	Gly	Trp	His	Asn	Cys	Lys	Lys	His	Gly	Pro	Thr	Ile	Gly	Trp	Cys	Cys	Lys	Gln
AP - C:	Pro	Ser	Gly	Trp	His	Asn	Cys	Lys	Ala	His	Gly	Pro	Thr	Ile	Gly	Trp	Cys	Cys	Lys	Gln
AP - A:	Pro	Ser	Gly	Trp	His	Asn	Cys	Lys	Ala	His	Gly	Pro	Thr	Ile	Gly	Trp	Cys	Cys	Lys	Gln

ATX I, II, III: Toxins from *Anemonia sulcata*
AP - A: Toxin from *Anthopleura xanthogrammica*
AP - C: Toxin from *Anthopleura elegantissima*

References:

ATX I: Wunderer, G. and Eulitz, M.: Eur. J. Biochem. (in press)
ATX II: Wunderer, G. et al.: Eur. J. Biochem. 68, 193-198 (1976)
ATX III: Beress, L. et al.: Hoppe Seyler's Z. physiol. Chem. 358, 985-988 (1977)
Martinez, G. et al.: FEBS Lett. 84, 247-252 (1977)
AP - A: Tanaka, M. et al.: Biochemistry 16, 204-208 (1977)
AP - C: Bennett, C. cited in: Norton, T.R. et al.: in: Drugs and Food from the Sea (P.N.Kaul, C.J.Sinderman, eds.), 37-50, University of Oklahoma Press 1977

With the exception of ATX III, the sea anemone toxins are structural homologous proteins. All contain three disulfide bridges which stabilize their structure. Neither the sequence nor the laser Raman spectroscopical studies have revealed any structural homology between ATX II and the other polypeptide neuro- and cardiotoxins isolated from snake or scorpion venoms (14).

Physiology of the sea anemone toxins

The effects of the three toxins from *Anemonia sulcata* on the neuromuscular transmission and nerve action potential have been investigated on the exposed open muscle of the isolated first walking leg of the crayfish *Astacus leptodactilus*. The toxins act presynaptically prolonging the duration of the action potential (16). The sea anemone toxin ATX II acts on the sodium channel by selectively slowing down the sodium inactivation (10, 21, for reviews see 12, 15).

Pharmacology of the sea anemone toxins

Pharmacological studies on sea anemone toxins (ATX I and ATX II) have been devoted to the investigation of the mode of their action on the mammalian heart muscle (1, 3, 17). Toxin ATX II evokes a dose-dependent positive inotropic effect in nanomolar (2-100 nM) range (1). However, in contrast to cardiac glycosides, ATX II does not inhibit the action of the enzyme Na/K-ATPase (2). The positive inotropic effect of ATX I on the isolated guinea pig auricles was found to be 100 times weaker than the effect of ATX II (3). Electromechanical studies with ATX II on mammalian cardiac muscle revealed that the positive inotropic effect was accompanied by a prolonged duration of the action potential (17). ATX II is also capable of depolarizing skeletal muscle membranes, and of prolonging the duration of the muscle fibre action potential (4). Both phenomena appear to be mediated via the fast sodium channel.

The small size of ATX II (Mw = 4935; G. Wunderer personal communication), its specific action on, both, the mammalian heart and sceletal muscle and on the neuromuscular transmission even in nanomolar concentrations, and its high affinity to the fast sodium channel system renders it as one of the most interesting substances for the study of the excitation phenomena of the heart muscle, nerve and muscle membranes and for the characterization of the sodium channel.

The method devised for the isolation of the toxins of the Anemonia sulcata (5, 7) has proved to be a generally applicable procedure for the isolation of biologically active polypeptides from marine organism(23,24)

References

1. Alsen, C., Béress, L., Fischer, K., Proppe, D., Reinberg, T. and Sattler, R.W.: Naunyn-Schmiedeberg's Arch. Pharmacol. 295, 55-62 (1976).
2. Alsen, C. and Reinberg, T.: Bull. Inst. Pasteur 74, 117-119 (1976).
3. Alsen, C., Béress, L. and Tesseraux, I.: Toxicon 16, 561-566 (1978).
4. Alsen, C., Harris, J.B. and Tesseraux, I.: Brit. J. Pharmacol. - Submitted (1981).
5. Béress, L. and Béress, R.: Kieler Meeresforschungen 27, 117-127 (1971).
6. Béress, L., Béress, R. and Wunderer, G.: FEBS-Letters 50, 311-314 (1975).
7. Béress, L., Béress, R. and Wunderer, G.: Toxicon 13, 359-364 (1975).
8. Béress, L. and Zwick, J.: Marine Chem. 8, 333-338 (1980).
9. Béress, R., Béress, L. and Wunderer, G.: Hoppe-Seyler's Z. Physiol. Chem. 357, 409-414 (1976).
10. Bergmann, C., Dubois, J.M., Rojas, E. and Rathmayer, W.: Biochim. Biophys. Acta 455, 173-184 (1976).
11. Blanquet, R.: Perspectives in Toxicology, ed. A.W. Bernheimer., John Wiley & Sons, N.Y. p. 149, 1977
12. Catterall, W.A.: Ann. Rev. Pharmacol. Toxicol. 20, 15-43 (1980).
13. Norton, T.R., Shibata, S., Kashivagi, M. and Bantley, J.: J. Pharm. Sci. 65, 1368-1374 (1976).
14. Prescott, B., Thomas, G.J., Jr., Béress, L., Wunderer, G. and Tu, A.T.: FEBS-Letters 65, 144-147 (1976).
15. Rathmayer, W.: Advances in Cytopharmacology, ed. B. Ceccarelli and F. Clementi, Raven Press, New York, Vol. 3, 335-344, 1979
16. Rathmayer, W. and Béress, L.: J. Comp. Physiol. 109, 373-382 (1976).
17. Ravens, U.: Naunyn-Schmiedeberg's Arch. Pharmacol. 296, 73-78 (1976).
18. Richet, C.: Compt. Rend. Soc. Biol. 55, 246-248 (1903).
19. Richet, C.: Arch. Physiol. (Pflügers) 108, 369-388 (1905).
20. Ritchie, J.M. and Rogart, R.B.: Rev. Physiol. Biochem. Pharmacol. 79, 1-49 (1977).
21. Romey, G., Abita, J.P., Schweitz, J., Wunderer, G. and Lazdunski, M.: Proc. Natl. Acad. Sci., USA, 73, 4055-4059 (1976).

22. Shibata, S., Norton, T.R., Izumi, T., Matsuo, T. and Katsuki, S.: J.Pharmacol.Exp.Ther. 199, 298-309 (1976).

23. Theede, H., Schneppenheim, R. and Béress,L.: Marine Biol. 36, 183-188 (1976).

24. Wunderer, G., Béress, L., Machleidt, W. and Fritz, H.: Methods in Enzymology 45, 881-888 (1976).

STRUCTURE AND PROPERTIES OF MASTOPARAN II - OLIGOPEPTIDE FROM THE VENOM OF HORNET *Vespa orientalis*

Igor Nasimov, Leona Snezhkova, Olga Reshetova, Anatoly Miroshnikov

Shemyakin Institute of Bioorganic Chemistry, USSR Academy of Sciences, Moscow, USSR

Introduction

The primary structure of mastoparan II - histamine-releasing peptide from the venom of hornet *Vespa orientalis* was determined. The amino acid sequence of the peptide consisting of 14 amino acids was studied by means of solid phase method as well as by mass spectrometry of the short peptide received upon restricted acid hydrolysis and tryptic hydrolysis. The physicochemical properties of mastoparan II were compared with those of another native histamine-releasing peptide, 22-membered MCD-peptide from the venom of honey bee *Apis mellifera*.

Results

The generally accepted molecular mechanism of allergen reaction is based on the assumption of antigen (allergen) binding by two molecules of immunoglobulin E, fixed by its F_c-fragments on the surface of mast cell membrane. Upon subsequent degranulation of mast cell the release of biogen amines (histamine) takes place the allergen reaction being the external display of this event (1).
Venoms of some insects contain polypeptides which can degranulate mast cells with histamine-releasing. 22-membered MCD-peptide with such properties was isolated from the venom of

Chemistry of Peptides and Proteins, Vol. 1

the honey bee *Apis mellifera* (2), 14-membered peptide (mastoparan), found in the venom of the wasp *Vespula lewisi* (3, 4). Histamine-releasing (HR) studies of the peptides differing in their chemical composition is of considerable interest as there are reasons to assume this effect to be determined by interaction of the positively charged amino acids with the negatively charged groups on membrane surface of target cells.
That is why the study of the primary structure and physicochemical properties of HR peptides is the substantial step for creation of a model of spatial arrangement of groups necessary for start of histamine secretion and, finally, for estimation of the nature of cell receptors, which release histamine selectively.
We have found the 14 membered peptide (mastoparan II) from the venom of the hornet, *Vespula orientalis*, to stimulate HR from mast cells, being quite different in chemical structure from the MCD-peptide (5). Amino acid composition and N-terminal amino acid of mastoparan II are shown in Table 1. C-terminal determi-

Table 1. Amino acid analysis of the mastoparan II.

Amino acid		
Alanine	3.1	(3)
Glycine	0.2	(0)
Isoleucine	1.9	(2)
Leucine	3.3	(3)
Lysine	2.9	(3)
Aspartic acid	0.9	(1)
Valine	1.9	(2)
Total	14	
N-terminus: Isoleucine		

nation of this peptide by means of CPA, CPB, CPC failed and allows to suppose the C-terminal amino acid to be α-amide. N-terminal amino acid sequence of mastoparan II, containing 3 Lys residues, was determined by the solid phase method. After DITC - coupling of the peptide to the γ-aminopropylglass and 14 cycles of Edman degradation the following amino acid sequence was determined:

$$\text{Ile-Asn-Leu-}X_1\text{-Ala-Ile-Ala-Ala-Leu-Val-}X_2\text{-}X_3\text{-}X_4\text{-}X_5\ .$$

On DNS analysis of the tryptic hydrolysate the presence of DNS-Ile, DNS-Ala, DNS-Leu, DNS-Val, as well as of α,ε-di-DNS-Lys was shown. Taking into account the amino acid composition of mastoparan II, and the results of solid phase sequence and N-terminal determination of tryptic hydrolysate one can see that residues X_1, X_2, X_3 can be Lys residues only.

For final reconstruction of the molecule the analysis of products of the restricted acid hydrolysis of mastoparan II

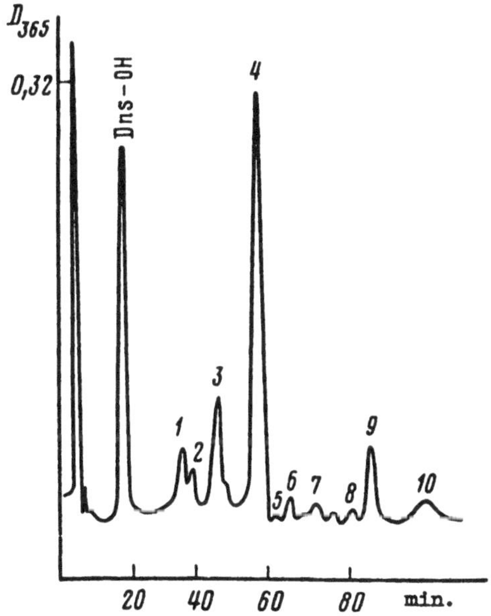

Fig. 1. Separation of mixture of methyl esters of dansylated tryptic hydrolysate of mastoparan II.

(5,7 N HCl, 110°, 45 min) by mass spectrometry was applied and peptide fragments were analysed as methyl esters of DNS-peptides.

After dansylation and esterification the hydrolysate was separated on RP column, eluate being monitored by absorption at 365 nm. Each of fractions 1-10 (Fig. 1) was evaporated, the residue was transfered into an ampule and placed into the ion source of mass spectrometer LKB 9000. Electron impact mass spectra were obtained at electron energy 70 eV, ion source temperature 230-250°, evaporator temperature 20-150°. MS analysis showed the presence of both individual peptides and the mixtures of 2-3 peptides (fractions 4-7, 9, 10). The mixed fractions were separated by thermal evaporation in a direct inlet system of the mass spectrometer.

Mass spectra of methyl esters of DNS-peptides contain intensive peaks of molecular ions $|M|^+$ as well as peaks of ions of the dansylated amine fragments (6, 7): $|\text{DNS-NH-CH(R}_n\text{)-CO-NH=CH(R}_{n+1}\text{)}|^{+\cdot}$.

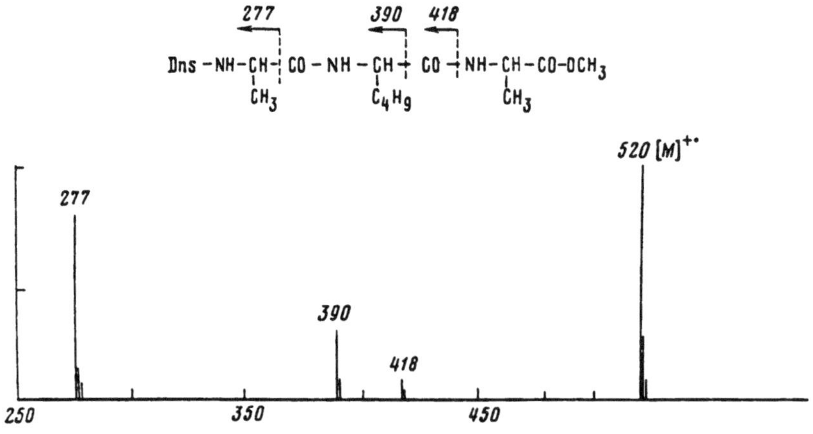

Fig. 2. Mass spectrum of DNS-Ala-Ile-Ala-OMe.

Peaks of ions derived due to elimination of the amino acid side chains were absent in mass spectra. The mass spectrum of DNS-Ala-Ile-Ala-OMe (fraction 7, Fig. 1) is presented in Fig.2 as the typical one. The characteristic features of the mass spectrum are the intensive peak of molecular ion (m/z 520) as well as peaks of ions of charged amine fragments (m/z 390, 277). The majority of peptides produced upon the restricted acid hydrolysis of mastoparan II were identified by the similar way. Lys-containing peptides were not found in the products separated by the RP chromatography. The analysis of the mass spectra of total dansylated and esterified restricted acid hydrolysate, obtained at evaporation temperatures of 150° showed the presence of DNS-Leu-Lys-OMe, DNS-Lys-Ala-OMe, DNS-Val-Lys-OMe, DNS-Lys-Val-OMe.

In mass spectra of modified products of nonspecific acid hydrolysis DNS-Val-Leu-amide was detected as well as DNS-Val-Leu-OMe. The former (amide) was coeluted with DNS-Ile-OMe and DNS-Ile-Ala-OMe upon RP chromatography. For reliable identification of C-terminal fragment of mastoparan II the tryptic hydrolysate of this peptide was dansylated and separated on RP column (Lichrosorb RP 18) in methanol-water with methanol gradient

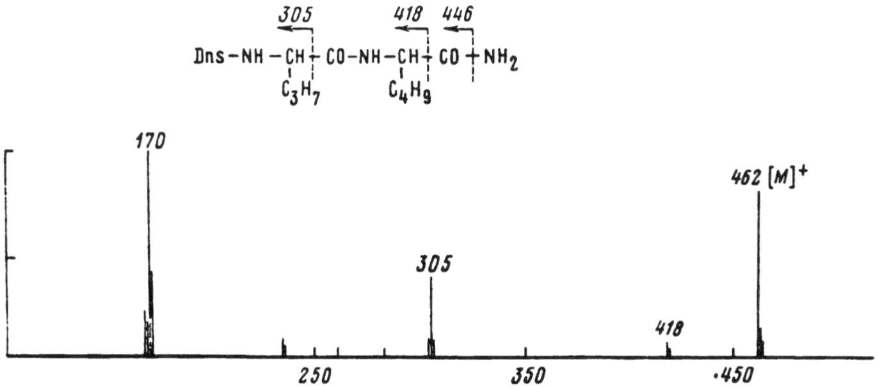

Fig. 3. Mass spectrum of DNS-Val-Leu-NH_2.

from 50 to 100%. As a result the DNS-Val-Leu-amide was isolated in a pure state. Mass spectrum of this compound (Fig. 3) contains an intensive peak of molecular ion (m/z 462) as well as peaks of amino fragments (m/z 418, 305) and the peak of dimethylaminonaphthalene (m/z 170).

Thus, the application of nonspecific acid hydrolysis for the mastoparan II, dansylation and esterification of short peptides followed by RP-HPLC, and mass spectrometry analysis of separated fractions in combination with MS analysis of DNS, OMe - derivatives as well as of DNS-Val-Leu-amide allowed us to determine the structure of this 14-membered peptide as

Ile-Asn
Ile-Asn-Leu
Leu-Lys
Lys-Ala
Ala-Ile
Ala-Ile-Ala
Ala-Ile-Ala-Ala
Ile-Ala
Ile-Ala-Ala
Ala-Leu
Ala-Leu-Val
Val-Lys
Lys-Val
Val-Leu-NH_2
Ile-Asn-Leu-Lys-Ala-Ile-Ala-Ala-Leu-Val-Lys-Lys-Val-Leu-NH_2 .

Peptides identified by the MS analysis of the DNS, OMe derivatives of tryptic hydrolysate are underlined.

Such an approach does not allow to distinguish neither Leu residue from Ile nor their relative positions in sequence (Leu-Ile or Ile-Leu). Thus their positions in the structure of mastoparan II were determined by sequenator analysis of the whole molecule of mastoparan II. On the other hand because of the presence of C-terminal α-amide it was not possible to attach

mastoparan II to the solid support through its C-terminus. So after DITC-coupling followed by sequenator analysis there was uncertainty concerning the C-terminal structure.
The joint use of sequenator and mass spectrometry analysis allowed to establish the amino acid sequence of mastoparan II. Taking into consideration the widespread of the C-terminal α-amides in the native toxic peptides we decided to develop the most reliable MS technique for determination of such derivatives as a constituent of the short peptides. We have compared the mass spectra of TFA derivatives of the Val-Leu-amide as well as of DNS derivatives of the same amide. This amide was treated with 20% trifluoroacetic anhydride in methylene chloride at 20° for one hour. Mass spectrum of the 5-TFA-oxazolyl-derivative showed a big set of nonintensive peaks of fragments of TFA derivative.
After the treatment of DNS-Val-Leu-NH_2 with trifluoroacetic anhydride a pronounced peak of molecular ion was obtained (m/z 540, Fig. 4).
From the comparison of the mass spectra of the different derivatives of the Val-Leu-NH_2 (obtained upon trifluoroacetylation, dansylation, dansylation plus trifluoroacetylation) it became

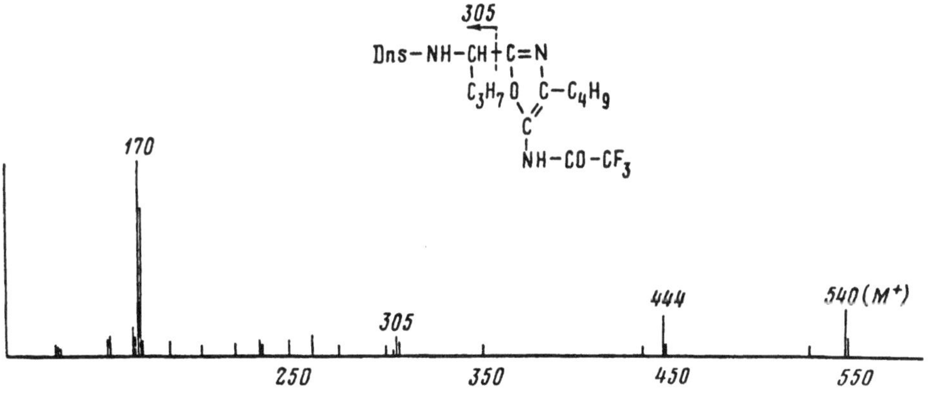

Fig. 4. Mass spectrum of trifluoroacetylated DNS-Val-Leu-NH_2.

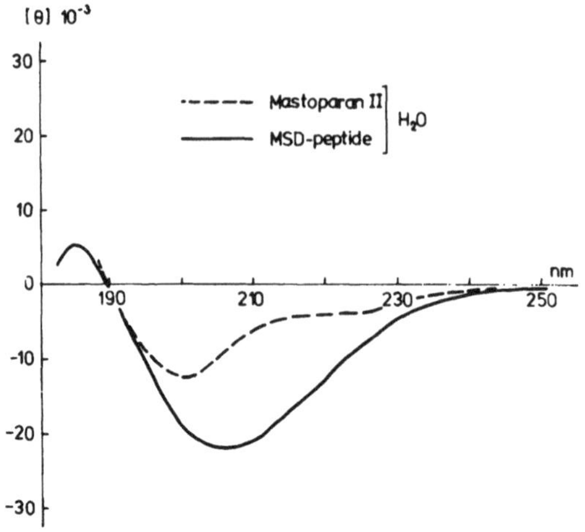

Fig. 5. CD spectra of mastoparan II and MCD-peptide in water.

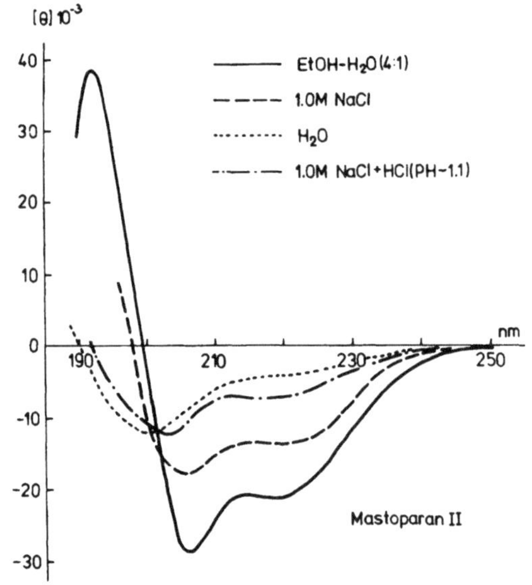

Fig. 6. CD spectra of mastoparan II in different solvents.

evident the second one to be the most convenient, as the derivatisation proceeds in one step, mass spectrum contains the abundant peak of molecular ion and does not contain fragment ions in high molecular weight region.

Thus we suppose the MS analysis of DNS-derivatives of the α-amidated short peptides to be a method of choice for quick reliable and sensitive determination of the α-amidated C-termini of the peptides and proteins.

The spatial structure of mastoparan II in solution was studied by means of circular dichroism. Whereas MCD-peptide, the most specific agent of HR from mast cells, possesses the rigid spatial structure provided by two disulfide bridges, linear polypeptide mastoparan II seems to have a set of conformations in solution as one can see from the CD spectra (Fig. 5).

It is interesting to note that when changing water for water-organic mixtures (Et-OH : H_2O=4:1) CD spectrum contains the positive band at 190 nm and the negative one at 208 nm with the shoulder at 220 nm (Fig. 6). Shape of the curve and the value of molecular ellipticity are similar to those of CD spectra of the classical helical polypeptides and proteins as well as to the CD spectra of MCD peptide. That indicates

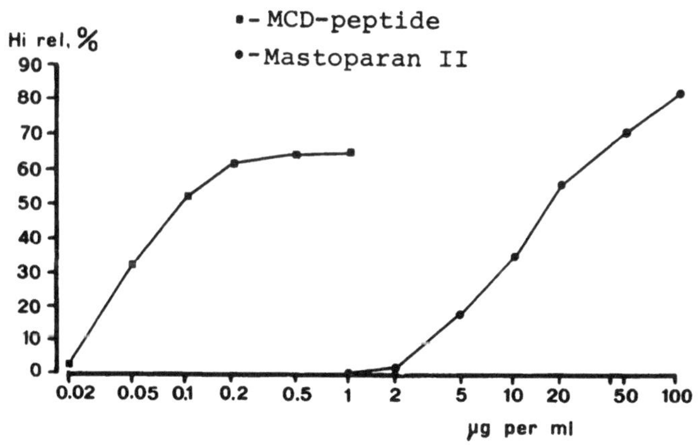

Fig. 7. Comparison of histamine-releasing action of MCD-peptide and mastoparan II on mast cells.

the spatial stability of bicyclic molecule of MCD-peptide consisting of 22 amino acid residues. It is possible that conformation of mastoparan II on the surface of membrane is similar to the conformation of MCD in solution.
When studing the HR activity of mastoparan II on mast cells it was shown that the biological action of the polypeptide from the venom of hornet is pronounced in higher concentrations than for MCD-peptide. Upon the action of MCD-peptide the HR reaches its maximum (70% of overall histamine content) at a concentration of about 0,2 mkg/ml, upon mastoparan II action the HR tends to increase progressively up to the total exhausting of histamine store in mast cell (Fig. 7).

References

1. Stanworth, D.R., Kings, M., Ray, P.D., Moran, J.M., Moran, D.M.: Biochemical J. 180, 665-668 (1979).
2. Habermann, E.: Science 177, 314-322 (1972).
3. Hirai, Y., Yashuhara, T., Yoshida, H., Nakajima, T., Fujino, M., Kitada, C.: Chem. Pharm. Bull. 27, N 8, 1942-1944 (1979).
4. Hirai, Y., Kawada, M., Yasuhara, T., Yoshida, H., Nakajima, T.: Chem. Pharm. Bull. 27, N 8, 1945-1946 (1979).
5. Miroshnikov, A.I., Snezhkova, L.P., Nasimov, I.V., Reshetova, O.I., Rosynov, B.V., Gushchin, I.S.: Bioorg. Khim. 7, N 10, in press.
6. Day, R.A., Falter, H., Lehman, J.P., Hamilton, R.E.: J. Org. Chem. 38, 782-788 (1973).
7. Marino, G., Malorni, A.: Adv. in Mass.Spectr. in Biochem. and Medicine, Spectrum Publications, Inc. 11, 249-255, 1976

ISOLATION, PHYSICO-CHEMICAL AND BIOLOGICAL PROPERTIES OF THE IMMUNITY POLYPEPTIDE BIOREGULATOR FROM THYMUS

O.A.Pisarev, V.G.Morosov, V.Kh.Khavinson, L.K.Shataeva, G.V. Samsonov

Institute of Macromolecular Compounds, Academy of Sciences of the USSR, Leningrad

Introduction

The main subject of modern immunology is to investigate the ways and means for the recovery of the disturbed immunological functions in the process of human vital activity. One of the hypotheses accounting for a T-lymphocyte formation is the assumption that in thymus a hormonal factor is produced, due to its action the marrow stem cells obtain T-lymphocyte properties and are further differentiated into T-cells participating in immunological reactions (1). Thymosin is a well-known isolated thymus factor (2). There are some reports on the thymic humoral factor isolation (3), thymopoietin (4) as well as the circulating blood thymus factor (5). This study had the aim of the isolation of a polypeptide fraction stimulating immunogenesis and the examination of its composition, physico-chemical and biological properties.

Results

The main stages of the compound production from thymus are represented in Fig.1. Before the stage of the ion-exchange chromatography more than 10 compounds of different origin were present. The multi-component extract was fractionated on Biocarb (6-8) into three fractions; electrophoresis on polyacrylamide gel showed that in fraction one there are nine com-

Chemistry of Peptides and Proteins, Vol. 1

pounds, in fraction 2 3 compounds and in fraction 3 there is 1 compound present. Fraction 1 eluates from the ion-exchange column at pH 3.0, fraction 2 at pH 6.5, fraction 3 at pH 9.5 - 10.0.

thymus tissue
acetone
(-4°, 48h)
homogenization
extraction
(6 x 3 % CH_3COOH + $ZnCl_2$, 72 h)
centrifugation
acetone
(-4°, 1 h)
dissolution in 5 % CH_3COOH
ion-exchange chromatography (Biocarb-T)
thymalin

Fig.1. Scheme of the isolation of thymalin.

According to gel chromatography on Sephadex G-25 and G-10 fraction 3 was also homogeneous. Investigations by ultracentrifigation demonstrated a symmetrical peak for fraction 3 with the sedimentation coefficient 0.7 Sw.

The percentages of each fraction, molecular masses, isoelectric points, tyrosine contents (by Lowry) and also the results of biological testing are shown in Figs. 2 and 3.

Fraction 1 has no biological activity. Fraction 2 stimulated immunological reactions 15-20 times stronger than compared to the extract. The biological activity of the extract is associated mainly with fraction 3. Intramuscular injection (0.5 mg/kg) of this fraction to the mouse increased the number of lymphocytes in blood by 50-70 %, and 0.1-0.5 mg/kg stimulated the formation of antibody forming cells threefold. At higher dose (25 mg/kg) this fraction had no stimulating effect and in some cases suppresed immunological reactions. Fraction 3 was named "thymalin" and was later used for clinical and biological tests.

Preparation	Dose mg/kg	Number of lymphocytes	Number of circulating antibody	Number of antibody forming cells	Phagocytic reaction	Activity of peroxidase
Fraction 1	25	-	-	-	-	-
	2.5	-	-	-	-	-
Fraction 2	25	±	±	-	±	±
	2.5	-	-	-	-	-
Fraction 3	25	+	±	-	±	±
	0.5	+	+	+	+	+
	0.1	+	±	+	±	+

Fig. 2. Effect of thymus preparations on the indices of immunologic reactivity.

Component	Content in extract(%)	Mol.	Isoelectr. points	Tyrosine content by Lowry %
Fraction 1	80	700	2.5	4.5
Fraction 2	15	5000	4.5	1.0
Fraction 3	5	5000	9.5	1.0

Fig. 3. Properties of thymic fractions.

Study of the thymalin amino acid composition confirmed its polypeptide structure. Fig.4 shows that thymalin is a basic polypeptide, consisting of 38 amino acid residues. In the thymalin structure there are 11 basic and 4 acid amino acids. N-terminal analysis performed according to (9) demonstrated that N-terminal amino acids are lysine and alanine.

Indices	Thymalin properties
Mol.	50 000
Isoelectric point	9.5
Number of amino acid residues	38
Amino acid composition	Asp_2, Thr_2, Ser_2, Glu_2, Trp_1, Pro_3, Gly_3, Ala_4, Val_1, Met_1, Ile_1, Leu_1, Tyr_1, Cys_1, Phe_1, His_1, Lys_8, Arg_3

Fig. 4. Physico-chemical properties of thymalin.

Conformational properties of thymalin were examined by the ORD method. Thymalin undergoes a cooperative conformational transition in the 8.5 - 10.5 range with $[\alpha]_D^{410°}$ changing from -230° to -165°. Estimates according to (10,11) reveal no ordered structure in the pH area of 4.0-8.5, whereas at pH 10.5 ca. 20 % α-helix is present.

The results obtained in the study of the thymalin effect on the immunity indices are given in Fig.5. The number of T- and B-lymphocytes was determined by E-POK and EAG-POK content in the lymphatic nodules and the spleen of the animals as described in (12). After thymectomy the number of caryocites, T-lymphocytes in the spleen,has decreased to a great extent. Cell composition in the lymphatic nodules has not changed. After thymalin injection the number of caryocites has completely recovered and the number of T- and B-lymphocytes has enhanced in the spleen. In the lymphatic nodules, the changes of caryocite content, T- and B-lymphocytes appeares less noticeable.

The effect of thymalin on T-cell population was confirmed by the investigation of its action in the culture of lymphocytes produced from thymectomyzed and pseudo-operated guinea pigs. The percentage of T-cells in the spleen months after the operation was 28.6 $\pm$ 2.9 for the pseudo-operated animals

Group of animals	Number of caryocites mln		Number of T-lymphocytes th/mg		Number of B-lymphocytes th/mg	
	Spleen	Lymphatic nodules	Spleen	Lymphatic nodules	Spleen	Lymphatic nodules
Pseudo-operated	58.5 ± 18.3	84.3 ± 7.2	158.4 ± 9.4	310.3 ± 19.9	342.5 ± 20.1	362.3 ± 27.3
Thymectomized	225.1 ± 16.2	69.1 ± 4.8	69.2 ± 12.0	278.1 ± 25.1	275.4 ± 23.2	372.2 ± 31.1
Thymectomized + thymalin	352.6 ± 16.8	85.2 ± 5.7	153.2 ± 7.6	295.5 ± 21.6	301.3 ± 25.8	385.2 ± 30.9

Fig. 5. Effect of the thymus factor on the indices of guinea pig immunity.

and 16.8 ± 20 for the thymectomized ones. When the thymus factor was applied the content of T-cells in the lymphocyte culture of thymectomized guinea pigs increased to the initial indices. The content of T-cells in the lymphocyte culture of the pseudooperated animals did not markedly change.

From the above data we assume that the decrease of the number of T-cells in the spleen after thymectomy occurs owing to T_1-population of lymphocytes in mature animals which are present mainly in thymus and spleen. Thymectomy of mature animals only slightly affects the cell composition of the lymphatic nodules containing mostly T_2 population of lymphocytes. Thus one can suggest that thymalin recovers the number of T-lymphocytes possessing a function of a helper in a certain stage of their development and thereby favours the definition of antigene and the induction of the cooperative immune response to thymus-depending antigenes. Due to the lack of thymus factor after thymectomy the number of T-lymphocytes decreases, this being the possible reason for supressed immunological reactivity of thymectomized animals.

Thymus-dependence of lymphocyte population as well as the helper activity of antigene-activated cells are evidently mediated by the presence of thymus factor on the surface of lymphocytes. Hence we may suggest that T-lymphocytes are the carriers of a thymus factor inducing a cooperative immune response to thymus-depending antigenes. The effect of the thymus factor on different physiological systems is likely to be associated with the function of T-cells which contain genetic information and they are the participants not only of immunologic but also of metabolic processes, such as the regeneration of tissues, the synthesis of protein and nucleic acids, carbohydrate and fat metabolism.

References

1. Burnet, F.M.: Lancet 1, 358-360 (1970).
2. Goldstein, A.L., Guna, A.: Proc.Natl.Acad.Sci. 69, 1800-1810 (1972).
3. Trainin, N., Small, M.: J.Exp.Med. 132, 885-897 (1970).
4. Basch, R., Goldstein, G.: Proc.Natl.Acad.Sci. 71, 1474-1478 (1974).
5. Bach, A.L., Dardenne, M., Pleau, M.A., Bach, M.A.: Ann.N.Y.Acad.Sci. 249, 186-210 (1975).
6. Morozov, V.G., Khavisnon, V.Kh., Pisarev, O.A.: Proc.Acad.Sci. USSR. 233, 491-494 (1977).
7. Pisarev, O.A., Shataeva, L.K., Samsonov, G.V.: Bull.Acad.Sci., USSR. 29(1), 37-42 (1980).
8. Pisarev, O.A., Shataeva, L.K., Samsonov, G.V.: J.Physical Chemistry, USSR. 54. 2391-2395 (1980).
9. Cray, W.R.: Methods in Enzymol.11, 469-475 (1967).
10. Moffit, W., Yang, J.T.: Proc.Natl.Acad.Sci.42, 596-600 (1956).
11. Shechter,E., Blout, E.: Proc.Natl.Acad.Sci. 51, 794-798 (1964).
12. Stadecker, M.J., Bishop, G., Wortiz, H.H.; J.Immunol.3, 1834-1837 (1973).

SOME RECENT ADVANCES IN THE METHODS OF PROTEIN SEQUENCE ANALYSIS

Agnes Henschen, Friedrich Lottspeich,
Max-Planck-Institut für Biochemie,
D-8033 Martinsried/München, Germany

Wolf Brandt and Claus von Holt
Department of Biochemistry, University of Cape Town,
Rondebosch 7700, South Africa

Introduction

The Edman degradation method (1) was first described more than 30 years ago(2). However, since then the method has perpetually been refined, improved and modified. The most important improvement was, obviously, the invertion of the protein sequenator by Edman and Begg 14 years ago (3). Two problems in protein sequence analysis have received much attention: how to obtain sequence information with ever less material and how to obtain ever longer direct sequences. The present communication deals with aspects of both problems. A novel high performance liquid chromatography system for the identification of the amino acid phenylthiohydantoin (PTH) derivatives is presented; the method is of considerable advantage in microsequence analysis. Furthermore, some recent contributions to the understanding of the chemistry of the phenylisothiocyanate degradation, including the chemistry of the major side reaction, are described.

Towards Microscale Sequence Analyses

The implication of "micro" in "microsequencing" is gradually changing, as more and more sensitive methods are being developed. At present, the term "microsequencing" is normally used when 10 nanomoles of material,

Chemistry of Peptides and Proteins, Vol. 1

or less, is used as starting material for the sequencial degradation. Scaling down the methods of protein structure elucidation creates problems on many levels. An obvious problem is to develop a sufficiently sensitive identification method for the end products of the sequence analysis, i.e. the PTH derivatives of the amino acids. Earlier, thin-layer chromatography was the most commonly used identification method, but this method suffers from being insufficiently sensitive as the detection limit is around 1 nanomole.

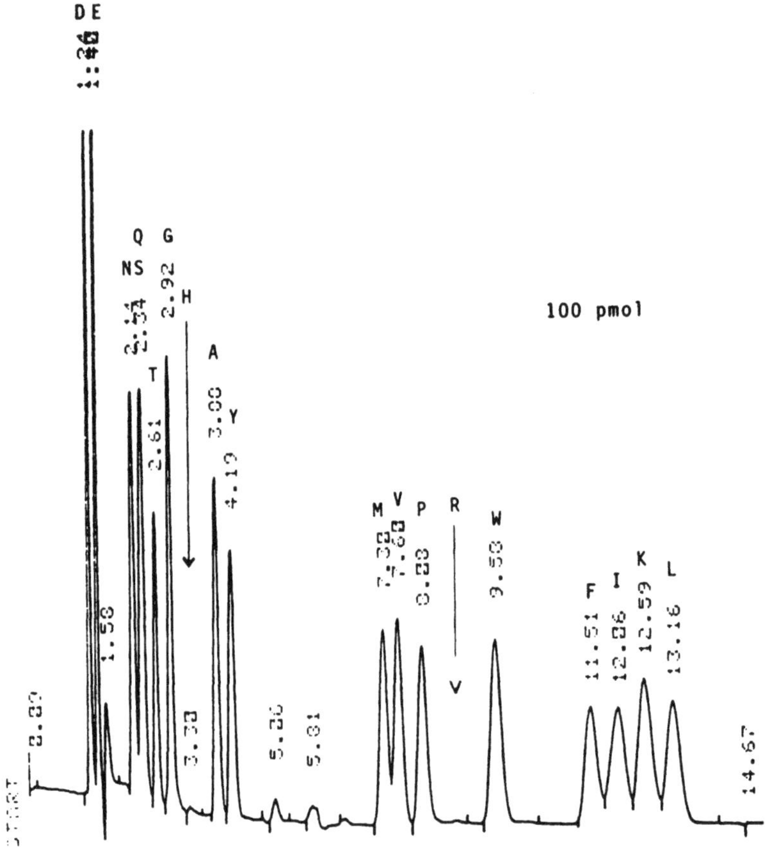

Fig. 1. High performance liquid chromatography of PTH derivatives of amino acids, 100 picomoles of each amino acid derivative being applied. Detection wave-length: 254 nm at 0.006 absorption units full scale. (From ref. 6.).

Since a few years, high performance liquid chromatography is steadily gaining in popularity as identification method (4). A newly developed, isocratic solvent system (5,6) offers special advantages at microscale sequencing, as isocratic elution gives a stable background ultraviolet absorption, allowing highly sensitive effluent monitoring. The chromatography conditions are given in the table. Chromatographies of PTH-amino acid reference mixtures at the 100 picomole and 5 picomole level are shown in Fig.s 1 and 2, respectively. In Fig. 2 the base-line obtained

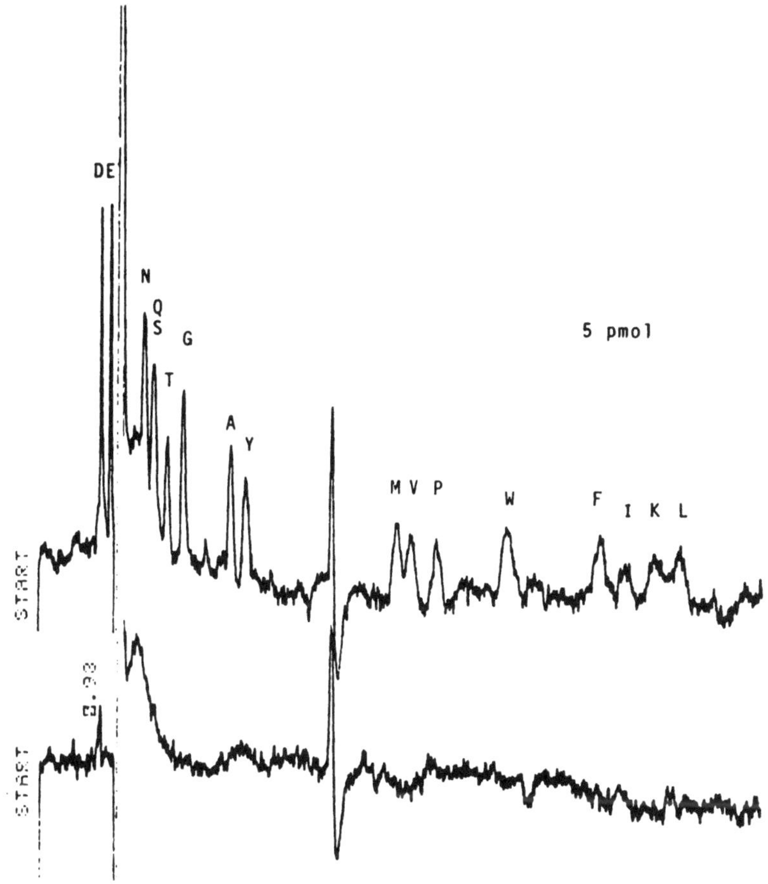

Fig. 2. High performance liquid chromatography of PTH derivatives of amino acids, 5 picomoles of each amino acid derivative (upper curve) or solvent only (lower curve) being applied. Detection wave-length: 254 nm at 0.0016 absorption units full scale. (From ref. 6.).

at solvent injection at the highest recording sensitivity is included. Such an even base line is difficult to obtain with gradient elution, even when the differences in absorption of the two components is corrected by, e.g. azide or acetone, or when the background is corrected by computerised blank-run subtraction. In addition to the PTH-amino acids shown in Fig. 1, the PTH derivatives of S-carboxymethylcysteine, methionine sulfone, hydroxyproline, hydroxylysine, aminobutyric acid, norleucine, the methyl esters of aspartic and glutamic acid and many others may be separated by the solvent system described (6). The system has now been successfully used during two years, for the identification of the out-put from sequenator runs on 0.5-5 nanomoles of protein or peptide material.

Towards Longer Direct Sequence Analyses

Long direct sequence analyses can only be obtained with the help of sequenators. The main difficulties encountered when long sequenator runs have to be interpreted are caused by: "overlap" of the amino acid derivatives from their appropriate cycle to the following one(s), gradual decrease in yield of the relevant amino acid derivatives and gradual increase in "background" of the non-relevant amino acid derivatives.

Table. High performance liquid chromatography conditions for the separation of PTH derivatives of amino acids.

Column	Hibar 250-4, LiChrosorb RP-18, 5µ (Merck, Darmstadt)
Buffer	68.5% 0.01 M sodium acetate, pH5.2 31.5% acetonitrile +0.5% dichloroethane
Flow	1.5 ml/min
Temperature	62° C
Analysis time	15 min
Detection limit	3 picomole

It has been noticed for some time that the most serious "overlaps" appear after the proline residues in the sequence (7). It could be shown that the "overlaps" were due to the slow and therefore incomplete release of the proline derivative during the acid cleavage. The different prolines along a sequence may cleave off at different rates, as exemplified in Fig. 3. However, with prolonged cleavage or higher cleavage temperature (Fig. 4) complete release can be achieved. Thus, once the positions of the prolines are known, it is possible to modify the cleavage conditions at the corresponding positions. In this way, longer direct sequences can be obtained.

Recently, the major cause of the decrease in yield and increase in "background" has been recognised (8). It was found that both were due to a hitherto unknown specific cleavage

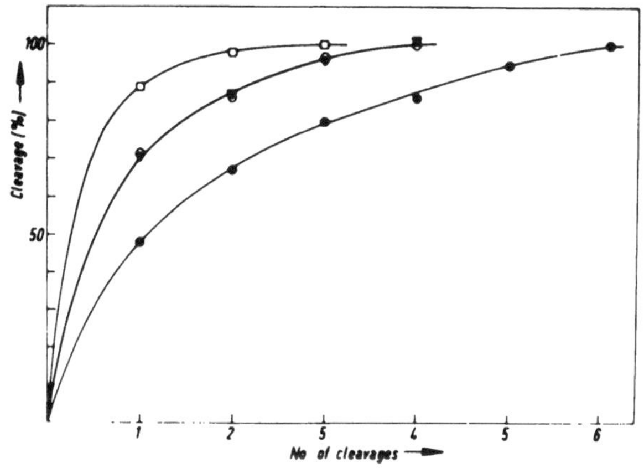

Fig. 3. Proline cleavages at sequenator degradation of the sequence P(1)-E(2)-P(3)-A(4)-K(5)-S(6)-A(7)-P(8)-A(9)-P(10)-K(11)- in chicken erythrocyte histone H2B. Cleavage course of P(1): o, P(3): □, P(8): ▼ and P(10): ●. (From ref. 7.).

at aspartic acid residues and furthermore to cleavage at threonine and serine residues in the peptide chain during the incubation with the anhydrous heptafluorobutyric acid. The "background" is obtained from the gradually appearing new N-terminal peptide chain ends and the decrease in yield is due to the washout of the small peptides formed at the cleavages.

The reaction was studied on a model peptide, Phe-Asp-Ala-Ser-Val (a generous gift of Professor E. Wünsch). Analysing the N-terminals formed at the incubation of this peptide with acid (Fig. 5, right side), it could be concluded that the cleavage occurred on the C-terminal side of aspartic acid, on the N-terminal side of serine (threonine) and on the N-terminal side of the C-terminal amino acid. The cleavage at serine and threonine is due to the well-known N → O shift mechanism shown in Fig. 6, lower scheme. The cleavage at

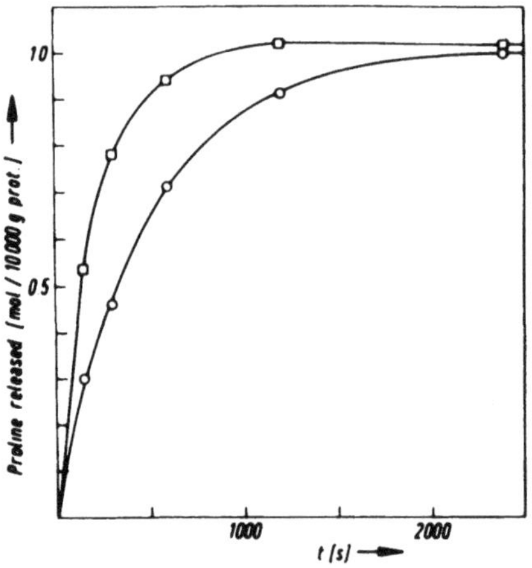

Fig. 4. Proline cleavage at degradation of the sequence P(1)-R(2)- in salmine, at 50°: o and 60°: □. (From ref. 7.)

aspartic acid differs from the one in dilute acid, where cleavage is obtained on both sides of aspartic acid. It is suggested that the reaction mechanism is the one indicated in Fig. 6, upper scheme. A similar mechanism could then be envisaged for the release of the C-terminal amino acid, both mechanisms involving the free carboxyl group.

The function of the carboxyl group in the cleavage reaction could be demonstrated by blocking it with an ester group (8). After esterification the corresponding peptide bonds were stable (Fig. 5, left side). The hydroxyl groups could also be stabilised by blocking. For this purpose acetylation was used (8).

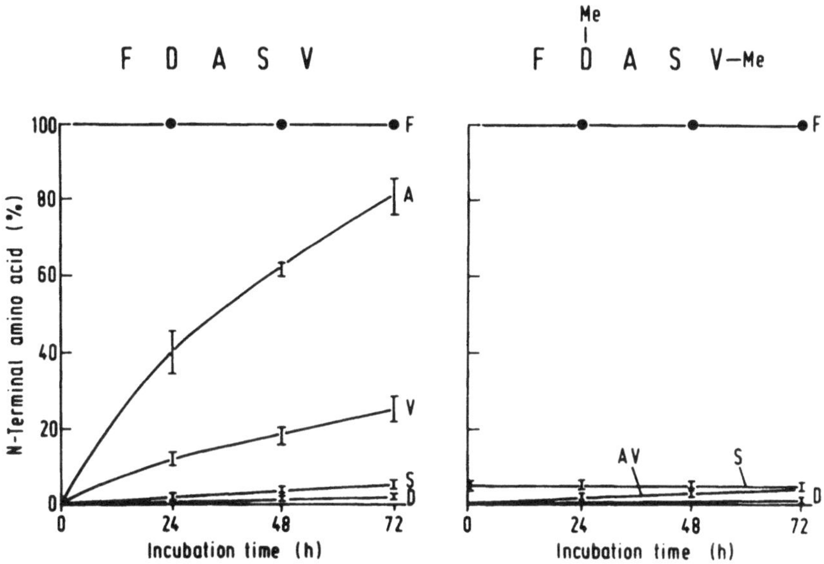

Fig. 5. Appearance of new N-terminal amino acids on incubation of the peptide Phe-Asp-Ala-Ser-Val with anhydrous heptafluorobutyric acid at 55°, before (left) and after (right) esterifying the carboxyl groups with methanol. The original N-terminal, Phe, is used as an internal standard. (From ref. 8.).

In proteins the carboxyl groups could be blocked by amidation with glycine methyl ester and the hydroxyl groups by acetylation. In this way the "background" was very considerably reduced. However, the acetylation of the hydroxyl groups gave the disadvantage of an O → N shift when the corresponding amino acid reached the N-terminal position, resulting in a blocked N-terminal. Thus, a different blocking group has to be

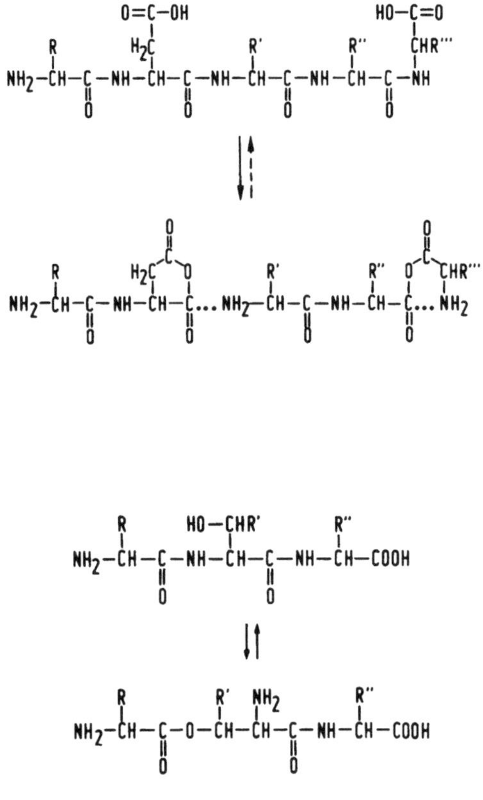

Fig. 6. Suggested rearrangements during incubation with anhydrous heptafluorobutyric acid leading to cleavage at aspartic acid and C-terminal amino acid (upper scheme) and at threonine or serine (lower scheme).

found so that the reaction may be optimised. It seems, however, reasonable that modifying the side chains contributing to the above-mentioned sequencing side reactions will enable us to obtain longer direct sequences on less material.

References

1. Edman, P., Henschen, A.: in Protein sequence determination (Needleman, S.B., ed.), 2nd edn., Springer, Berlin, pp.232-279, 1975
2. Edman, P.: Arch. Biochem. 22, 475-476 (1949).
3. Edman, P., Begg, G.: Eur. J. Biochem. 1, 80-91 (1967).
4. Zimmerman, C.L., Appella, E., Pisano, J.J.: Anal. Biochem. 77, 569-573 (1977).
5. Henschen-Edman, A., Lottspeich, F.: in Methods in peptide and protein sequence analysis (Birr, C., ed.), Elsevier/North Holland, Amsterdam, pp. 105-114, 1980
6. Lottspeich, F.: Hoppe-Seyler's Z. Physiol. Chem. 361, 1829-1834 (1980).
7. Brandt, W.F., Edman, P., Henschen, A., von Holt, C.: Hoppe-Seyler's Z. Physiol. Chem. 357, 1505-1508 (1976).
8. Brandt, W. F., Henschen, A., von Holt, C.: Hoppe-Seyler's Z. Physiol. Chem. 361, 943-952 (1980).

FIELD DESORPTION MASS SPECTROMETRY OF OLIGOPEPTIDES

Wolfgang Voelter

Abteilung für Organische und Physikalische Biochemie des Physiologisch-chemischen Instituts der Universität Tübingen, Hoppe-Seyler-Straße 1, D-7400 Tübingen-1,FRG

Michael Przybylski

Institut für Organische Chemie, Universität Mainz, Johann-Joachim-Becher-Weg 18-20, D-6500 Mainz, FRG

Introduction

Electron impact (e.i.) mass spectrometric analysis of oligopeptides has been greatly hampered by their involatility. Therefore, several methods have been developed to enhance the volatilization by derivatization and thus enable mass spectroscopic investigation of peptides by conventional electron impact techniques (1,2). However, only up to penta/hexapeptide derivatives are investigable by this approach. Especially for investigation of highly polar and involatile compounds field desorption mass spectrometry has become a useful tool(3). Although still limited, the present results demonstrate that at least partly useful sequence information can be obtained from f.d. fragmentation of peptides.

Experimental conditions

Since the initial development of f.d. mass spectrometry (4) a lot of progress was achieved experimentally (5,6). In most laboratories the field ion emitters are prepared according to reference 7 by high temperature activation of 10 µm tung-

Chemistry of Peptides and Proteins, Vol. 1

sten wires with benzonitrile. By this procedure microneedles of 20-30 μm length are produced on the metal surface. The sample is brought on the emitter by the syringe (8,9) or dipping technique (5). For the syringe technique the emitter is loaded with 10-20 μg of peptide while for the dipping procedure concentrations of 50-100 μg/20 μl are needed. The peptides are dissolved in DMSO, DMSO/H_2O or DMSO/H_3O^+. All spectra discussed in this report are recorded with a Varian MAT CH 711 double focussing spectrometer equipped with an e.i. (electron impact)/f.i. (field ionization)/f.d. (field desorption) combination ion source. The emitter after being introduced into the ion source is heated gradually in about 3 to 4 minutes from ambient temperature to a heating current of 25 mA. Conditions achieving the best anode temperature (BAT) are: 1000 (10 %) resolution, 11 kV potential between field emitter anode and cathode, 180-220°C source filament temperature and 2.5 kV electron multiplier voltage. E.i. spectra of high-boiling perfluorokerosine are used for mass calibration.

Field desorption mass spectra of protected oligopeptides

For the synthetic peptide chemist purity and structural control of the intermediates by different chromatographic and spectroscopic methods is one of the most urgent demands. Concerning the spectroscopic methods besides ^{1}H-NMR and ^{13}C NMR (10-19) mass spectra (20-24) yield the most valuable structural information of protected peptide samples. Different mass spectrometric methods for the structure analysis of oligopeptides were developed recently (1,2,25). Especially derivatization reactions to increase the volatility of the samples were investigated extensively (26). According to the method of Biemann after multiple derivatization steps up to penta-/hexapeptide derivatives are investigable by gas chromatography-electron impact mass spectrometry (27). This approach is also suitable for the sequence determination of protected or unprotected peptide chains with up to 50 amino acid residues (28). During the necessary hydrolysis steps, however, the information about the protective groups is lost. Protecting groups used by the synthetic chemist in most cases do not improve the volatility of the peptide because these residues are labile and are often of high molecular weight. In connection with our studies in the field of hypothalamic hormones f.d. mass spectra of protected oligopeptide sequences used for the synthesis of thyrotropin-releasing hormone and somatostatin respectively their derivatives were recorded and interpretated (6,29-34).

As so far only few f.d. mass spectra of protected peptides were investigated the spectra of some amino acid derivatives with commonly used protecting groups should demonstrate the characteristic fragmentation patterns.

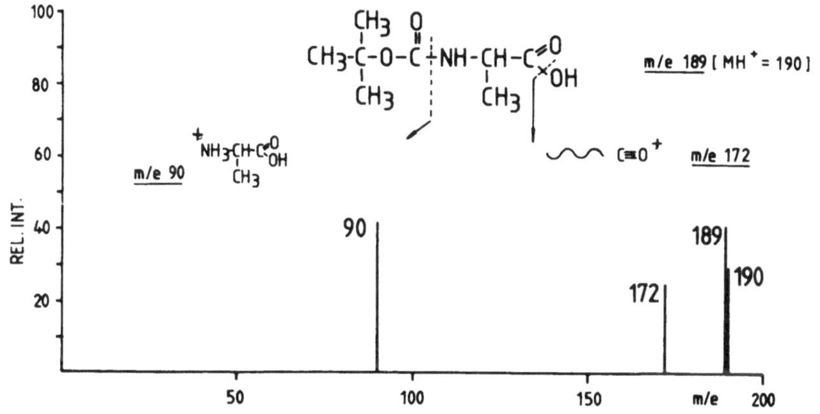

Figure 1. F.d. mass spectrum and interpreted fragmentation of Boc-Ala-OH at 8 mA e.h.c. (emitter heating current); solvent: DMSO.

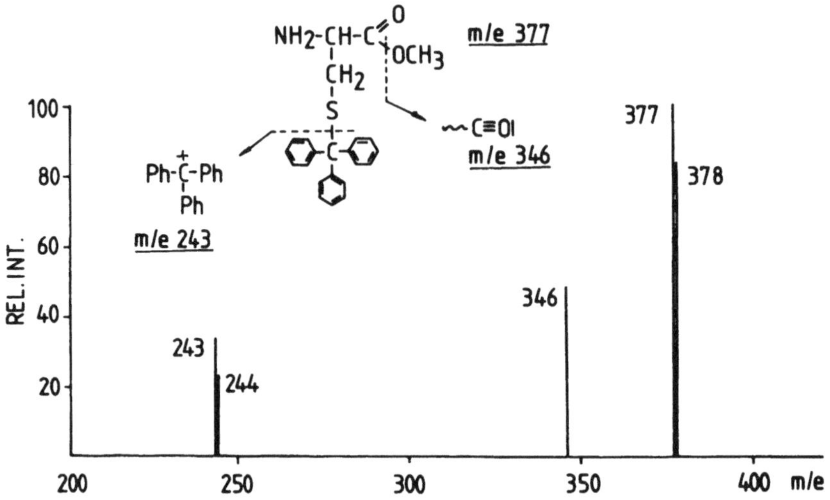

Figure 2. F.d. mass spectrum of H-Cys(Trt)-OCH$_3$ and interpreted fragmentation at 6 mA e.h.c.; solvent: DMSO.

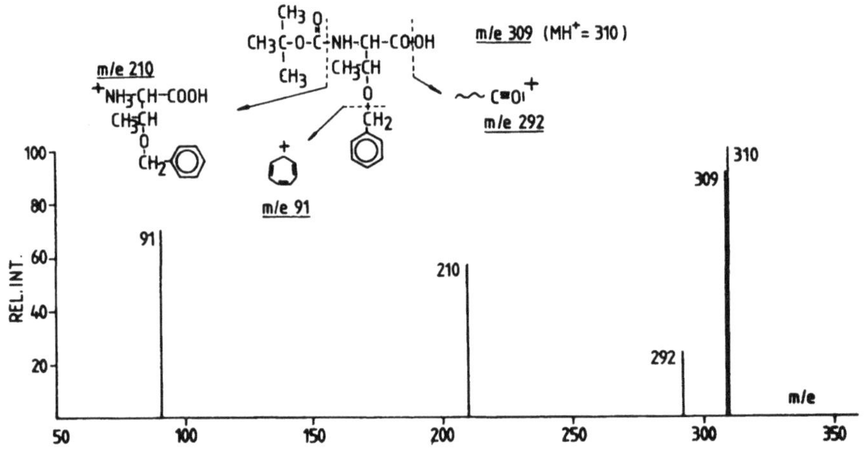

Figure 3. F.d. mass spectrum of Boc-Thr(Bzl)-OH and interpreted fragmentation at 13 mA e.h.c.; solvent:DMSO.

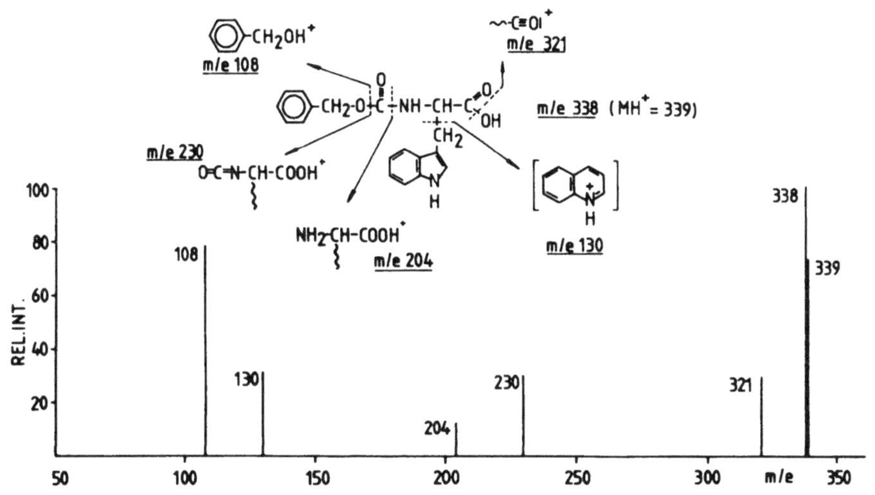

Figure 4. F.d. mass spectrum of Z-Trp-OH and interpreted fragmentation at 12 mA e.h.c.; solvent: DMSO.

From figures 1 to 4 the following conclusions can be drawn: The presence of protecting groups, even if thermolabile, increases the desorption rate from the surface of the field emitter anode, the molecular ion peaks show increased intensity and the fragments arise by characteristic cleavage of the residues of the protecting groups, e.g. M-Boc, M-Tpm, M-Bzl, M-Z. These facts support former f.d. mass spectroscopic investigations on peptides, protectded with the dimethoxybenzhydryl (Mbh) group, Z-Glu(Mbh)-Ser(Bzl)-Gly-NH_2 and Z-Glu(Mbh)-Phe-Pro-NH_2, in which intense molecular ions at m/e 740 respectively 750 are observed. However, the fragmentation of the peptide backbone is greatly reduced and only two typical fragments arising from the cleavage of an anisole moiety is observable (35).

During our synthetic approaches to somatostatin (36-39) we used f.d. mass spectra intensively for the characterization and structural information of the intermediates(40). The peptide Boc-Ala-Gly-Cys(Trt)-OMe with the natural sequence of somatostatin was synthesized according the following scheme:

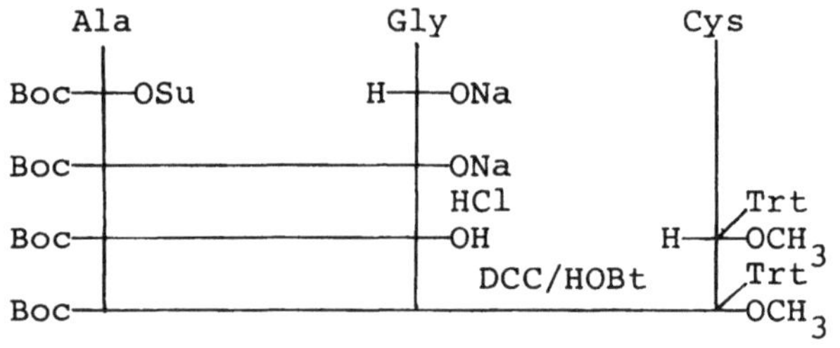

Figure 5. Synthesis of Boc-Ala-Gly-Cys(Trt)-OMe.

Figure 6 shows the field desorption mass spectrum of the protected tripeptide. Similar to the fragmentation pattern of protected amino acids characteristic ions arise from the t-butyloxycarbonyl and trityl groups but besides also a cleavage of the peptide backbone is observed.

The f.d. mass spectrum (figure 8) of the intermediate Boc-Ala-Gly-Cys(Trt)-Lys(Boc)-Asn-Phe-Phe-OMe, the synthesis of which was achieved according to the scheme of figure 7, again shows characteristic fragments arising from the cleavage of the Boc and trityl groups.

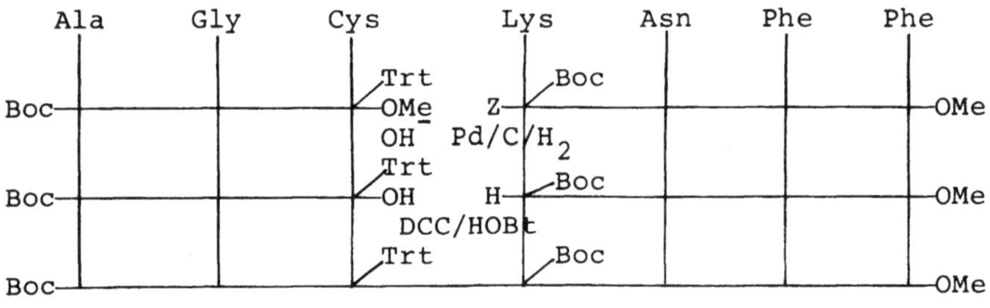

Figure 7. Synthesis of Boc-Ala-Gly-Cys(Trt)-Lys(Boc)-Asn-Phe-Phe-OMe.

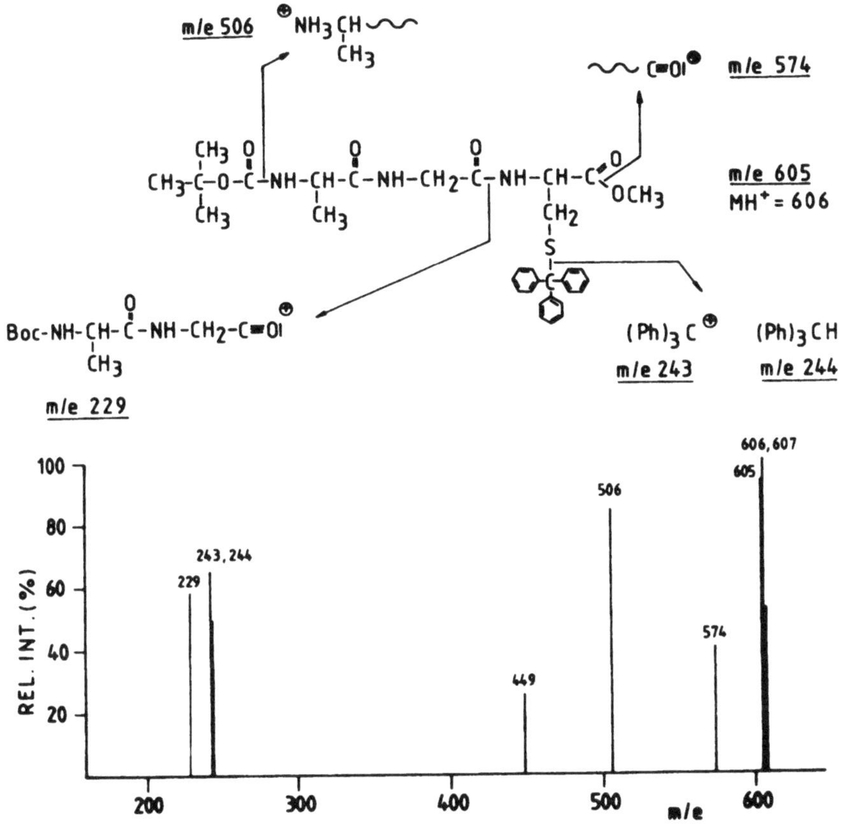

Figure 6. F.d. mass spectrum of Boc-Ala-Gly-Cys(Trt)-OMe and interpreted fragmentation at 10 mA e.h.c.; solvent: DMSO.

The base peak at m/e 243 is caused by the triphenylmethyl ion; the corresponding fragmentary peak of the molecule at m/e 1000/1001 (M-243, M-244) is also of relative high intensity. The peak at m/e 969 arises from the cleavage of the OCH_3 group.

The loss of one molecule of water (most probably from the Asn side chain) from the M-100 fragment is indicated by the peaks at m/e 1125/1126. Under the experimental conditions only the N^{α}-Boc group of the alanine residue is cleaved off, the N^{ε}-Boc group remains intact.

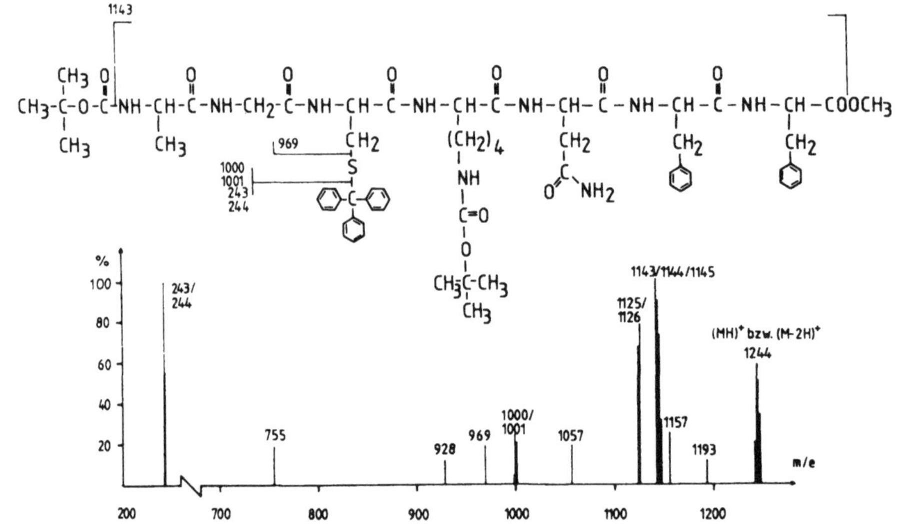

Figure 8. F.d. mass spectrum of Boc-Ala-Gly-Cys(Trt)-Lys(Boc)-Asn-Phe-Phe-OCH_3 at 15 mA e.h.c. and interpretation of characteristic fragments; solvent: DMSO.

In one of the strategies used for the total synthesis of somatostatin (36-40) the protected fragment with the natural sequence 1-11 of somatostatin was produced according to the scheme in figure 9. The protected tetrapeptide methylester was synthesized via its benzyloxycarbonyl derivative Z-Trp-Lys(Boc)-Thr(Bzl)-Phe-OMe; its structural proof was also achieved by f.d. mass spectrometry (figure 10). Characteristic fragments arise not only from the cleavages of the protecting groups (Z and Bzl) but also from those of the peptide backbone; as observed for the f.d. mass spectroscopic properties of Boc-Ala-Gly-Cys(Trt)-Lys(Boc)-Asn-Phe-Phe-OCH_3 the N^{ε}-Boc group is not subjected to cleavage under the experimental conditions mentioned in Figure 10. Besides the fragmentation symbolized by dashed lines in figure 10 further peaks are interpretable: at m/e 903: $(MH-H_2O)^+$, m/e 842: $(M-C_6H_6)^+$, m/e 642: $(Z\text{-Trp-Lys-Thr(Bzl)-C}{\equiv}O)^+$. The f.d. mass spectrum of the synthetic protected undecapeptide methylester (figure 9) Boc-Ala-Gly-Cys(Trt)-

Lys(Boc)-Asn-Phe-Phe-Trp-Lys(Boc)-Thr(Bzl)-Phe-OCH_3 shows the following characteristic fragments (figure 11):

m/e	fragment
107	$(Ph-CH_2-O)^+$
148	$(H_3N-Lys-OH.H)^+$
242	$(^+C(Ph)_3)$
243	$(HC(Ph)_3)^+$
244	$(HC(Ph)_3.H)^+$
276	$(Thr-Phe-OCH_3)^+$
315	$(Trp-Lys\ C{\equiv}O)^+$
413	$(H_3-N-Asn-Phe-Phe\ C{\equiv}O)^+$
693	$(Trp-Lys(Boc)-Thr-Phe-OCH_3)^+$
784	$(Trp-Lys(Boc)-Thr(Bzl)-Phe-OCH_3)^+$
983	$(Boc-Ala-Gly-Cys-Lys(Boc)-Asn-Phe-Phe-C{\equiv}O)^+$

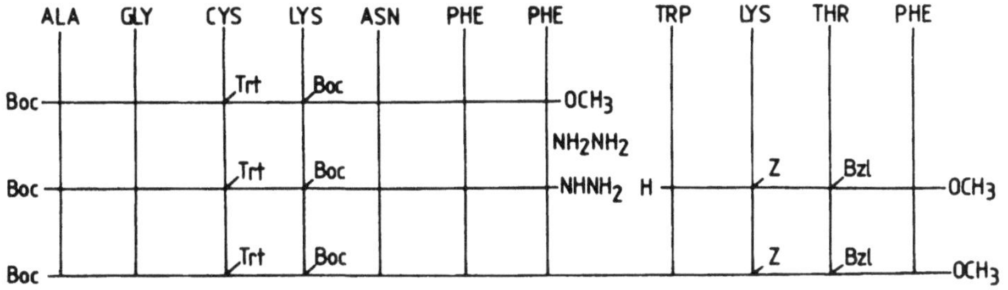

Figure 9. Synthesis of the protected undecapeptide methylester with the natural sequence of somatostatin.

In further reactions from the protected undecapeptide methylester Boc-L-Ala-Gly-L-Lys(Trt)-L-Cys(Boc)-L-Asn-L-Phe-L-Phe-L-Trp-L-Lys(Boc)-L-Thr(Bzl)-L-Phe-OCH_3 the azide was formed via the hydrazide and then a coupling was tried with the protected tripeptide (C-terminal region of somatostatin) H-Thr(Bzl)-Ser(Bzl)-Lys(Trt)-OH. However, even using different concentrations in isoamyl nitrite and nitrosyl

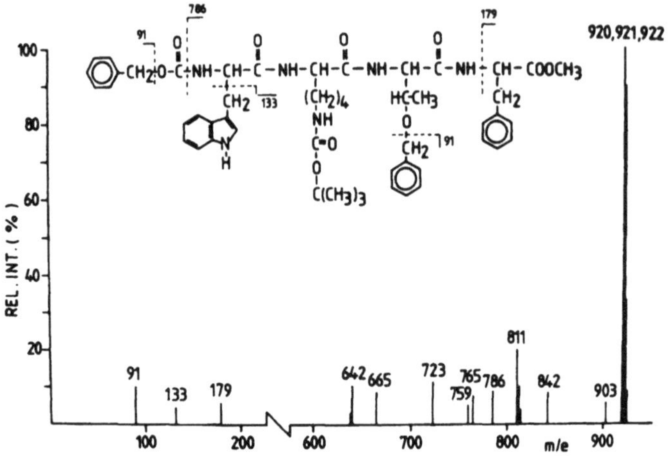

Figure 10. F.d. mass spectrum of Z-Trp-Lys(Boc)-Thr(Bzl)-Phe- OMe and interpreted fragmentation at 15 mA e.h.c.; solvent: DMSO.

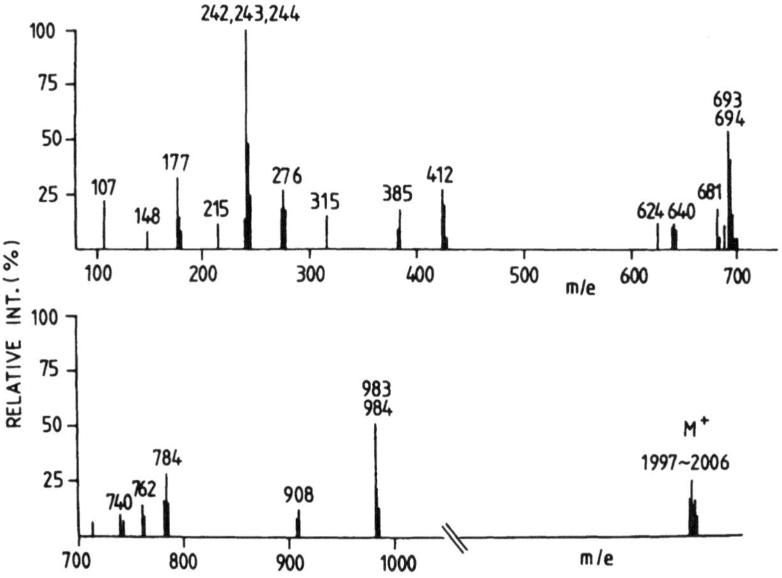

Figure 11. F.d. mass spectrum of Boc-L-Ala-Gly-L-Cys(Trt)-L-Lys(Boc)-L-Asn-L-Phe-L-Phe-L-Trp-L-Lys(Boc)-L-Thr(Bzl)-L-Phe-OCH_3 at 23 mA e.h.c.; solvent: DMSO.

chloride no coupling could be achieved. For the control of this reaction with a negative result f.d. mass spectrometry gave valuable information and finally a new strategy for the synthesis of this hormone had to be developed (38-40). For this new strategy a new amino protecting group was applied for the first time, the 1-(1-adamantyl)-1-methylethoxycarbonyl(Adpoc) group. The advantages of this group soon became obvious: It can be removed 10^3 times faster than the Boc group under very mild acidolytic conditions (e.g. 3 % trifluoroacetic acid in CH_2Cl_2) and thus allows its cleavage selectively in presence of protecting residues of the t.-butyl type, benzyloxycarbonyl, benzyl or trityl groups. The Adpoc amino acids are completely stable towards alkali and sunlight and the crystalline compounds do not decompose for months at room temperature. Adpoc-protected peptides show increased solubility because of the strong lipophilic character of the residue.

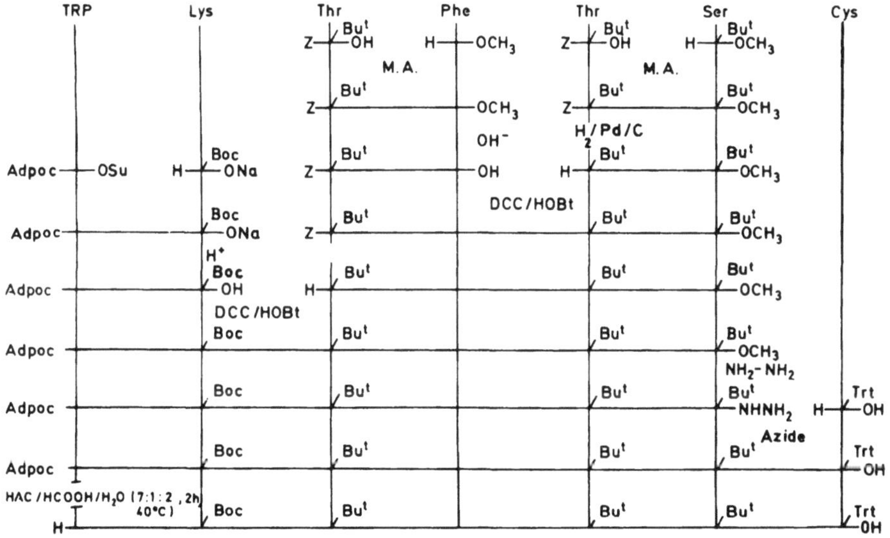

Figure 12. Synthesis of the Adpoc-protected C-terminal heptapeptide with the natural sequence of somatostatin.

In the new developed strategy for the synthesis of somatostatin the C-terminal heptapeptide H-Trp-Lys(Boc)-Thr(Bu^t)-Phe-Thr(Bu^t)-Ser(Bu^t)-Cys(Trt)-OH was synthesized first (figure 12). The side chains of Ser and Thr were protected by t-butyl, that of lysine by the Boc and the one of Cys by the Trt residues. The α-amino residue of Trp was blocked by the Adpoc group which could be removed selectively in presence of the Boc,Bu^t and Trt residues. The control of the synthesis of this protected heptapeptide was performed by thin-layer chromatography, amino acid analysis, NMR spectroscopy and especially f.d. mass spectrometry. The f.d. mass spectrum of the Adpoc-protected hexapeptide with the natural sequence of somatostatin is shown in figure 13. The f.d. mass spectrum of the protected hexapeptide methylester shows at elevated emitter temperature of 76°C besides the molecular ion only a $(M-220)^+$ peak which is caused by cleavage of the Adpoc grup.

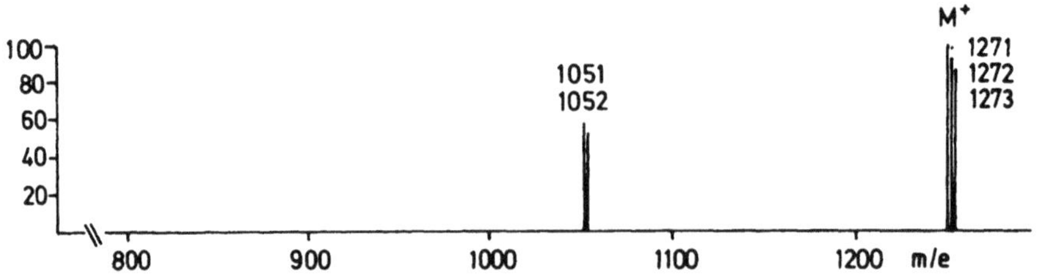

Figure 13. F.d. mass spectrum of Adpoc-Trp-Lys(Boc)-Thr(Bu^t)-Phe-Thr(Bu^t)-Ser(Bu^t)-OCH_3 at an emitter temperature of 76°C; solvent: DMSO.

Field desorption mass spectra of unprotected oligopeptides

TRH and its derivatives were investigated extensively by f.d. mass spectrometry as models for peptide hormones of low molecular weight (6,31,32). From a comparison of e.i. and f.d.mass spectra of TRH the following differences of the two spectroscopid methods become obvious (figure 14): By e.i. mass spectrometry no molecular ion is detectable; some structural information is received only from fragment ions caused mainly by cleavages of the peptide bonds (41). The f.d. mass spectrum

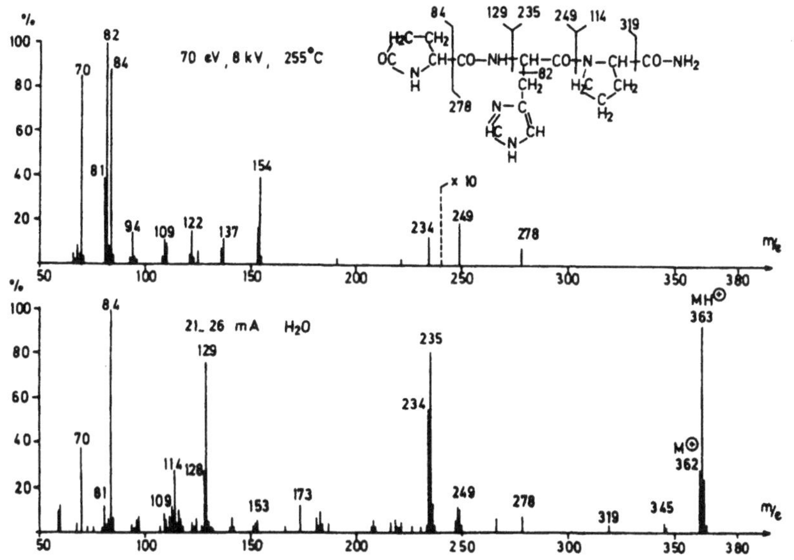

Figure 14. Electron impact and field desorption mass spectrum of TRH.

shows an intense protonated molecular ion peak at m/e 363 (MH^+) caused by a proton transfer reaction at the field emitter surface and therefore the mass units of the fragment ions can be determined often only with a certainty of plus or minus one. If an emitter heating current of 10-14 mA is used the f.d. spectra show almost only molecular ion peaks. By raising the emitter heating current to 25 mA fragments appear by fission of the peptide chain which allow partial assignment of the sequence. The fragments at m/e 249 and 114 arise by the cleavage of the His-Pro peptide bond. The fragment m/e 278 is attributed to the His-Pro-NH_2 structure. The peaks m/e 84,129 arise by the Pyr and Pyr-NH_2 fragments. Several N-alkylated TRH derivatives (42-49) show a similar fragmentation pattern in their f.d. mass spectra. The f.d. mass spectrum of Pyr-His-Pro-NH-CH_3 shows the following m/e values: 377(MH^+), 376 (M^+), 358, 343, 292, 277, 264, 250, 249, 234, 221, 177, 162, 158, 146, 129, 128, 122, 121, 110, 95, 93, 84, 82, 70, the one of Pyr-His-Pro-NH-C_4H_9 gives signals at m/e 419(MH^+), 418(M^+), 333, 301, 277, 210, 209, 201, 192, 172, 170, 169, 134, 128, 84, 70. Fragments including the Pro-NH_2 residue are shifted to higher masses by 14 amu for Pyr-His-Pro-NH-CH_3 (m/e 292, 249, 128) and by 56 amu for Pyr-His-Pro-NH-C_4H_9 (m/e 333, 169). Identical Pyr- and His-containing fragments appear at m/e 249,84. A similar pattern was also found for the f.d. spectrum of the TRH analogue Pyr-Phe-Pro-NH_2 (50-52) (m/e: 372(M^+), 288, 245, 129 (Pyr-NH_2), 84 (Pyr)). For the related dipeptides Pro-Glu-OH, Phe-Pro-OH, Pro-Phe-OH abundant molecular ions were found (49). In Figure 15 the f.d. spectra of Pro-His-Glu-OH, Glu-Pro-His-OH and Glu-His-Pro-OH (recorded at BAT) are shown. The f.d. mass spectra of the derivatives with an N-terminal Glu residue show no molecular ion peak. The most abundant $(M-H_2O)^+$ ions result from the formation of a Pyr residue.

In the meantime molecular ions of peptides up to several thousands daltons were observed (6,34,53,54). Under the conditions of reduced ion accelerating voltage the f.d. spectrum of somatostatin (recorded at 22 mA) showed an MH^+ peak of highest abundance at m/e 1638 (6,34).

In a systematic f.d. mass spectrometric investigation on peptides with the natural sequence of gastrine and glutathione it was further demonstrated that partial sequence information may be received also for peptides of higher molecular weight (6,33). Molecular ions of high intensity (M^+ at m/e 596; MH^+ at 597) were received for the C-terminal tetrapeptide of gastrine (Trp-Met-Asp-Phe-NH_2) at emitter anode currents of $>$ 10 mA. From the peaks at m/e 149,164 respectively 432,449 the C-terminal Phe-NH_2 residue can be determined while those at 279,263 respectively 317/318 are caused by the fragments of the Asp-Phe-NH_2 respectively N-terminal Trp-Met residues (figure 16).

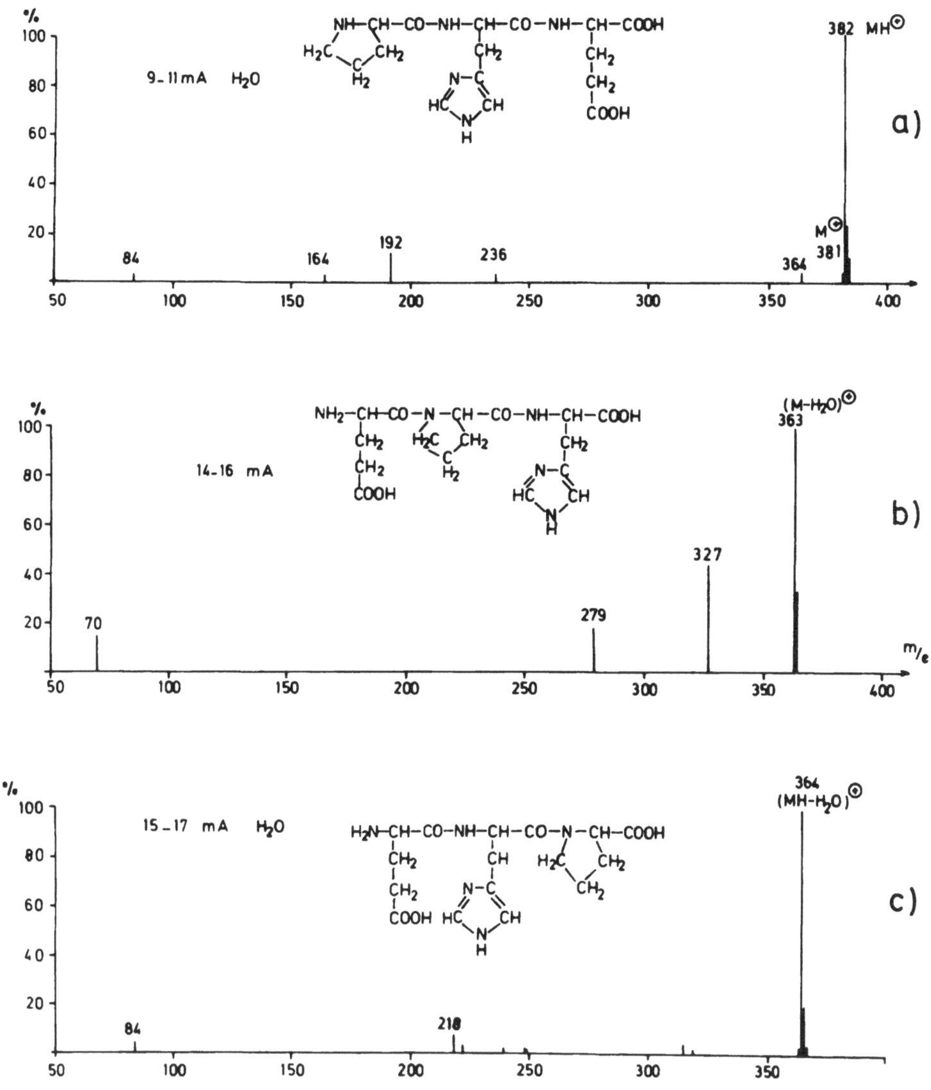

Figure 1 5. F.d. mass spectra of Pro-His-Glu-OH (a), Glu-Pro-His-OH (b) and Glu-His-Pro-OH (c).

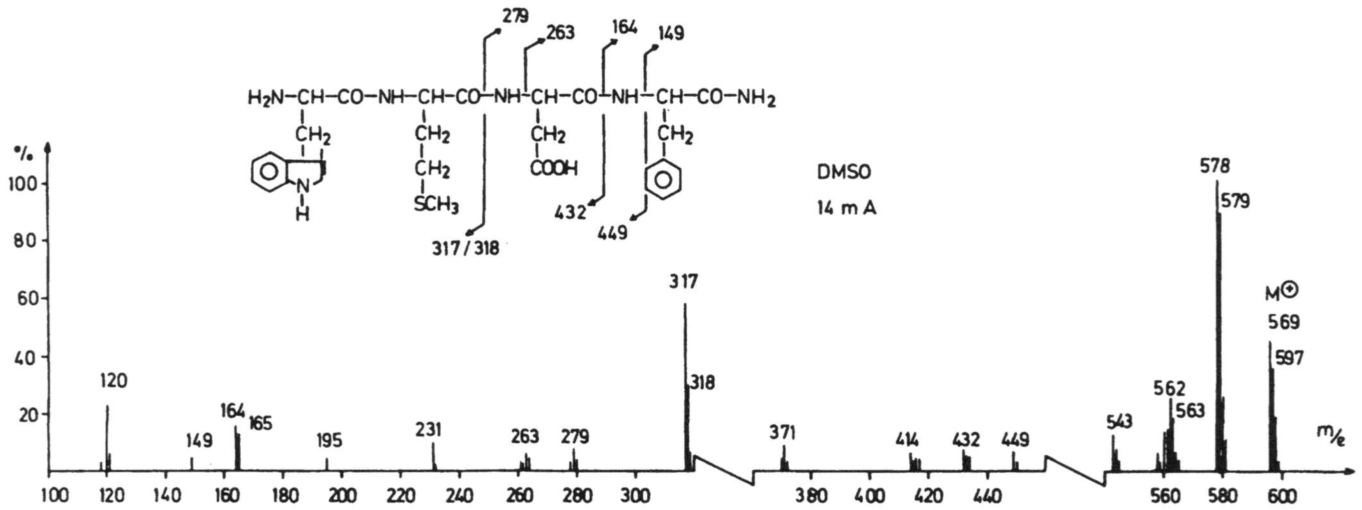

Figure 16. F.d. mass spectrum of Trp-Met-Asp-Phe-NH_2 and interpreted fragmentation at 14 mA e.h.c.; solvent: DMSO.

References

1. Morris, H.R.: in: Pain, R.H., Smith, B.J. (eds): New Techniques in Biophysics and Cell Biology, p. 149, Wiley-Interscience, New York, 1973

2. Biemann,K.: in: Waller, G.R.(ed): Biochemical Applications of Mass Spectrometry, p. 405, Wiley-Interscience, New York, 1972

3. Beckey, H.D.: Int.J.Mass Spectrometry Ion Phys.2, 500 (1969).

4. Beckey,H.D., Hoffmann,G., Maurer, K.H., Winkler,H.U.: Adv. Mass Spectrometry 5, 626 (1971).

5. Beckey, H.D., Schulten,H.R.: Angew.Chem. 87, 425 (1975).

6. Przybylski,M., Voelter,W.:Molecular Weight and Structure Analysis of Oligopeptide Hormones and Peptide Conjugates by Field Desorption Mass Spectrometry, in: Voelter,W.,Weitzel, G.(eds): Structure and Activity of Natural Peptides, p.141, Walter de Gruyter, Berlin, New York, 1981

7. Schulten,H.R.,Beckey,H.D.: Org.Mass Spectrometry 6, 885 (1972).

8. Schulten,H.R.: Int.J.Mass Spectrometry Ion Phys.32, 97 (1979).

9. Olson,K.L.,Cook.J.,Rinehart,K.L.: Biomed.Mass Spectrometry 1, 358 (1974).

10.Pople,J.A., Schneider,W.G., Bernstein,H.J.: High-Resolution Nuclear Magnetic Resonance, McGraw-Hill, New York, 1959

11.Suhr,H.: Anwendung der kernmagnetischen Resonanz in der organischen Chemie, Springer-Verlag, Berlin, 1965

12.Emseley,J.W., Feeney,J., Sutcliffe,L.H.: High Resolution Nuclear Magnetic Resonance Spectroscopy, Pergamon Press, Oxford, 1966

13.Mooney,E.F.: Annual Review of NMR Spectroscopy, Academic Press, London, 1968, 1969

14.Casy,A.E.: PMR Spectroscopy in Medicinal and Biological Chemistry, Academic Press, London, 1971

15.The Sadtler Standard Spectra, Sadtler Research Laboratories, Philadelphia

16.Levy, G.C., Nelson, G.L.: Carbon-13 Nuclear Magnetic Resonance for Organic Chemists, Wiley-Interscience, New York, 1972

17. Stothers,J.B.: Carbon-13 NMR Spectroscopy, Academic Press, New York, 1972

18. Farrar, T.C., Becker,E.D.: Pulse and Fourier Transform NMR, Academic Press, New York, 1971

19. Breitmaier, E., Voelter, W.: ^{13}C NMR Spectroscopy, 2nd edition, Verlag Chemie, Weinheim, 1978

20. Biemann,K.: Mass Spectrometry, Organic Chemical Applications, McGraw-Hill,New York, 1962

21. McLafferty, F.W.: Interpretation of Mass Spectra, W.A.Benjamin, New York, 1966

22. Spiteller,G.: Massenspektrometrische Strukturanalyse organischer Verbindungen, Verlag Chemie, Weinheim, 1966

23. Budzikiewicz, H., Djerassi, C., Williams, D.H.: Mass Spectrometry of Organic Compounds, Holden-Day, San Francisco, 1967

24. Kienitz,H.: Massenspektrometrie, Verlag Chemie, Weinheim, 1968

25. Arpino,P.J., McLafferty, F.W.: in: Nachod, F.C., Zuckerman, J.J., Randall,W.E.(eds): Determination of Organic Structures by Physical Methods, vol.6, p.1, Academic Press,New York, 1976

26. Das,B.C., Lederer, E.: in: Niederweiser,A., Pataki,E.(eds): New Techniques in Amino Acids, Peptide and Protein Analysis, p. 175, Ann Arbor Sci. Publ., Ann Arbor, 1971

27. Nau,H., Förster, H.J., Kelley,J.A., Biemann,K.: Biomed. Mass Spectrom.2, 326 (1976).

28. Nau,H.: Angew.Chem. 88, 74 (1976).

29. Gupta,D., Voelter, W.: Hypothalamic Hormones - Structure, Synthesis and Biological Activity, Verlag Chemie,Weinheim, 1975

30. Voelter,W.: Hypothalamus-Regulationshormone, in: Zechmeister,L., Herz,W., Grisebach,H., Kirby,G.W.(eds): Fortschritte der Chemie organischer Naturstoffe, Progress in the Chemistry of Organic Natural Products, vol.34, p.439, Springer-Verlag, Wien, New York, 1977

31. Gupta,D., Voelter, W.: Hypothalamic Hormones - Chemistry, Physiology and Clinical Applications, Verlag Chemie, Weinheim, New York, 1978

32. Lüderwald,I., Przybylski,M., Ringsdorf,H., Silberhorn,D., Kalbacher,H., Voelter,W.: Z.Naturforsch.33b, 809 (1978).

33. Przybylski,M., Lüderwald,I., Kraas,E., Voelter,W.: Z. Naturforsch.34b, 736 (1979).

34. Voelter,W., Kalbacher,H., Przybylski,M.: in preparation

35. Lüderwald,I., Przybylski,M., Ringsdorf,H., Silberhorn,D., Zech,K., Voelter,W.: Adv.Mass Spectrometry 7B, 1513 (1978).

36. Kalbacher,H., Bürvenich,C., Fuchs,S., Horn,H., Klingler, W., Pietrzik,E., Zech,K., Voelter,W.: Z.Naturforsch. 31b, 1702 (1976).

37. Voelter, W., Kalbacher,H.: 1-(1-Adamantyl)-1-methylethoxycarbonyl(ADPOC): A New Group for Amino Protection in Peptide Synthesis with Advantageous Properties, in: Gross,E., Meienhofer,J. (eds): Peptides, Structure and Biological Function, p. 325, Pierce Chemical Company, Rockford, Illinois, 1979

38. Voelter,W.: Advances in and Comparison of Protective Groups for Peptide Synthesis, in: Beers,R.F., Bassett,E.G. (eds): Polypeptide Hormones, p. 135, Raven Press, New York, 1980

39. Kalbacher,H., Voelter,W.: 1-(1-Adamantyl)-1-methylethoxycarbonyl (ADPOC): A New Useful Amino Protecting Group, in: Voelter,W., Weitzel, G.: Structure and Activity of Natural Peptides, p. 383, Walter de Gruyter, Berlin, New York, 1981

40. Kalbacher,H.: Dissertation, Universität Tübingen, 1980

41. Desiderio, D.M., Burgus, R., Dunn, T.F., Vale, W., Guillemin R., Ward, D.N.: Org. Mass Spectrom. 5, 221 (1972).

42. Kalbacher, H., Zech, K., Voelter, W.: Chemiker-Ztg. 99, 460 (1975).

43. Kalbacher, H., König, W.A., Voelter, W.; Hoppe-Seyler's Z. Physiol. Chem. 356, 1827 (1975).

44. Klingler, W., Kalbacher, H., Göbel, P., Seif, F.J., Zech, K., Voelter, W.: Hoppe-Seyler's Z.Physiol. Chem. 357, 269 (1976).

45. Kalbacher, H., Voelter, W., Vajda, M.: Acta Chim.Acad.Sci. Hung. 95, 67 (1977).

46. Voelter, W., Kalbacher, H., Zeller, K.-P.: Chemiker-Ztg. 100, 238 (1976).

47. Fuchs, St., Klingler, W., Voelter, W., Göbel, P.: Liebigs Ann. Chem. 1977, 602.

48. Voelter, W., Bauer, H., Haas, G., Zech, K., Fuchs, St.: Kémiai Közlemények 49, 245 (1978).

49. Lüderwald, I., Przybylski, M., Ringsdorf, H., Silberhorn, D., Zech, K., Voelter, W.: Field Desorption Mass Spectrometry of TRH Derivatives and Related Oligopeptides, in: Gupta, D., Voelter, W. (eds): Hypothalamic Hormones - Chemistry, Physiology and Clinical Applications, p. 85. Verlag Chemie, Weinheim, New York, 1978

50. Göbel, P., Klingler, W., Horn, H., Voelter, W.: Klin.Wschr. 52, 1128 (1974).

51. Zech, K., Horn, H., Voelter, W.: Chemiker-Ztg. 98, 209 (1974).

52. Voelter, W.: Chemiker-Ztg. 98, 544 (1974).

53. Pandey, R.C., Cook Jr., C.J., Rinehart,K.L.: J.Am.Chem. Soc. 99, 8469 (1977).

54. Pandey, R.C., Cook Jr., J.C., Rinehart, K.L.: J.Antibiotics 31, 241 (1978).

SEQUENCE ANALYSIS OF MEMBRANE-MODIFYING PEPTIDE ANTIBIOTICS BY GAS CHROMATOGRAPHY AND MASS SPECTROMETRY

Wilfried A. König and Mitat Aydin
Institut für Organische Chemie und Biochemie, Universität Hamburg, D-2000 Hamburg, Germany

Introduction

Peptide sequence analysis is a challenging task in the field of structure elucidation of natural products. Efficient and sensitive chemical procedures have been elaborated for this purpose.
As an alternative method combined gas chromatography and mass spectrometry may be used, as long as peptides of moderate size are concerned. This is the case for peptide antibiotics, which usually do not exceed a chain length of 20 amino acids.
The occurrance of unusual amino acids, blocked end groups and constituents which are not amino acids makes structure identification of peptide antibiotics a difficult job.
For a successfull application of GC/MS it is necessary to optimize both the separation and identification part of the technique.
a) Gas chromatography
Highly volatile and thermally stable derivatives as well as very well deactivated gas chromatographic capillary columns are necessary for a good separation.
b) Mass spectrometry
High sensitivity of the instrument guarantees unambiguous information of the structure, amino acid sequence and molecular weight from the mass spectrum.

Chemistry of Peptides and Proteins, Vol. 1

Results

1 - 2 mg samples of peptide antibiotics are partially hydrolyzed in conc. hydrochloric acid at 37°C. The resulting peptide fragments (di- to hexapeptides) are converted to volatile derivatives.
Separation on glass capillaries and identification of partial sequences from their mass spectra proceeds in one experiment (Scheme 1).

Scheme 1.

Peptide Antibiotic

↓ *conc. HCl / 37°C / 12 -24h*

Peptide Fragments (2-6 amino acids)

↓ *volatile derivatives*

Separation on Glass Capillary Column
EI and CI Mass Spectra

Only two types of peptide derivatives proved to be adequate for GC/MS investigations. The N-trifluoroacetyl-peptide methyl esters as introduced by Weygand and Prox (1,2) are sufficiently volatile at least up to tetrapeptides to be eluted from glass capillary columns (Scheme 2).
A further increase in volatility can be achieved by reducing all carbonyl groups in N-acetylated peptides with diborane according to Biemann and coworkers (3,4,5). The resulting O-trimethylsilylated polyamino alcohols are sufficiently volatile at least up to hexapeptides for GC/MS. These methods have been

applied for the analysis of membrane modifying peptide antibiotics trichotoxin A 40 (6) and alamethicin (7).

Scheme 2.

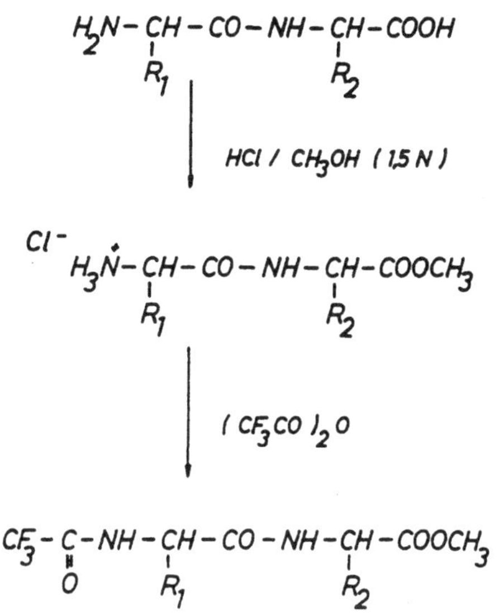

<u>Trichotoxin A 40</u>

The structure of trichotoxin A 40 is given in Scheme 3.

Scheme 3

1 2 3 4 5 6 7 8 9 10 11 12 13 14 15 16 17 18
Ac-Aib-Gly-Aib-Leu-Aib-Glx-Aib-Aib-Aib-Ala-Aib-Aib-Pro-Leu-Aib-Ival-Glx-Valinol
(Ala) (Ala) ↑ (Aib)

Trichotoxin A-40

Peptid I	Ac-Aib-Gly-Aib-Leu-Aib-Glx-Aib-Aib-Ala-Ala-Aib-Aib-OH
Peptid II	Ac-Aib-Gly-Aib-Leu-Aib-Glx-Aib-Aib-Aib-Ala-Aib-Aib-OH
Peptid III	Ac-Aib-Ala-Aib-Leu-Aib-Glx-Aib-Aib-Aib-Ala-Aib-Aib-OH
Peptid IV	Pro-Leu-Aib-Aib-Glx-Valinol
Peptid V	Pro-Leu-Aib-Ival-Glx-Valinol

In positions 2, 9 and 16 amino acids may be exchanged by homologues. By mild hydrolysis in trifluoroacetic acid at 37°C trichotoxin could be cleaved into three homologous N-acetyl dodecapeptides and two homologous pseudohexapeptides with a C-terminal valinol moiety and one α-aminoisobutyric acid (aib) being replaced by isovaline (ival) (Scheme 3). After preparative separation of the peptide fragments by countercurrent distribution the dodecapeptides were partially hydrolyzed according to Scheme 1 and converted to volatile derivatives according to Scheme 2. The gas chromatographic separation of the mixture of di-, tri- and tetrapeptides is presented in Figure 1.

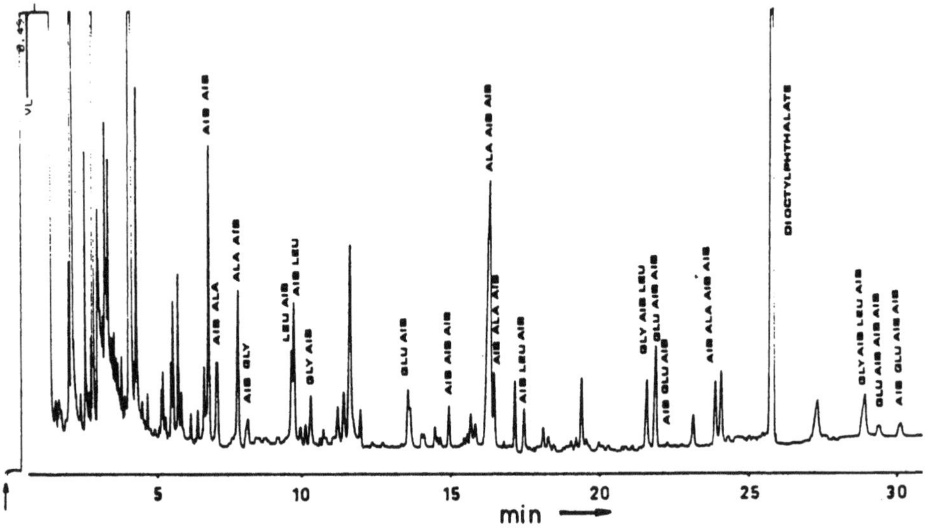

Figure 1. Glass capillary gas chromatogram of N-TFA-peptide methyl esters from a partial hydrolyzate of the N-terminal dodecapeptide fragment of trichotoxin A 40.

From the overlapping partial sequences as identified by electron impact (EI) and isobutane chemical ionization (CI) mass spectra the complete dodecapeptide sequences could be derived

(Scheme 4).

Scheme 4.

```
aib-gly
    gly-aib
    gly-aib-leu
    gly-aib-leu-aib
        aib-leu
        aib-leu-aib
                aib-glu-aib
                aib-glu-aib-aib
                    glu-aib
                    glu-aib-aib
                    glu-aib-aib-aib
                        aib-aib
                        aib-aib-aib
                                aib-ala
                                aib-ala-aib
                                aib-ala-aib-aib
                                    ala-aib
                                    ala-aib-aib
----------------------------------------------------
(Ac)-aib-gly-aib-leu-aib-glu-aib-aib-aib-ala-aib-aib
```

In the same way the pseudohexapeptides were treated. The gas chromatogram is shown in Figure 2. Again the complete sequence could be reconstructed by overlapping fragments.

Mass spectrometric identification of the amino acid sequence of peptide derivatives is facilitated by sequence specific fragment ions resulting from α-cleavage next to the carbonyl groups of the peptide bonds (ions at m/z = 166, 194, 279, 307, 364, 392 in Figure 3).

In the EI mass spectrum the molecular ion intensity may be so low, that the C-terminal amino acid can not be detected. The CI mass spectra with isobutane as reactant gas, however, show very abundant protonated molecular ions and are quite suitable to complement the mass spectrometric information.

Alamethicin

For the investigation of alamethicin the reduction method according to Biemann was applied. The resulting O-trimethyl-

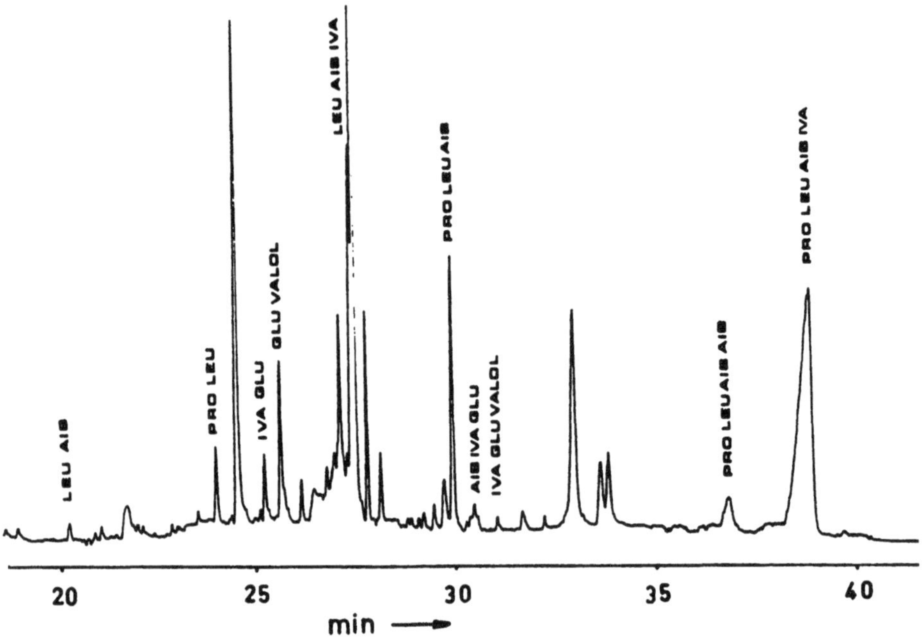

Figure 2. Glass capillary gas chromatogram of N-TFA-peptide methyl esters from a partial hydrolyzate of the C-terminal pseudohexapeptide fragment of trichotoxin A 40.

silylated polyamino alcohols are particularly volatile and can be eluted from 25 m glass capillary columns at least up to hexapeptides.

From alamethicin 33 peptide fragments were obtained including several analogous sequences (Figure 4 and Scheme 5). Again most of the original sequence could be reconstructed by overlapping fragments. The mass spectra of polyamino alcohol derivatives also exhibit pronounced sequence specific ions originating from α-cleavage processes induced by a positive charge on the nitrogen atoms. In addition CI mass spectra may be helpful to detect the molecular ion (Figure 5).

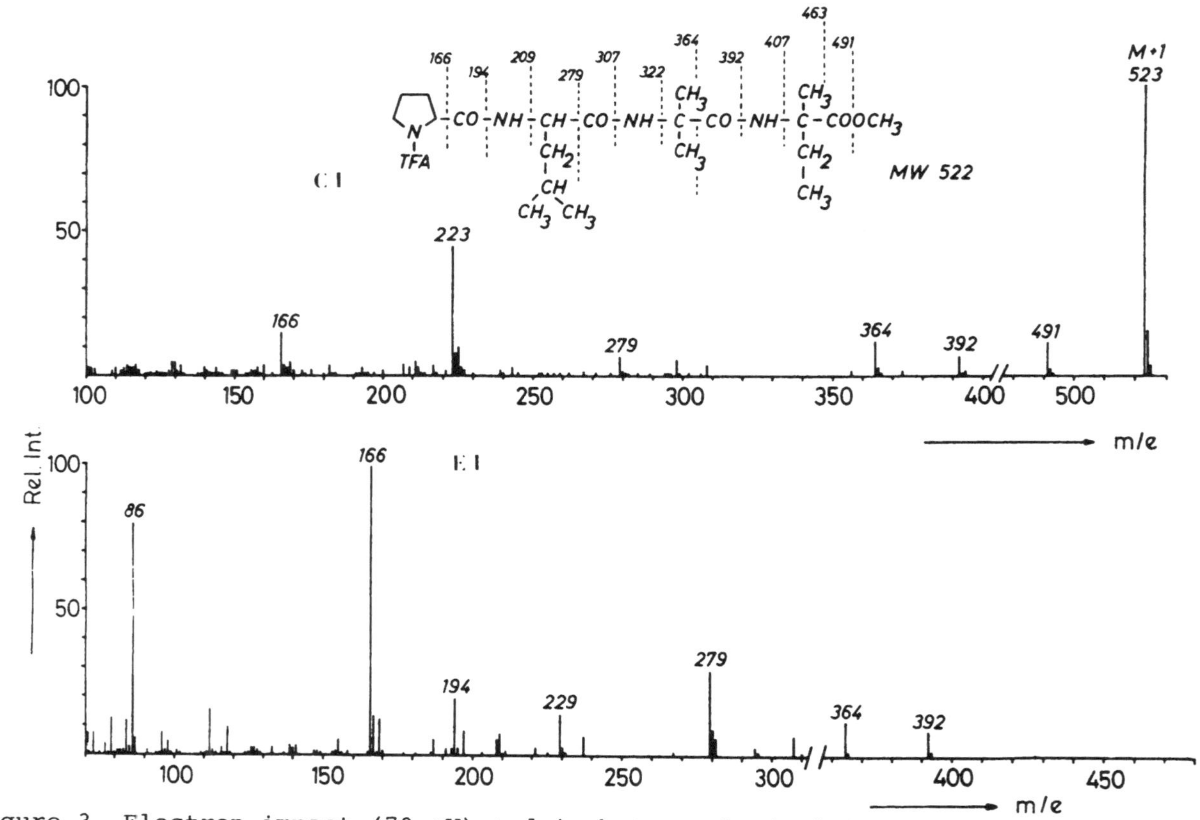

Figure 3. Electron impact (70 eV) and isobutane chemical ionization mass spectra of N-TFA-pro-leu-aib-iva-0-methyl ester from a partial hydrolyzate of the C-terminal hexapeptide of trichotoxin A 40.

SCHEME 5.

Partial sequences from a hydrolyzate of alamethicin .

```
    pro-aib(4)
    pro-aib-ala(13)
    pro-aib-ala-aib(18)
    pro-aib-ala-aib-ala(26)  pro-aib-ala-aib-aib(27)  pro-aib-aib-aib-ala(31)  pro-aib-aib-aib-aib(32)
        aib-ala-aib(9)                ala-aib-aib(8)
            ala-aib(1)
                    ala-gln-aib(17)
                        gln-aib(7)
                        glu-aib-val(21)
                        glu-aib-val-aib(28)  ala-gln-aib-aib-aib(25)
                                val-aib(2)
                                val-aib-gly(10)
                                val-aib-gly-leu-aib(30)
                                        gly-leu-aib(12)
                                            leu-aib(5)
                                            leu-aib-pro-val(29)
                                                    pro-val-aib(15)
                                                    pro-val-aib-aib(22)  pro-val-aib-val(23)  pro-val-val-aib(24)
                                                        val-aib-aib(11)      val-aib-val(14)
                                                            aib-aib(3)
                                                                    glu-gln(19)
                                                                    glu-gln-phol(33)
                                                                            phol(6)
------------------------------------------------------------------------------------------------------------------
      1    2   3   4   5   6   7   8   9  10  11  12  13  14  15  16  17  18  19  20
(Ac-aib-)pro-aib-ala-aib-ala-gln-aib-val-aib-gly-leu-aib-pro-val-aib-aib-glu-gln-phol
                 aib ala aib         aib                         val val
```

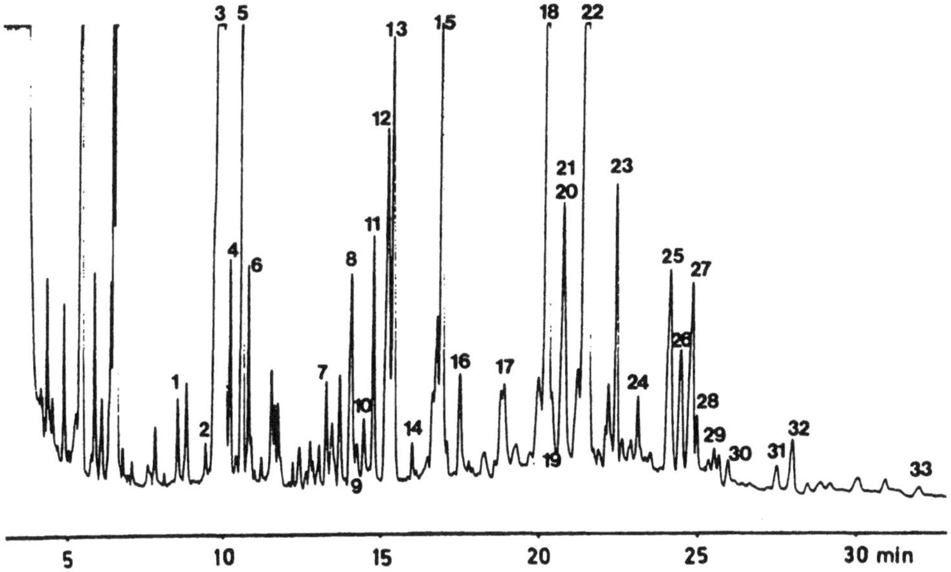

Figure 4. Glass capillary gas chromatogram of N-ethyl,O-trimethylsilyl-polyamino alcohol derivatives of a partial hydrolyzate of alamethicin.

Identification of amide side chains

A particularly difficult problem in peptide analysis is the determination of amide side chains. Glutamine and asparagine are usually hydrolyzed under acidic conditions to the corresponding acids. We have found, that during 12 - 24 h of partial hydrolysis in conc. hydrochloric acid a substantial amount of the amide function is retained. In the subsequent formation of derivatives the carboxyl and amide functions can be distinguished from each other as demonstrated in Scheme 6. Figure 6 represents the mass spectrum of derivatized glu-gln from the partial hydrolyzate of alamethicin.

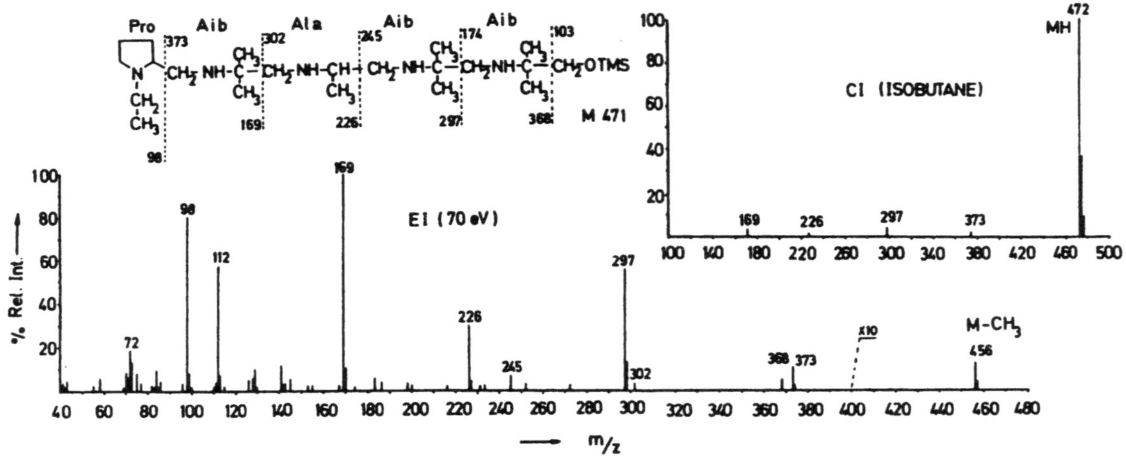

Figure 5. Electron impact (70 eV) and isobutane chemical ionization mass spectra of N-ethyl,O-trimethylsilyl-polyamino alcohol derivative of a pentapeptide fragment from the partial hydrolyzate of alamethicin.

Scheme 6

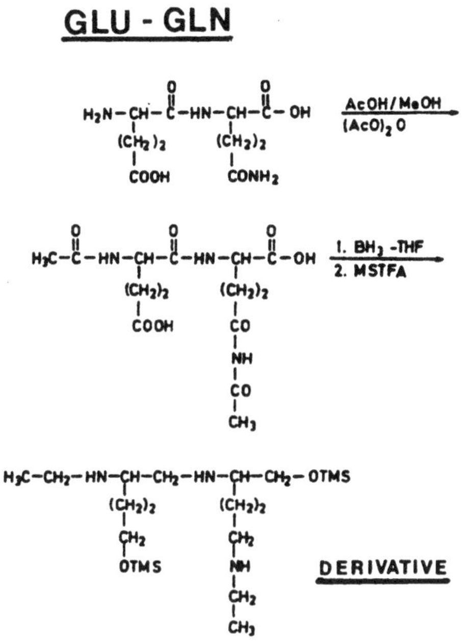

Conclusions

It may be concluded from these results, that capillary gas chromatography is a highly efficient and sensitive method for separating volatile peptide derivatives. The N-ethyl,O-trimethylsilyl polyamino alcohols as obtained by N-trimethylsilylation are superior to trifluoroacetylated peptide esters in volatility. Both types of derivatives yield excellent mass spectra and are very well suited for deriving the amino acid sequence. The chemical ionization mass spectra with isobutane as reactant gas complement the electron impact spectra by prominent protonated molecular ions.

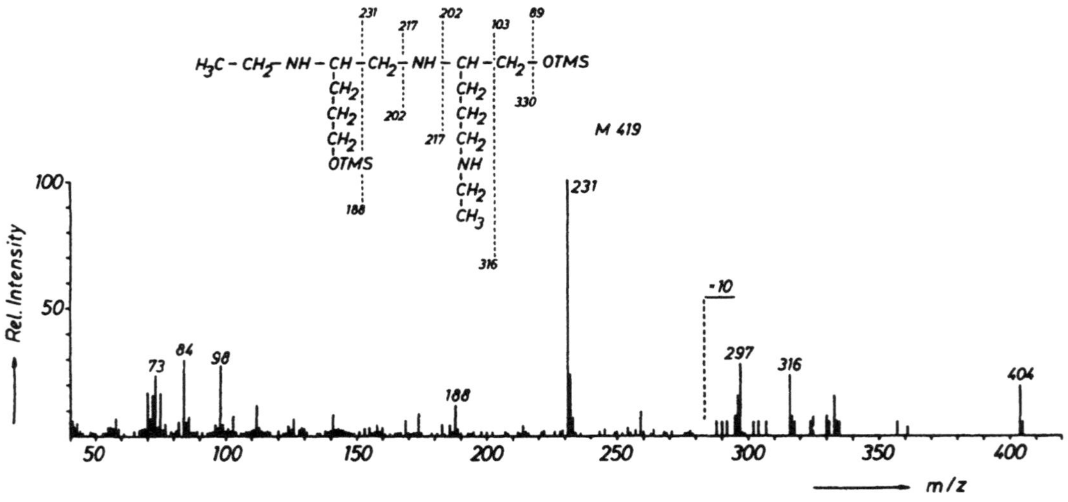

Figure 6. Electron impact (70 eV) mass spectrum of Glu-Gln derivative from partial hydrolysate of alamethicin.

References

1. Weygand, F., Prox, A., Fessel, H.H., KunSun, K.: Z. Naturforsch. 20 b, 1169 (1965).
2. König, W.A., Engelfried, C., Hagenmaier, H., Kneifel, H.: Liebigs Ann. Chem. 1976, 2011-2020.
3. Kelley, J.A., Nau, H., Förster, H.-J., Biemann, K.: Biomed. Mass Spectrom. 2, 313 (1975).
4. Frank, H., Desiderio, D.M.: Anal. Biochem. 90, 413 (1978).
5. Carr, S.A., Herlihy, W.C., Biemann, K.: Biomed. Mass Spectrom. 8, 51-61 (1981).
6. Brückner, H., Greiner, M., König, W.A., Jung, G.: Angew. Chem. 91, 508-509 (1979); Angew. Chem. Int. Ed. Engl. 18, 476-477 (1979).
7. Pandey, R.C., Cook, J.C., Rinehart, K.L.: J. Amer. Chem. Soc. 99, 8469 (1977).

STRUCTURE ELUCIDATION AND PROPERTIES OF NIKKOMYCINS, A NEW CLASS OF NUCLEOSIDE - PEPTIDE ANTIBIOTICS

Hanspaul Hagenmaier
Institut für Organische Chemie der Universität Tübingen
D-7400 Tübingen

Wilfried A. König
Institut für Organische Chemie der Universität Hamburg
D-2000 Hamburg 13

Nikkomycins are nucleoside-peptide antibiotics exhibiting fungicide and insecticide activity, which is due to an inhibition of chitin biosynthesis. From the culture filtrate of Streptomyces tendae (Tü 901) was first isolated nikkomycin Z (1). Its structure was elucidated by mass spectrometry of the products obtained after chemical degradation: uracil, an amino hexuronic acid and the new amino acid 2-amino-4-hydroxy-4-(5-hydroxy-2-pyridyl)-3-methylbutyric acid.

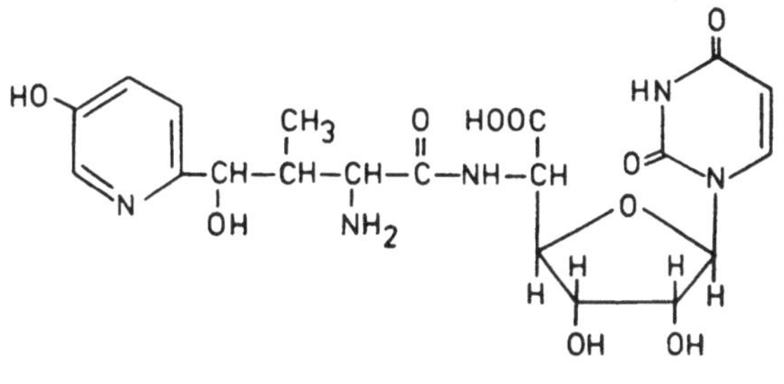

Fig. 1. Structure of nikkomycin Z

The amino hexuronic acid of nikkomycin Z was found to be identical with that one of the structurally related group of antibiotics, the polyoxins; it therefore is 5-amino-5-deoxy-ß-D-allo-furanuronic acid. The configuration of the amino

Chemistry of Peptides and Proteins, Vol. 1

acid had yet to be determined. For this purpose the N-ethylthio(thiocarbonyl)-derivative of the amino acid was prepared (Fig.2).

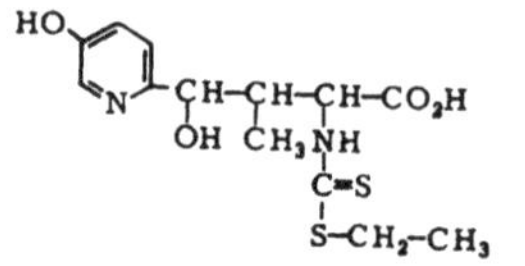

Fig.2. N-ethylthio(thiocarbonyl)-derivative of 2-amino-4-hydroxy-4-(5-hydroxy-2-pyridyl)-methylbutyric acid.

According to Ishikawa et al. (2) all tested naturally occurring L-α-amino acids show a positive Cotton effect as cyclohexylammonium salt of the N-ethylthio(thiocarbonyl)-derivative. The cyclohexylammonium salt of the amino acid derivative did show a positive Cotton effect. It was therefore concluded that the amino acid in nikkomycin Z belongs to the L-series, which means S-configuration for the α-C-atom.

Attempts to deduce the relative configuration at the C-3 and C-4 atom of the amino acid by ^{1}H-NMR-spectroscopy were not unequivocally successful. Therefore an X-ray analysis of the crystalline amino acid was carried out. Details of this analysis have been published (3). The ORTEP-representation of the spatial orientation of the atoms is shown in Fig. 3.

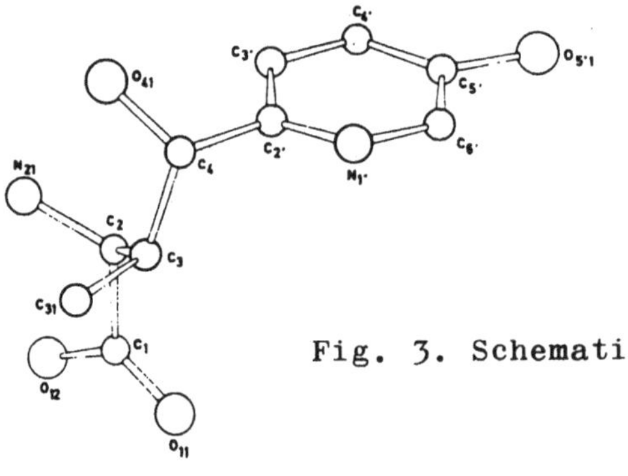

Fig. 3. Schematic ORTEP-representation.

The peptide bond in nikkomycin was shown to be susceptible towards hydrolytic cleavage, even in aqueous solution at room temperature. The pH dependency of the peptide bond hydrolysis is shown in Fig. 4. PO_4^{3-} ions have a catalytic effect on the velocity of hydrolysis, an effect which is so far unexplained.

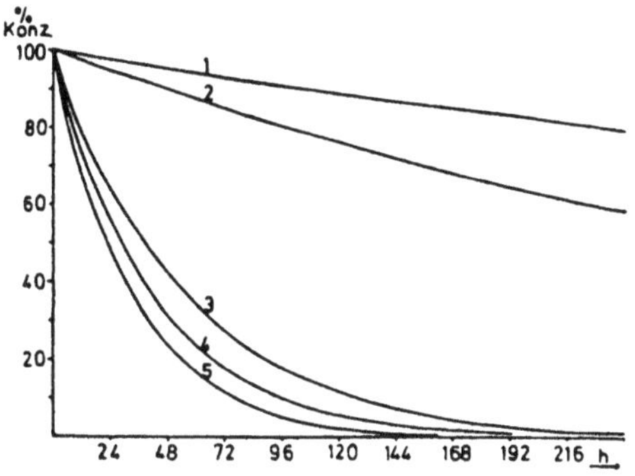

Fig. 4. Rate of hydrolysis of nikkomycin Z. 1: pH 3,0; 2: pH 4,0; 3: pH 6,8; 4: pH 8,0; 5: 0.01M PO_4^{3-} pH 6,8.

When crude nikkomycin preparations were hydrolyzed in phosphate buffer on a preparative scale, not only the amino acid and the corresponding nucleoside of nikkomycin were isolated from the reaction mixture, but also a number of structurally related compounds. Besides the nucleoside of nikkomycin three more nucleosides were identified and their structure elucidated by NMR-spectroscopy and mass spectrometry, two of which are peptide-nucleosides without biological activity (Fig. 5).

Corresponding peptide-nucleosides with the amino acid, found in nikkomycin Z, were subsequently isolated from the culture filtrate of streptomyces tendae (nikkomycin X and nikkomycins I and J), which show similar activity as nikkomycin Z (4).

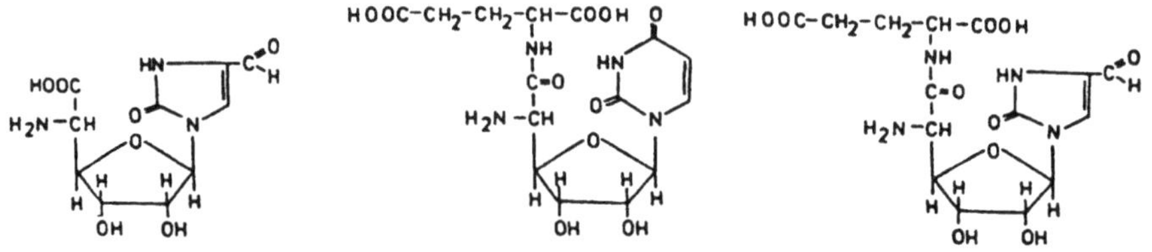

I II III

Fig. 5. I: nucleoside of nikkomycin X; II: peptide-nucleoside (nikkomycin N) of nikkomycin J; III: peptide-nucleoside (nikkomycin M) of nikkomycin I.

Finally nikkomycin B was isolated (5) in which the pyridine ring of nikkomycin X is replaced by a phenyl ring. The fungicide activity of this compound is rather low compared to nikkomycin X.

Since nikkomycin X and Z are produced in amounts up to 4g/l culture filtrate and due to their high fungicide and insecticide activity they are promising compounds for practical application in pest control.

References

1. Dähn, U., Hagenmaier, H., Höhne, H., König, W.A., Wolf,G., Zähner,H.: Arch. Microbiol. 107, 143 (1976).
2. Ishikawa, K., Achiwa, K., Yamada, S.I.: Chem. Pharm. Bull. 19, 926 (1971).
3. König, W.A., Pfaff, K.P., Bartsch, H.H., Schmalle, H., Hagenmaier, H.: Liebigs Ann. Chem. 1980, 1728.
4. Hagenmaier, H., Keckeisen, A., Dehler, W., Fiedler, H.P., Zähner, H., König, W.A.: Liebigs Ann. Chem. 1981, 1018.
5. König, W.A., Hass, W., Dehler, W., Fiedler, H.P., Zähner, H.: Liebigs Ann. Chem. 1980, 622.

II Peptide Syntheses. Biological Activity and Analytical Problems of Synthetic Peptides

ELECTROCHEMICAL INTRODUCTION AND SELECTIVE REMOVAL OF A NEW TYPE OF AMINO- AND CARBOXY-PROTECTING GROUP FOR PEPTIDE SYNTHESIS

Günther Jung, Mohamed Hassen Khalifa and Anton Rieker
Institut für Organische Chemie der Universität Tübingen, D-7400 Tübingen, Germany

Introduction of protecting groups

The electrochemical oxidation of sterically hindered 2,6-di-*tert*-butyl-4-aryl(or alkyl)-phenols 1 (1,2) yields phenoxenium cations 4. These aryloxenium ions are formed via a homogeneous redox reaction between the intermediate cation radical 2 and the neutral radical 3, respectively via a two-electron process including the deprotonation of cation 2. Nucleophiles (NuH) react with the phenoxenium cation 4 yielding para-substituted quinol derivatives 5 (R=aryl or alkyl).

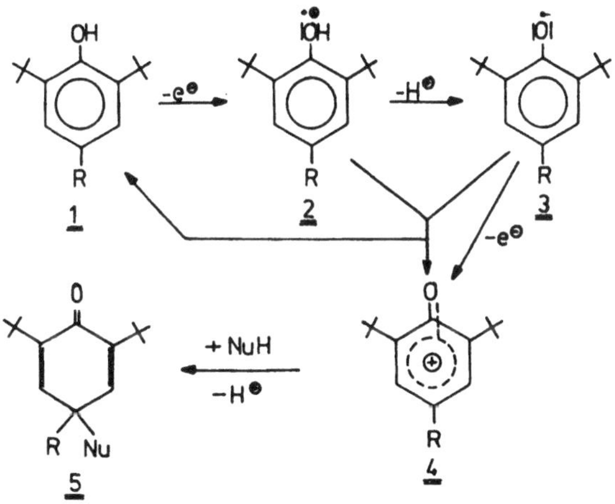

Chemistry of Peptides and Proteins, Vol. 1

This reaction can be carried out in neutral electrolyte systems in the presence of amino acid derivatives as nucleophiles yielding either N-protected or C-protected intermediates useful for peptide synthesis (3,4,5).

a) Amino acid esters as nucleophiles

The anodic oxidation of 2,6-di-tert-butyl-4-phenyl-phenol (3,5-di-tert-butyl-4-biphenylol) 6 (1,2) in the presence of amino acid esters in dichloromethane followed by a saponification step gives N-[3,5-di-tert-butyl-4-oxo-1-phenyl-2,5-cyclohexadienyl] amino acids 7 (R^3=H) in 80-95% preparative yields (3,4).

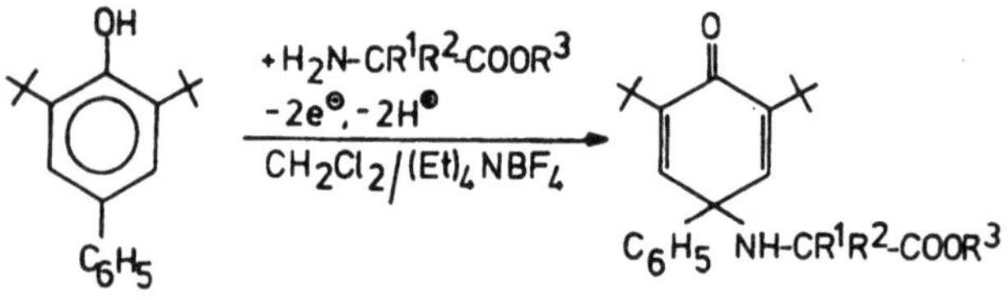

6 7

As shown on more than 10 different derivatives synthesized so far N-PChd amino acids are readily crystallizable compounds free of racemate.

b) N-protected amino acids as nucleophiles

Another application of the anodic oxidation of sterically hindered phenols is the preparation of quinol esters 8 from 2,6-di-tert-butyl-4-phenyl-phenol 6 and Z-resp.Boc-amino acids as nucleophiles. Quinol esters, e.g. Z-amino acid-[3,5-di-tert-butyl-4-oxo-1-phenyl-1,5-cyclohexadienyl] esters 8 or corresponding peptide derivatives constitute a new class of carboxy protected intermediates. Via CF_3COOH catalysis cleaving one of the tert-butyl groups they can be transformed into the ether soluble pyrocatechol esters 9. These isolable activated esters 9 couple to amino groups of amino acid and peptide esters often in more than 80 % yield.

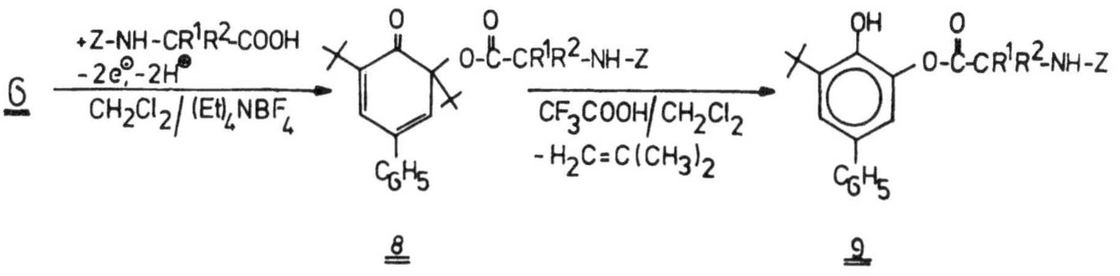

Removal of N-PChd group

The cleavage conditions for the N-PChd group are either mild acidolysis, e.g. 10-50 % CF_3COOH in CH_2Cl_2 within about 15 min, or hydrogenolysis, e.g. H_2/Pd-C (10 %) in CH_3OH/HCl within ca.1h. Therefore the PChd protection is selectively applicable in combination with the Z, Bzl and the Boc, Bu^t groups.

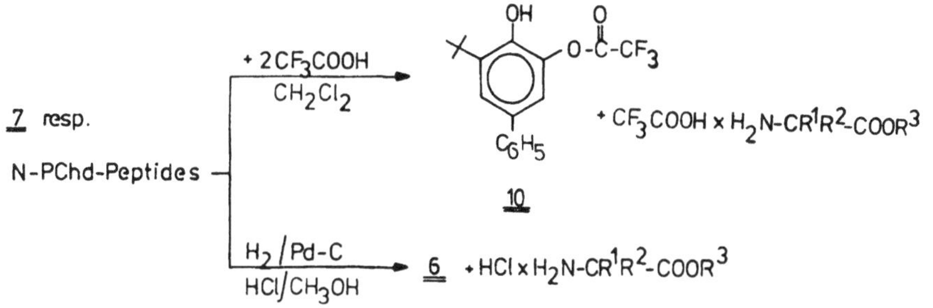

The pyrocatechol trifluoroacetate 10 is easily separated from the peptide trifluoroacetates by extraction with diethyl ether. In the same way the regenerated phenol 6 can be isolated from peptide hydrochlorides, and it is reusable. Recent experiments have shown the electrochemical reduction to be a third alternative way for deblocking N-PChd protected peptides. The N-PChd group is completely stable towards the alkaline conditions used for saponification of peptide esters and the aminolytic conditions used for removal of the Fmoc group. Due to the simultaneous hydrogenolytic and acidolytic sensitivity of the PChd group and its electrochemical oxidative intro-

duction and reductive removal a variety of orthogonal combinations of protecting groups is possible in particular, when we have side chain protected PChd amino acids available. Furthermore, the different sensitivities of the PChd group can still be modified, e.g. by introducing suitable substituents on the phenyl ring. Moreover, by substituting the 4-phenyl for the 4-*tert*-butyl group on phenol *6* one can obtain N-(1,3,5-tri-*tert*-butyl-4-oxo-2,5-cyclohexadienyl) amino acids, which can be transformed into further interesting derivatives (6).
It is interesting that the rate of trifluoroacetolysis of N-PChd amino acids is much faster for α,α-dialkyl than for the protein amino acids. Thus, PChd-Aib-OH is cleaved quantitatively within 1 min and PChd-Ala(Gly or Leu)-OH within 5-10 min to about 70 % using 3 % trifluoroacetic acid in dichloromethane.

Peptide synthesis with N-PChd protection
The applicability of this new amino and carboxy protection and the acid-catalyzed carboxy activation is shown so far on the synthesis of a number of oligopeptides containing only difunctional residues. The extension of the new protective principle to combinations of protecting groups for the trifunctional amino acids is in progress. Representative examples of synthesis revealed readily crystallizable N-PChd oligopeptides with excellent solubilities in common organic solvents, complete retention of configuration and the possibility of monitoring reactions by UV detection.
So far we prepared more than 20 N-PChd-protected oligopeptide esters, amides and acids, including 1 tetrapeptide, 1 hexapeptide and 1 heptapeptide. As a first example the well-known C-terminal tripeptide of oxytocin, H-L-Pro-Leu-Gly-NH_2 was synthesized. In a second test synthesis we built up by fragment condensation the lipophilic heptapeptide segment 14-20 of human lymphoblastoid interferon HuIFN-α(Ly) Ala-Leu-Ile-Leu-Leu-Ala-Gln. The same peptide had been prepared in our laboratory using Z- and Boc-protection.

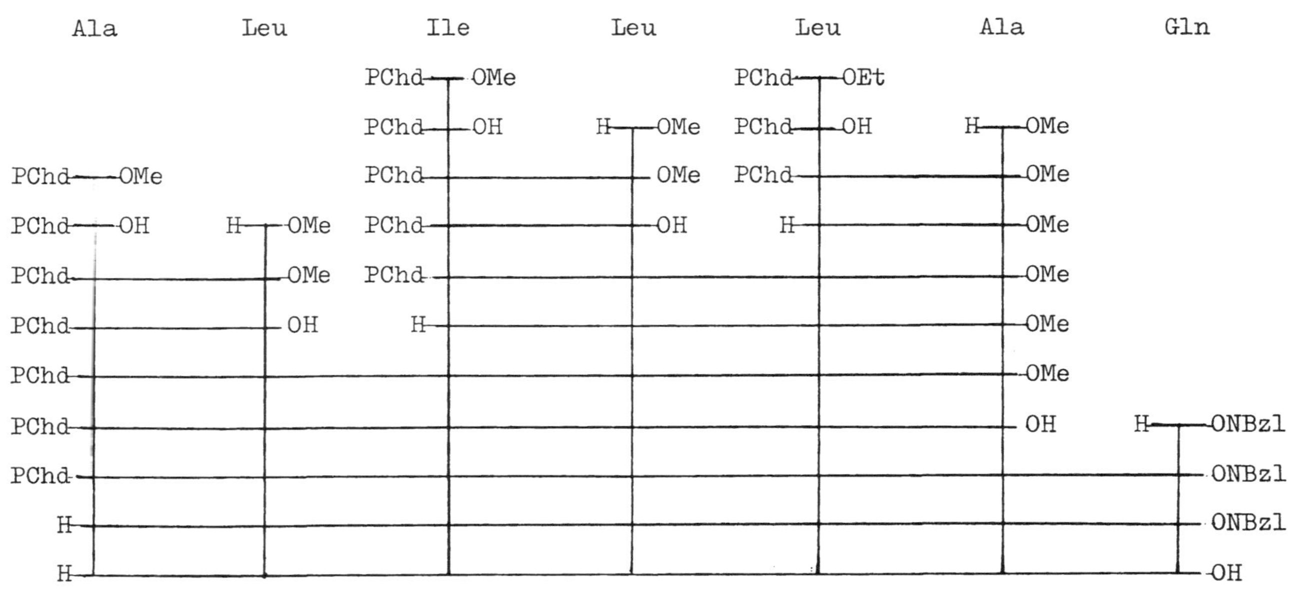

Fig.´. Synthesis of the heptapeptide segment HuIFN-α (Ly) 14-20 <u>via</u> fragment condensation and N^{α}-PChd protection.

The coupling steps were carried out according to the scheme in Fig.1 using dicyclohexylcarbodiimide/1-hydroxybenzotriazole. The average preparative yields were mostly higher than 80%. The N-PChd heptapeptide ester is still good soluble. After removal (Sephadex LH 20, Methanol) of traces of dicyclohexylurea and an impurity introduced via glutamine nitrobenzyl ester the N-PChd hexapeptide ester was found to be pure in five TLC systems, elemental and amino acid analysis and ^{13}C-NMR spectroscopy. After trifluoroacetolytic removal of the PChd group the resulting heptapeptide ester showed identical R_F values as a sample obtained via a different route. The analytical data of the N- and C-deprotected heptapeptide ester obtained by hydrogenation were, too, in full agreement with the expected values. The gas chromatographic analysis of the N-pentafluoropropionyl-amino acid-propyl esters of a total hydrolysate on the chiral phase Chirasil-Val (N-propionyl-L-valine tert-butyl-amide polysiloxane) gave no evidence for racemization. However, during a 24 h test coupling of PChd-Ala-Leu-OH to H-Ile-Leu-Leu-Ala-OMe using DCC and one mole HOBt we observed 6 % D-Leu. This racemization has nothing to do with the use of the PChd group. Upon shortening the coupling time to 3 h and using two moles HOBt the amount of racemization decreased beyond the limits of detectability.

As an example for the signal assignments of the new protecting group the ^{13}C NMR spectrum of N-PChd-Pro-Leu-Gly-NH_2 is shown in Fig.2. The diastereotopic splitting of quinonoidal carbon signals C-3,5, C-2,6 and the <u>tert</u>-butyl signals due to the chiral peptide moiety is significant.

Selected experimental procedures (3-6)

Synthesis of N-PChd L-alanine

8.46 g (30 mmol) 2,6-di-<u>tert</u>-butyl-4-phenyl-phenol, 6.19 g (60 mmol) free alanine methyl ester, and 6.42 g (60 mmol) 2,6-dimethyl-pyridine in 300 ml dry 0.1 M $(Et)_4NBF$/dichloromethane are electrolyzed under N_2 in an undivided cell, equipped with Pt net electrodes (anode potential ca. +1000 mV <u>vs</u> Ag/

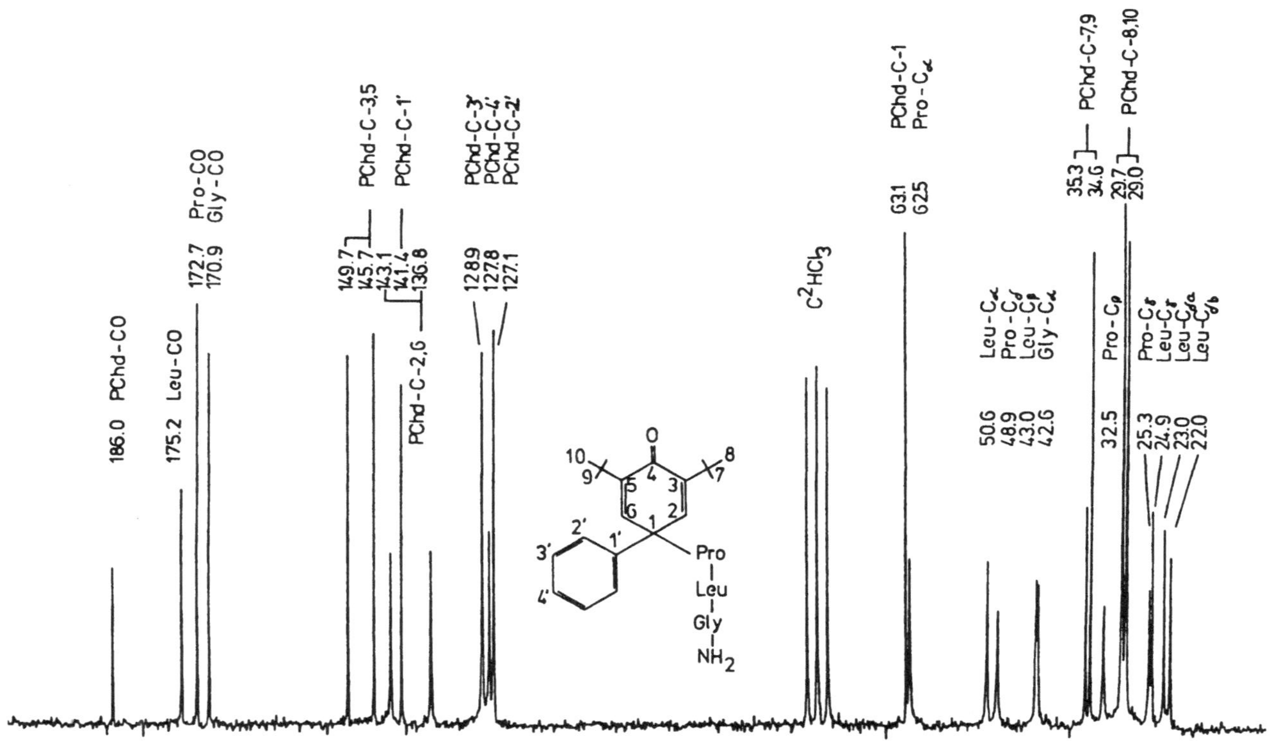

Fig.2. ^{13}C NMR spectrum (20.116 MHz) of PChd-L-Pro-Leu-Gly-NH_2 (concentration: 100 g/l, $^2HCCl_3$, t = 25°C).

$Ag^{\oplus}$; current < 300 mA, increasing potential to +1300 mV on decreasing of current). The electrolysis is terminated after about 8 h (at this point the current is about 50 mA and almost complete consumption of the phenol is observed by TLC). The solvent is evaporated and the residue extracted with 5 x 50 ml ether. The insoluble electrolyte is filtered off, and the ether is evaporated. The residual oil containing PChd-Ala-OMe is saponified with 2 equivalents 1N NaOH in 50 ml methanol within about 2 h (TLC). After adjusting to pH 1-2 by 1 N HCl in the cold the solution is extracted with 3 x 50 ml ethyl acetate. After evaporation of the solvent the residual PChd-Ala-OH is recrystallized from light petroleum (30 - 50°C). Yield 9.7 g (86 %, based on the phenol); m.p. 181-183°C, R_F = 0.45 ($CHCl_3/CH_3OH$ = 8:1, silica gel plates, Merck No. 5554).

Synthesis of PChd-Ile-Leu-Leu-Ala-OMe

5.27 g (10 mmol) PChd-Ile-Leu-OH , 2.53 g (10 mmol) H-Leu-Ala-OMe x HCl, 1,15 g (10 mmol) N-methylmorpholine and 1.52 g (10 mmol) 1-hydroxybenzotriazole are dissolved in 30 ml dichloromethane. After addition of 10 ml 1 N dicyclohexylcarbodiimide/dichloromethane at 0°C the solution is stirred 30 min at 0°C and 12 h at 23°C. After addition of 0.5 ml acetic acid the dicyclohexylurea is filtered off and the solvent is removed in vauo. The residue is dissolved in 50 ml ethyl acetate and washed 3x each with 30 ml 5 % citric acid/water, 5 % bicarbonate/water and water. After drying over sodium sulfate the tetrapeptide is crystallized by addition of light petroleum. Yield 6.1 g (84 %); m.p. 182°C, R_F = 0.65 ($CHCl_3/CH_3OH$ = 8:1, silica gel plates, Merck No. 5554).

Hydrogenolytic removal of PChd-group

5.0 g (7 mmol) PChd-Ile-Leu-Leu-Ala-OMe in 25 ml 1 N HCl/methanol are hydrogenated in the presence of 1 g Pd/C (10 %). After about 4 h the solvent is evaporated, and the residue is extracted with 5 x 25 ml diethyl ether.(This extract gives after evaporation and recrystallization a recovery of 1.9 g (96 %) 2,6-di-tert-butyl-4-phenyl-phenol). The ether insoluble tetrapeptide ester hydrochloride is recrystallized from methanol/ether. Yield 3.05 g (89 %); m.p. 249-252°C; R_F = 0.75 (1-butanol/acetic acid/water = 4:1:1, silica gel plates, Merck No. 5554).

Trifluoroacetolytic removal of PChd group

1.15 g (2.5 mmol) PChd-Leu-Gly-NH_2 in 10 ml trifluoroacetic acid/dichloromethane (1:1) are stirred 15 min at room temperature. The solvents are removed in vacuo and the residue is extracted with 3 x 25 ml diethyl ether,which dissolves the pyrocatechol monotrifluoroacetate. CF_3COOH x H-Leu-Gly-NH_2 is filtered off and recrystallized from dry methanol/ether. Yield 360 mg (96 %).

References

1. Speiser, B., Rieker,A.: J.Chem.Research (S), 314-315 (1977), (M) 3601-3618 (1977); J.Electroanal.Chem.102, 373-395(1979).

2. Rieker, A., Dreher, E.-L., Geisel, H., Khalifa, M.H.: Synthesis 1978, 851-855.

3. Khalifa, M.H., Jung, G., Rieker, A.: Angew.Chem. 92, 739-740 (1980); Angew.Chem.Int.Ed.Engl.19, 712-713 (1980).

4. Rieker, A., Khalifa, M.H., Jung, G.: in: K.Brundfeldt (ed.): Peptides 1980, p. 116-120, Scriptor, Copenhagen 1981

5. Khalifa, M.H., Jung, G., Rieker, A.: manuscript submitted.

6. Khalifa, M.H., Thesis, University of Tübingen 1980

NEW ASPECTS OF PEPTIDE SYNTHESIS BY FOUR COMPONENT CONDENSATIONS

Ivar Ugi, Wolfgang Breuer, Peter Bukall, Samir Falou, Gerhard Giesemann, Rudolf Herrmann, Gerhard Hübener, Dieter Marquarding, Peter Seidel, Reinhard Urban

Organisch-Chemisches Institut der Techn. Universität München
D-8046 Garching, Germany

The recent syntheses of tetravaline derivatives [1], glutathione [2], and alamethicin segments [3,4] served as model experiments in order to demonstrate that stereochemically uniform peptide segments can be synthesized by four component condensations (4CC) with productive stereoselectivity, followed by partial cleavage with destructive stereoselectivity [1 - 7]. There is also some evidence for the usefulness of 4CC for peptide segment coupling [4,7].

The solution to the problems that are outlined below were needed for fully exploiting the preparative potential of 4CC for peptide synthesis.

The chiral α-ferrocenyl alkylamines (4) [8] are needed as the chiral amine components of peptide segment syntheses by stereoselective 4CC. Hitherto these amines (4) were not readily available [9]. Recently we found that they can be made by a one-pot condensation [10] according to Scheme 1. Ferrocene (1) and aldehydes (2) condense in acidic media to form the carbenium ion (3) which can be quenched by ammonia, yielding (4).

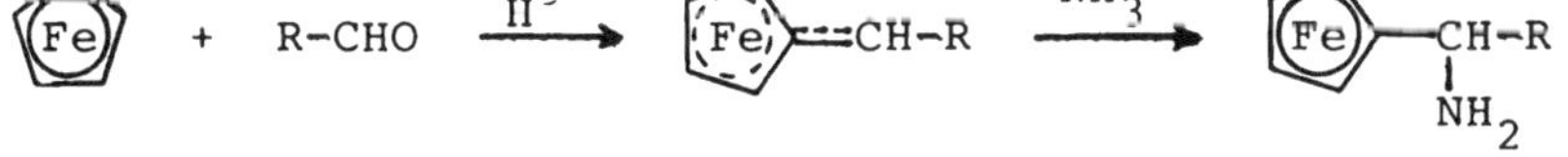

(1) (2) (3) (4)

Scheme 1

The chiral α-isocyano esters which are needed as isocyano components of some 4CC peptide syntheses can now be made from the corresponding α-formylamino esters by treatment with one mole equivalent carbonyl diimidazole and two mole equivalents methane sulfonic acid in HMPA. Thus, for example, chiral methyl α-isocyano-ß-phenyl-propionate can be made from N-formyl phenylalanine methyl ester [11].

Since the synthesis of peptide segments by 4CC is often accompanied by the formation of malonamide derivatives as by-products [12], we have conducted a search for reaction conditions under which this side reaction does not take place. We found that the formation of malonamide derivatives competes with the 4CC to some substantial extent in anhydrous methanol, and that it can be suppressed quite effectively by the addition of small amounts of water and of hydrogen chloride [13].

Recently, our model experiments in the field of peptide segment coupling by 4CC [4,7] have led to some significant results, in particular with 4-pyridine aldehyde [14,15] and chloral [16] as the aldehyde components.

As an aldehyde component of 4CC peptide segment coupling according to Scheme 2 4-pyridine aldehyde (7) is particularly interesting, because its 4CC products (9) with N-protected amino acid derivatives (5) and C-protected amino acid derivatives (6) and isocyanides (8) undergo selective cleavage to form peptide derivatives (10) under a great variety of conditions.

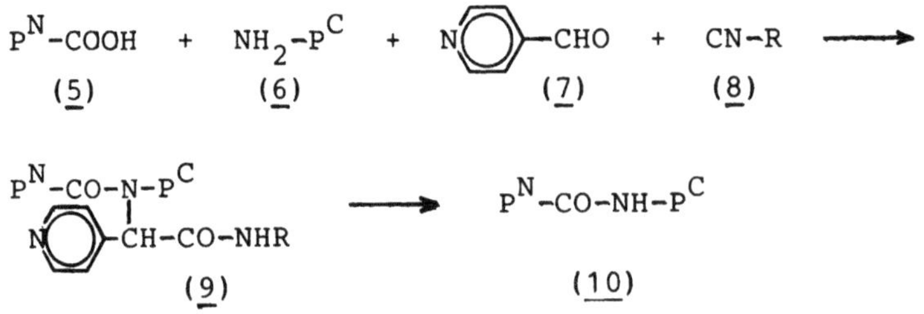

Scheme 2

For example, the 4CC products (9) are cleaved by electrolytic reduction [14], by formic acid in the presence of imidazole [17], by acetanhydride, photolysis [15a], or by autoxidation [15b]. Our recent experiments with (7) aimed at optimizing its 4CC yieldwise and also determining its behaviour with respect to racemization [17]. We found that the 4CC with (7) proceed particularly well, if (6) and (7) are precondensed in the presence of a molecular sieve to form an imine which subsequently reacts with (5) and (8). The cleavage (9) ⟶ (10) is most conveniently accomplished by autoxidation and removal of the accompanying pyridine derivatives from (10) with the acid of an ion exchange resin (amberlite 15, deactivated by 3-bromopyridine) [15b,17].

The synthesis of several peptide derivatives according to Scheme 2 was studied with respect to racemization [17,18]. Not only the C-terminal amino acid unit of (5) was found to racemize considerably (up to 10%) under unfavorable conditions, but it was also found that the amino component may also racemize, if amino acid esters are used as (6). With peptide esters as (6) no racemization of the amino component is observed. The degree of racemization in (5) depends strongly on the reaction conditions [17], in particular the solvent. With methanol or isopropanol as the solvent it is possible to carry out such 4CC peptide segment couplings without any detectable (limits of error: ±1%) racemization [17].

Among the aldehydes which may be used as the aldehyde component of 4CC peptide segment coupling [4,7,16], chloral excels by its availability, its reactivity, and the stability of its 4CC products vs. acidolysis. We believe that 4CC according to Scheme 3 have a good chance to become practically useful in peptide synthesis.

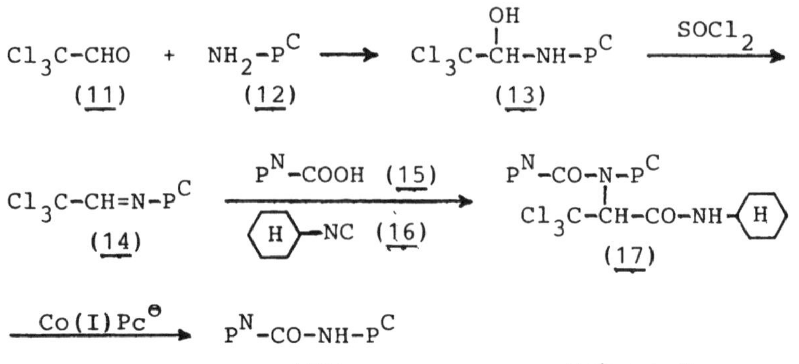

Scheme 3

The imines (14) are formed by treating with $SOCl_2$ the adducts (13) of chloral (11) and C-protected amino acid derivatives (12). In ~1 M DMF solution these imines (14) react smoothly with isocyanides and N-terminally protected or protonated peptides to form the 4CC products (17) (80- 100% crude yield) [19]. The peptide derivatives (18) are obtained from the 4CC products (17) by treatment with cobalt(I)phthalocyanine anion (Co(I)$Pc^{\ominus}$) [20]. For example, H-Val-OMe, H-Leu-OMe, H-Leu-OBzl, or H-Phe-OMe were successfully used as (12) in combination with BOC-Phe-OH, (H-Gly-Val-OH)·$MeSO_3H$, or (H-Gly-Leu-OH)·$MeSO_3H$ as (15) in 4CC with cyclohexyl isocyanide according to Scheme 3.

In order to simplify the search for reaction conditions under which 4CC peptide syntheses proceed with minimum racemization, we developed a particularly convenient racemization test. Mosher's acid [21] seemed to be a most promising auxiliary material for racemization tests on the basis of differences in the ^{19}F-NMR of the stereoisomers of (19) and its derivatives like (20).

In model experiments it was found that the synthesis of (20) from (19 RS) and H-Phe-OMe proceeds without any significant racemization within the limitations of the method (±2-3%), if MAI + 2 HOSu [22], or DCCI + 2 HOSu, or the azide method, or 4CC with (7) are used to couple (19) and H-Phe-OMe, whereas DCCI without HOSu leads to 19% racemization. When carbonyldiimidazole is used as the coupling reagent, 10% racemization are observed. The mixed anhydride of (19) and ethyl chlorocarbonate in the presence of N-methylmorpholine gives rise to 26% racemization, while the mixed anhydride with isobutyl chlorocarbonate produces 5% racemization [23].

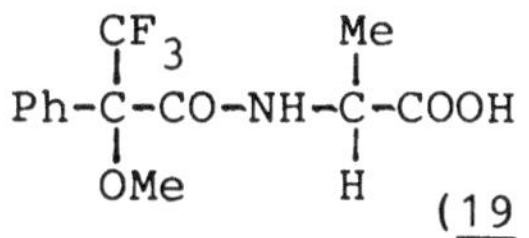

(19 RS) (= M-Ala-OH) $\delta(CF_3)$ = -9.4 ppm ($CDCl_3$; TFA extern)

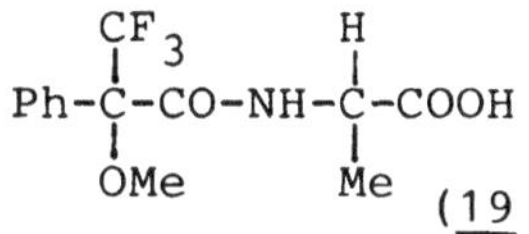

(19 RR) $\delta(CF_3)$ = -9.8 ppm

M-Ala-Phe-OMe $\delta(CF_3)$ = -9.65 ppm
(20 ŔSS)

M-D-Ala-Phe-OMe $\delta(CF_3)$ = -10.1 ppm
(20 RRS)

Scheme 4

References

1. Urban, R., Marquarding, D., Ugi, I.: Hoppe Seyler's Z. Physiol. Chem. 359, 1541 - 1552 (1978).
2. Urban, R.: Tetrahedron 35, 1841 - 1843 (1979).
3. Urban, R. (unpubl. results)

   see also: Urban, R., Marquarding, D., Ugi, I.: Proceedings of the American Peptide Symposium 1979.
4. Ugi, I., Marquarding, D., Urban, R., in: Chemistry and Biochemistry of Amino Acids, Peptides and Proteines. Vol. 6 (Weinstein, B., edit.). Marcel Dekker, New York (1981, in print).
5. Brandt, J., Jochum, C., Ugi, I., Jochum, P.: Tetrahedron 33, 1353 - 1363 (1977).

6. Dugundji, J., Kopp, R., Marquarding, D., Ugi, I.: Top. Curr. Chem. 75, 165 - 180 (1978).

7. Ugi, I., in: The Peptides. Vol. 2 (Meienhofer, J., edit.). Academic Press, New York 1979, p. 365 - 381.

8. Marquarding, D., Burghard, H., Ugi, I., Urban, R., Klusacek, H.: J. Chem. Res. (S), 82 - 83 (1977); J. Chem. Res. (M), 0915 - 0958 (1977).

9. Eberle, G., Lagerlund, I., Ugi, I., Urban, R.: Tetrahedron 34, 977 - 980 (1978).

10. Herrmann, R., Ugi, I.: Tetrahedron 37, 1001 - 1009 (1981).

11. Giesemann, G., v. Hinrichs, E., Ugi, I. (unpubl. results).

12. Gieren, A., Dederer, B., George, G., Marquarding, D., Ugi, I.: Tetrahedron Lett., 1503 - 1506 (1977).

13. Falou, S., Ugi, I. (unpubl. results; see also ref. 4).

14. Waki, M., Meienhofer, J.: J. Amer. Chem. Soc. 99, 6075 - 6082 (1977).

Waki, M., Minematsu, Y., Meienhofer, J., Izumiya, N.: Chem. Lett., 823 - 824 (1979).

15. (a) Bukall, P., Ugi, I.: Heterocycles (Sendai) 11, 467 - 470 (1978).

(b) Bukall, P., Ugi, I.: Heterocycles (Sendai) 15, 381 - 390 (1981).

16. v. Zychlinski, Ugi, I., Marquarding, D.: Angew. Chem. 86, 517 - 518 (1974); Angew. Chem. Internat. Ed. 13, 473 - 474 (1974).

17. Bukall, P.: Dissertation, T. U. München (1980).

18. Gil-Av, E., Charles, R., Bukall, P., Ugi, I. (unpubl. results).

19. Giesemann, G.: Dissertation, T. U. München (1981, in preparation).

20. Eckert, H., Breuer, W., Geller, J., Lagerlund, I., Listl, M., Marquarding, D., Seidel, P., Stüber, S., Ugi, I., Zahr, S., v. Zychlinski, H.: Pure Appl. Chem. 51, 1219 - 1233 (1979).

21. Sullivan, G. R., Dale, J. A., Mosher, H. S.: J. Org. Chem. 38, 2143 - 2147 (1973).

22. Aigner, H., Koch, G., Marquarding, D., in: Chemistry of Peptides and Proteins. III USSR - FRG Symposium. Makhachkala, October 2 - 6, 1980. (Voelter, W., Wünsch, E., Ovchinnikov, Yu. A., Ivanov, V. T., edit.). Walter de Gruyter, Berlin, New York 1981.

23. Breuer, W.: Dissertation, T. U. München (1981, in preparation).

ISOCYANIDES AS ACTIVATING REAGENTS IN PEPTIDE SYNTHESIS

Helmut Aigner, Gertraud Koch and Dieter Marquarding
Organisch-Chemisches Institut der Technischen Universität München, D-8046 Garching, Germany

Dicyclohexylcarbodiimide (DCCI) plays a major role as reagent for a one-pot coupling of properly protected amino acids yielding peptide derivatives[1]. The wide application of this reagent is due to the fact that it is readily available and easy to apply. The real breakthrough in its use, however, originates from the discovery that racemization during the coupling in many cases can be totally suppressed, or at least remarkably decreased, if the carbodiimide is employed in combination with certain additives, such as hydroxysuccinimide (HOSu) or hydroxybenzotriazole (HOBt)[1].

The effectiveness of DCCI in amino acid coupling has been the impulse for numerous attempts to use in an analogous manner other compounds containing a similar "reactive moiety" as carbodiimides, namely the C=N double bond. These investigations have been undertaken not only to get informations about the behaviour of those compounds in condensation reactions, but also possibly to find a reagent which could replace DCCI, i. e. to find a coupling reagent that avoids the two major shortcomings of DCCI, namely its skin irritating property and the low solubility of its product dicyclohexylurea, which very often makes the purification of the desired peptide product so difficult. Up to now, however, in practice none of the compounds in the literature seems to be able to compete with DCCI.

Attempts to introduce simple alkyl and aryl isocyanides as coupling reagents have not been satisfactory with respect to the

Chemistry of Peptides and Proteins, Vol. 1

yields of the desired peptides[2,3,4]. The yields are particularly low, when as the acid component amino acids are used whose amino group is protected through a t-butoxycarbonyl or benzyloxycarbonyl group, i. e. those protecting groups which are usually employed in peptide chemistry[3].

It is noteworthy, however, that after the normal work-up procedure the desired peptide derivatives usually are obtained in analytically pure form, even in those cases in which the yields are extremely low.

When DCCI is used as coupling reagent, very often the yield of the condensation product is remarkably increased, if HOSu is added to the reaction mixture[1]. We have found that addition of HOSu, or HOBt, has almost no influence on the yields of the peptides, if simple alkyl or aryl isocyanides serve as condensation reagents. However, a large effect is observed, when isocyanides with tertiary amino groups are used. The best results have been obtained in the case of those isocyanides, in which the distance between the tertiary amino group and the isocyanide group is determined by an ethyl or propyl moiety. A very effective condensation reagent is N-(2-isocyanoethyl)-morpholine (MAI) which has not yet been described in the literature. It is prepared by well-known procedures in two steps from the inexpensive chemical N-(2-aminoethyl)-morpholine. Most important with respect to the application of MAI is the fact that it does not have the customary smell of isocyanides.

<u>Preparation of MAI</u>

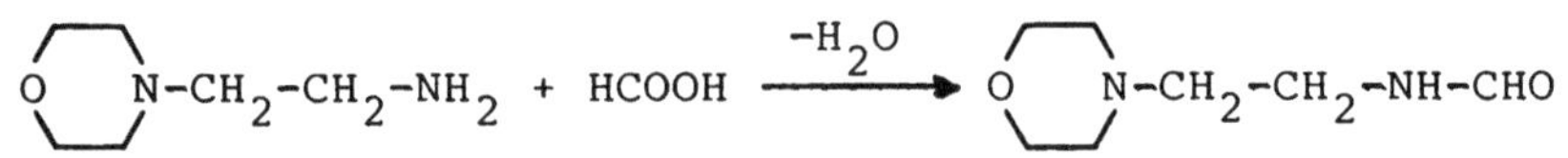

0.7 mole of N-(2-aminoethyl)-morpholine and 30 ml of formic acid in 300 ml of toluene are refluxed with a water separator. After the removal of the water the mixture is concentrated i. vac. and subsequently dissolved in dichloromethane. The solution is satu-

rated with ammonia (gas), filtered, and again evaporated i. vac.. The residue is distilled i. vac., N-(2-N'-formylaminoethyl)-morpholine (b. p. 122 - 124°, 0.05 torr) is obtained in 90% yield.
H-NMR ($CDCl_3$): -CHO 8.0 ppm (s) [TMS as internal standard]
IR (neat): C=O 1670 cm^{-1}.

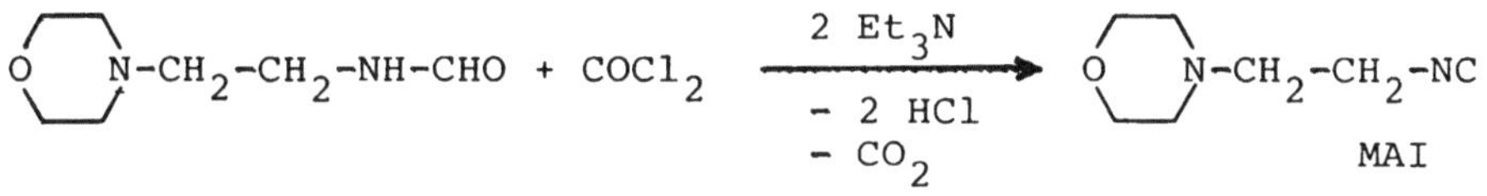

1 mole of N-(2-N'-formylaminoethyl)-morpholine, 320 ml of triethyl amine and 450 ml of dichloromethane are the starting materials. 1 mole of phosgene is introduced while stirring and cooling with ice. The mixture is allowed to reach room temperature and is stirred for 3 hours. Within 1 hour 40 g of ammonia (gas) are introduced (until saturation) while cooling with ice, and the mixture is filtered and concentrated. The residue is distilled i. vac., MAI (b. p. 70°, 0.05 torr) is obtained in 76% yield.
IR (neat): isocyanide 2160 cm^{-1}

From numerous experiments the following general procedure for the preparation of peptide derivatives has emerged.

Procedure for preparing peptide derivatives

The acid component (10 mmole), together with the additive, HOSu (20 mmole) or HOBt (10 mmole), is dissolved or suspended by stirring in a suitable solvent (dichloromethane, ethyl acetate, dioxane, isopropanol, t-butanol, tetrahydrofurane, or dimethyl formamide). After adding MAI (12 mmole) the reaction mixture is stirred for half an hour, then the amine component as hydrochloride (10 mmole) and a suitable tertiary amine (10 mmole) are added (triethyl amine, 4-dimethylaminopyridine (DMAP)).

In general, the reaction mixture is stirred for 24 hours and then after evaporation i. vac. the residue is distributed between ethyl acetate/1 N hydrochloric acid. The ethyl acetate phase is washed with 1 N hydrochloric acid, water, twice with 7.5% $NaHCO_3$-solution, and once again with water. After drying over sodium sulfate the solution is evaporated i. vac.

Note that it is necessary to let the acid component, the additive, and the isocyanide react over a suitable period of time before the amine component is added.

In Table 1 the yields of a number of peptide derivatives are given, which have been prepared according to the above procedure. It contains di-, tri-, and tetra-peptides. However, larger peptide segments have also been combined successfully[5].

Table 1: Peptide derivatives which have been synthesized according to the above general procedure.

Peptide derivative	additive	base	solvent	yield [%]
Z-Val*Gly-OMe	HOSu	Et_3N	AcOEt	81
Z-Val*Gly-OMe	HOBt	DMAP	t-BuOH	82
BOC-Aib*Aib-OMe	HOBt	DMAP	C_6H_6 +)	60
TCBOC-Val*Aib-OBzl	HOBt	DMAP	CH_2Cl_2	81
TCBOC-Aib*Pro-OBzl	--	Et_3N	CH_2Cl_2	69
TCBOC-Aib*Ala-OBzl	HOSu	DMAP	CH_2Cl_2	80
BOC-Gly-Ala*Val-OtBu	HOSu	--	CH_2Cl_2	95
BOC-Gly-Ala*Leu-OEt	HOSu	Et_3N	AcOEt	93
TCBOC-Aib*Val-Aib-OBzl	HOSu	DMAP	CH_2Cl_2	64
BOC-Gly-Leu*Ala-Gly-OEt	HOSu	Et_3N	CH_2Cl_2	72
BOC-Gly-Ala*Leu-Gly-OEt	HOSu	Et_3N	CH_2Cl_2	70
TCBOC-Aib-Ala*Aib-Ala-OBzl	HOBt	DMAP	CH_2Cl_2	55
TCBOC-Aib-Pro*Val-Aib-OBzl	HOSu	Et_3N	CH_2Cl_2	61

+) boiling after 3 h of reaction time

The critical issue in peptide synthesis is the racemization of the acid component during the coupling step. MAI as well as DCCI have been applied in the synthesis of TFA-Phe-Phe-OtBu from the racemization prone TFA-Phe-OH and H-Phe-OtBu. The racemization of the phenylalanine moieties has been measured according to the method developed by Gil-Av et al.[6] Preliminary results[7] are most promising (see Table 2). It is very surprising that the additive HOSu suppresses racemization only in those cases completely in which MAI is employed as coupling reagent.

Table 2: Racemization of TFA-Phe-OH [1 mmole] during the formation of TFA-Phe-Phe-OtBu.

additive [mmole]	coupling reagent [mmole]	phenylalanine D and L [%]	remarks
---	---	D 0 L 100	starting material H-Phe-OH
HOSu [2]	MAI [1.2]	D 0 L 100	---
HOSu [2]	MAI [1.2]	D 0 L 100	addition of a catalytical amount of DMAP
HOSu [2]	DCCI [1.1]	D 3 L 97	---
---	MAI [1.2]	D 17 L 83	---

If the condensation reaction with isocyanides is performed in absence of any additive, then the mechanism of the amide bond formation may be described by a scheme which is given below. The main step in this scheme is the α-addition of the acid component at a formally divalent isocyanide carbon. All of the consecutive steps represent nucleophilic substitutions, in which the various nucleophiles that are present in the reaction mixture attack the carbonyl carbon of the acid component. The ability of isocyanides to undergo α-additions is well-documented in the literature[8]. The intermediate products VII and VIII have been detected; the diacylamine derivative IX, which originates from a

Mumm rearrangement[9], has, however, not yet been found. Compound IX corresponds to the diacylurea derivative which is often a by-product in condensation reactions performed with carbodiimides in the absence of HOSu. A nucleophilic attack at the carbon atom of the imine leads ultimately to compounds which already belong to the scheme.

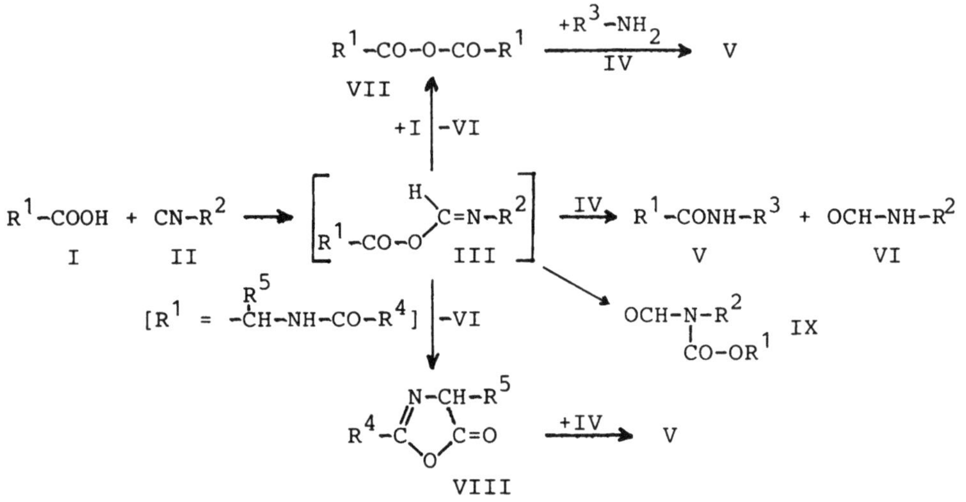

The presence of additives such as HOSu may change the reaction mechanism considerably, in particular, when MAI is the condensation reagent. We have observed that in a solution of 1.2 moles of MAI and 2 moles of HOSu in dichloromethane no isocyanide is detectable any more according to IR. This ratio depends strongly on the nature of the solvent as well as on the acidity of the additive. For example, in order to achieve this effect in DMF, a larger amount of HOSu must be added. In the case of HOBt, which is more acidic than HOSu, the corresponding ratio is given by 1.2 moles HOBt and 1 mole MAI in dichloromethane.

A protonation of MAI according to the next scheme explains these observations. Isocyanides are known to form strong hydrogen bonds[10]. In the case of unsubstituted isocyanides a very large excess of HOSu is needed, in order to protonate the isocyanide such that it cannot be detected in the infrared. Note that the propyl analog of MAI is also not as easy to protonate as MAI itself.

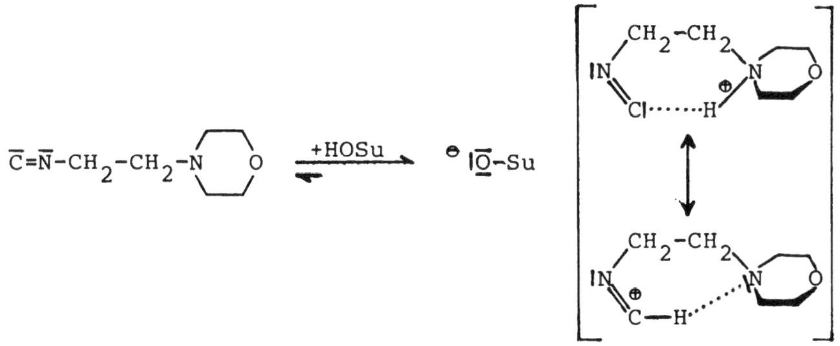

Furthermore, when an excess of MAI (or a basic solvent such as pyridine) is used in the reaction, then very often the HOSu or HOBt ester, resp., of the acid component can be isolated in high yields. It is noteworthy that this is possible even in the presence of the amine component!

The formation of these so-called "active" esters can be rationalized according to the following reaction scheme. Here the hydroxysuccinimide anion serves as the nucleophile which reacts with the α-adduct of the isocyanide and the acid component.

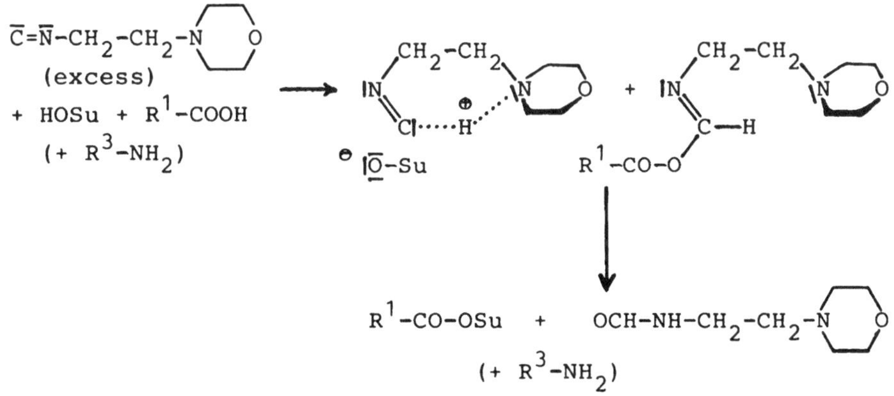

It has been assumed that in condensation reactions with DCCI the additive HOSu reacts with the DCCI-adduct of the acid component to form the HOSu ester of the latter. This reaction is supposed to take place before the DCCI-adduct can decompose into the racemization prone oxazolinone derivative of the acid component[11]. Our observations are evidence against such a reaction mechanism.

References

1. Wünsch, E.(edit.): Synthese von Peptiden. Teil II. Methoden der organischen Chemie. (Müller, E., edit.). Band 15/2. Georg Thieme Verlag, Stuttgart 1974, p. 103 - 119.
2. Yasuda, N., Ariyoshi, Y., Toi, K.: United States Patent, 3,933,783. Jan. 20, 1976.
3. Aigner, H., Marquarding, D.: Tetrahedron Lett., 3325 - 3326 (1978).
4. Wackerle, L.: Synthesis, 197 - 198 (1979).
5. Wünsch, E., Moroder, L. et al. (unpubl. results).
6. Gil-Av, E., Feibush, B., Charles-Sigler, R., in: Gas Chromatography. (Littlewood, A. B., edit.). Institute of Petroleum, London 1967, p. 227.

   Gil-Av, E., Feibush, B., Charles-Sigler, R.: Tetrahedron Lett., 1009 - 1015 (1966).
7. Charles, R., Gil-Av, E., Aigner, H., Koch, G., Marquarding, D. (unpubl. results).
8. Ugi, I. (edit.): Isonitrile Chemistry. Academic Press, New York and London 1971, p. 65 - 92.
9. Mumm, O., Hesse, H., Volquartz, H.: Chem. Ber. 48, 379 - 391 (1915).
10. Schleyer, P. von R., Allerhand, A.: J. Amer. Chem. Soc. 84, 1322 - 1323 (1962).

    Ferstandig, L. L.: J. Amer. Chem. Soc. 84, 1323 - 1324 (1962).

    Ferstandig, L. L.: J. Amer. Chem. Soc. 84, 3553 - 3557 (1962).

    Allerhand, A., Schleyer, P. von R.: J. Amer. Chem. Soc. 85, 866 - 870 (1965).
11. Rich, D. H., Singh, J., in: The Peptides. Analysis, Synthesis, Biology. (Gross, E., Meienhofer, J., edit.). Vol. 1. Academic Press, New York 1979, p. 249.

IMPACT OF CONFORMATION ON THE SYNTHETIC STRATEGIES FOR PEPTIDE SEQUENÇES

Manfred Mutter, Hermann Anzinger, Karsten Bode, Franz Maser and V.N. Rajasekharan Pillai (+)
Institute of Organic Chemistry, University of Mainz, D-6500 Mainz, Germany

Introduction

The physicochemical properties such as the solubility and the rate of coupling play an extremely important role in the stepwise and segment condensation approaches for the synthesis of peptides. The chain length of the peptide has a significant influence upon the solubility and coupling rates during stepwise synthesis. In a growing peptide chain, these properties are drastically influenced by conformational transitions. The primary sequence, nature of the involved protecting groups and the physical environment decide the conformational preferences of the concerned peptides. A thorough knowledge of the interdependence of these factors and their influence on the physicochemical properties of the peptides would permit an effective planning of synthetic strategies for the desired peptide and protein sequences. In order to delineate these aspects, we synthesized a number of model oligopeptides and investigated their conformational preferences with respect to composition, sequence, chain length and solvation on secondary structure. The observed correlation between conformation, solubility and coupling kinetics of the growing

(+) On leave from the Department of Chemistry, University of Calicut, Kerala-67635, India

Chemistry of Peptides and Proteins, Vol. 1

peptide chain from these studies helped to delineate the specific experimental strategies to be followed for a successful synthesis of short to medium-sized peptides.

For the synthesis of the oligopeptides the liquid phase method[1] proved to be highly suitable and the conformational analysis was carried out using CD, IR and NMR measurements on the polyoxyethylene (POE)-bound peptides under different conditions. The experimental techniques for the conformational analysis of peptides could be directly applied to the POE-bound peptides since the POE has been shown to have no significant influence on the conformations of the bound peptides[2]. This property of POE is due to its conformational behaviour characterized by a statistical coil structure with low density and high segmental mobility.

Chain Elongation and Physicochemical Properties of Homooligopeptides

The stepwise synthesis and subsequent conformational analysis of homooligopeptides on the soluble polymeric support, polyoxyethylene, reveals a drastic influence of the conformational transitions of the growing peptide on its physicochemical properties. The solubility of the peptides and the reactivity of the terminal amino group are significantly reduced in oligopeptides exhibiting β-structures. Thus in the POE-bound homooligomers of valine and isoleucine, the tendency to aggregate at very short chain lengths is evident from the drastic decrease in solubility with growing peptide chain. For example, when dissolved in methylene chloride the tetrapeptide derivatives yield highly viscous solutions at concentrations of 10% (w/v), and only at very low concentrations ($<$ 1%) does the viscosity decrease to normal values. Obviously, a network-like structure is formed by intermolecular interactions between the peptide chains. By adding polar solvents such as DMF or DMSO, these aggregates are disrupted and the viscosity decreases. With growing peptide chain, increasing amounts of DMF and DMSO had to be used to obtain homogeneous solutions. The highest POE-bound oligomers (n = 7,8) were sparingly soluble in mixtures of DMF and DMSO and resulted in gel-like solutions in CH_2Cl_2 even at concentrations of ca 1%. Whereas the low solubility of these oligopeptides represent the most severe obstacle in the stepwise elongation, quantitative yields are also difficult to achieve because of the strong steric hindrance of the coupling components.

The above described gradual decrease in solubility and coupling rates have their origin in the conformational transitions which occur in the stepwise elongation process. In the case of POE-bound homooligoisoleucines the intensity of the Cotton effect in the regions 210-220 nm and 190-200 nm gradually increases with increasing chain length (n = 2-5)[3]. The characteristic CD data point to the onset of a partial ß-structure, which is very much concentration-dependent, at the 7-peptide level. As in the case of $(Ala)_n$ and $(Val)_n$, the spectra of the 7-peptide and 8-peptide are completely different from those of the lower oligomers. In contrast to the POE-bound oligoalanine series, here the random-coil ⟶ ß-structure transition starts at shorter chain lengths. This indicates a much higher stability for the ß-structure in the oligoisoleucines. Even concentrations as low as 0.02 mg peptide/ml did not result in a significant disruption of the ß-sheets for $(Ile)_8$. This unusual stability of the 8-peptide may be explained by the possibility of the formation of an intramolecular ß-structure (cross ß-structure). Such a model should result in the bending of the peptide (ß-bend) to form intramolecular hydrogen bridges which are not destroyed by dilution.

Oligopeptides with tendencies to adopt α-helical or unordered structures showed no pronounced change in the solubility or coupling kinetics during chain elongation. For example, the oligomethionine, $\underline{L}$-$(Met)_{15}$-NH-POE, exhibits excellent solubility in a number of solvents (Table I). The physicochemical properties of this protected oligomer are strongly dominated by the C-terminal POE group. The kinetics of the coupling step showed no significant dependence of the chain length. Obviously, the transition from random coil to ∝-helical structure, as observed by CD measurements in organic solvents in the range n = 6-9, was not reflected in a drastic change of the concerned physicochemical properties of this peptide[4].

In the case of POE-bound $[GLu(Bzl)]_n$, the development of ∝-helical conformation starts at n = 7 in helix-promoting solvents. During the stepwise synthesis, a unique dependence between physiochemical properties and chain length was observed. The solubility of the oligomers in methylene chloride decreased considerably with the growing peptide chain and showed minimum solubility in the range n = 6-8. With further elongation of the peptide chain, a continuous increase in the solubility was observed, and for $n > 12$, the oligopeptide exhibited the same solubility as that of the lowest oligomers ($n < 5$). Obviously, this change in solubility with increasing chain length is paralleled by conformational transitions of the type random coil ($n \leq 5$) $\longrightarrow$ aggregated ß-like structures, ($5 < n \leq 8$) $\longrightarrow$ ∝-helical conformation ($n > 8$). This dependence of the solubility of the oligopeptides on the chain length and the involved conformational transitions is depicted in Figure 1. The occurence of minimum solubility of medium-sized peptides appears to be a general feature and seems to reflect the tendency of certain peptides to form intermolecularly hydrogen-bonded structures in this range.

The kinetics of the peptide forming step in the $[Glu(Bzl)]_n$ series also reflects the conformational features outlined above. For example, a considerable decrease in the reaction rates in methylene chloride was observed in the range $3 < n < 9$, indicating a reduced reactivity of the terminal amino group of the oligomers in the ß-conformations. On proceeding to longer chain lengths, transition to ∝-helical conformations occured

Table I

Dependence of the Solubility of POE-Bound Oligopeptides on their Conformation

POE[a]-bound peptide	Preferred conformation in trifluoro-ethanol	Solubility[b] in TFE	CH_2Cl_2	DMF
t-Boc-(L-Met)$_n$-NHPOE; n >10	α	++	+	+
(L-Met)$_5$-Pro-(L-Met)$_5$-NHPOE	r.c.	+	o	+
[L-Glu(OBzl)]$_8$-OPOE	β	o	-	-
[L-Glu(OBzl)]$_{14}$-OPOE	α	++	+	++
[L-Glu(OBzl)]$_7$-Pro-[L-Glu(OBzl)]$_7$-OPOE	r.c.(α)	++	+	++
[L-Glu(OBzl)]$_2$-Pro-[L-Glu(OBzl)]$_{12}$-OPOE	α	++	+	++
[L-Glu(OBzl)]$_{20}$-Gly-OPOE	α	++	+	++
(L-Val)$_8$-Gly-OPOE(M)	β	o	--	-
(L-Ile)$_n$-OPOE; n>6	β	-	--	--
(L-Ile)$_4$-Gly-(L-Ile)$_4$-NHPOE	r.c.(β)	+	o	+
(L-Ile)$_4$-Pro-(L-Ile)$_4$-NHPOE	r.c.(β)	+	o	+
(L-Ile)$_n$-X-OPOE; (X=Gly, Pro); n>6	β	-	--	--
(L-Ala)$_n$-OPOE; n>7	β	-	--	-
(L-Ala)$_5$-Pro-(L-Ala)$_4$-OPOE	r.c.	+	o	+

a Molecular weight 10,000; b ++, very good; +, good; -, low; --, very low; o, medium and a significant increase in the reaction rate was registered.

Co-oligopeptides: the Host-Guest Technique

A better understanding of the interdependence of the physico-chemical properties and conformational transitions of oligo-peptides could be obtained by the application of the so-called host-guest technique in the liquid phase method of peptide synthesis. The host-guest technique was originally introduced to obtain detailed information regarding the thermodynamic and conformational parameters of single guest amino acids in

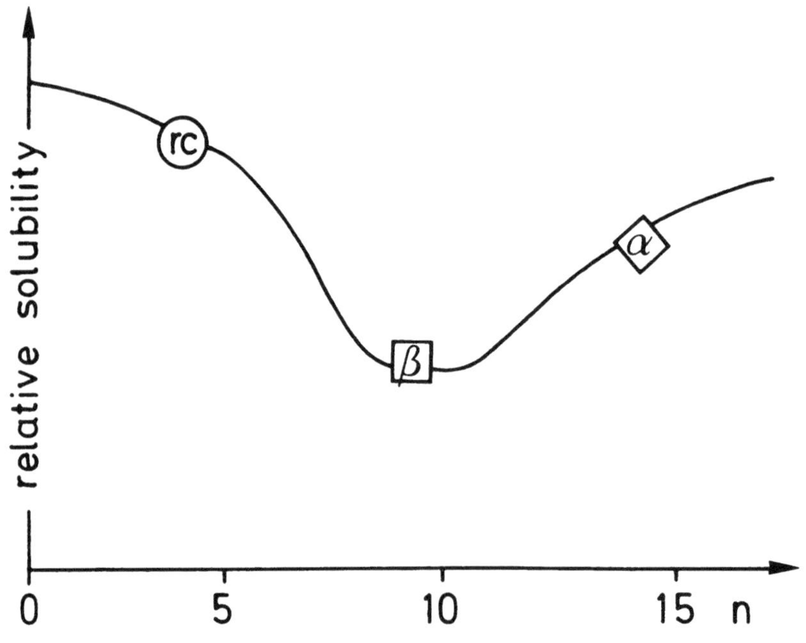

Fig. 1. Dependence of the relative solubility of oligo-peptides on the chain length (n) and the involved conformational transitions, illustrated for the case of POE-bound $[Glu(OBzl)]_n$.

various positions of a host peptide chain[5,6]. We adopted a variation of this technique to delineate the conformational effects upon the physicochemical properties of a peptide.

Considering conformational energies, the insertion of a Gly or Pro residue should result in a substantial disruption of ordered conformations. Most notably, due to steric interactions with its neighbouring residues, the proline residue is predicted to interrupt the growth of an α-helix. Moreover, the Pro residue cannot accomodate itself as part of a regular ß-structure. Similarly, the glycine residue is expected to destabilize ordered conformations, both the α-helix and ß-structure, because of its high conformational flexibility.

The above considerations become evident on the examination of the physicochemical properties of a number of cooligopeptides. Thus the solubilities of $(Ile)_4$-Gly-$(Ile)_4$ and $(Ile)_4$-Pro-$(Ile)_4$ differ drastically from the corresponding homooligomer of Ile. The Ile-Pro "host-guest peptide" is soluble in CH_2Cl_2, DMF or AcOH, whereas the homooligomer is practically insoluble in these solvents. The corresponding glycine substituted co-oligomer shows somewhat lower solubility compared to the proline substituted one ;but here also the difference in the physicochemical properties compared to the homooligopeptide is still striking.
The coupling rates also remain practically constant throughout the stages of chain elongation in the case of the host-guest peptide. For example, the reaction conditions for obtaining quantitative yields for the steps $(Ile)_5 \dashrightarrow (Ile)_8$ did not differ significantly from those in the steps $(Ile)_1 \dashrightarrow (Ile)_4$ in the case of $(Ile)_4$-Gly-$(Ile)_4$ or $(Ile)_4$-Pro-$(Ile)_4$. This is in sharp contrast to that observed in the case of the homo-oligopeptide. This difference in the physicochemical behaviour of these oligopeptides is purely of conformational origin. This is evidenced from the CD studies. In the case of the co-oligopeptides, random conformations appear to be predominant. The formation of extended structures, which is probably a

prerequisite for the aggregation of these short peptides, seems to be prevented by the guest residues. Thus, these guest residues, in a way, act as a flexible link between the two $(Ile)_4$ blocks.

Circular dichroism studies in the case of the Met series indicate that the α-helix content in trifluoroethanol of polyoxyethylene-bound Boc(Met)$_5$-Pro-(Met)$_5$ (28%) approaches that of Boc-(Met)$_7$ (30%) and is much lower than that of Boc-(Met)$_{11}$

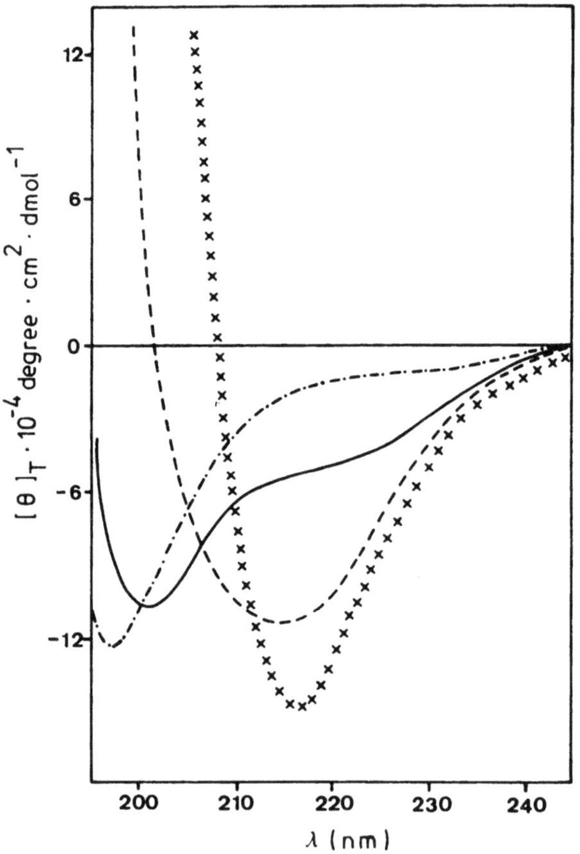

Fig.2. CD spectra of POE-bound oligoalanines. (xxxx): Boc-(Ala)$_{10}$ in water; (-·-·-): Boc(Ala)$_5$-Pro-(Ala)$_4$ in water; (---): Boc-(Ala)$_{10}$ in trifluoroethanol; (——): Boc(Ala)$_5$-Pro-(Ala)$_4$ in trifluoroethanol.

(80%). This observation shows that proline interupts the α-helix and it induces a helix-inducing effect (asymmetric nucleation) upon the C-terminal pentapeptide. Similarly, the CD spectra of $(Ala)_{10}$-OPOE are indicative of a ß-structure, whereas $(Ala)_5$-Pro-$(Ala)_4$-OPOE gives CD spectra characteristic of mostly unordered conformations (Figure 2)[7,8].
The drastic effect of a proline residue on ordered conformations is also demonstrated for the case of POE-bound $[Boc\text{-}Glu(OBzl)]_n$ co-oligopeptides in the solid state and in solution (Table II).

Table II

Influence of a L-Pro Residue on the α-Helix Content of POE-Bound Oligo-$[L\text{-}Glu(OBzl)]_n$.

Oligomer		α-Helix-content percentage
$[L\text{-}Glu(OBzl)]_n$;	n = 8	30
	n = 12	48
	n = 15	55
$[L\text{-}Glu(OBzl)]_n$-Pro-$[L\text{-}Glu(OBzl)]_7$;	n = 0	32
	n = 7	23
$[L\text{-}Glu(OBzl)]_n$-Pro-$[L\text{-}Glu(OBzl)]_{12}$;	n = 0	52
	n = 2	45

Similar destabilizing effects of ordered conformations were observed for the case of a glycine residue as guest amino acid also.

Implications in Synthetic Strategies for Peptides

The above-described investigations on the interdependence of chain length, physicochemical properties and conformational transitions of homooligopeptides and cooligopeptides help to delineate, to some extent, the experimental strategies to be followed for a successful synthesis of peptides. The present results suggest that the elongation of a peptide chain must not necessarily be paralleled by a decrease in solubility. The observation that medium-sized peptides show a minimum in solubility and reactivity is related to the tendency of these short segments to form ß-like structures and this behaviour appears quite general. The examples described in this paper show that this critical stage in peptide synthesis can be overcome either by (a) further elongation of the peptide chain (i.e. in cases where structures other than ß-conformation become more favourable at longer chain lengths), most notably by segment condensation, or by (b) insertion of a residue, which prevents the formation of a ß-structure, in the middle of a medium-sized peptide chain.

These findings can be of considerable impact for the conventional strategies of peptide synthesis. The segment condensation of medium-sized segments with low solubility to larger peptide chains may result in the increased solubility of the reaction product due to conformational transitions. Whenever it is possible, amino acid residues with low tendencies for ß-structure formation should be inserted in the middle of the peptide segments. Considering these conformational aspects, an effective synthetic strategy for larger peptides can be envisioned as follows: First, determine the conformational preferences (nature of the secondary structure) for the various possible segments following empirical prediction rules[9,10]. Then, arrange the sequence in such a way as to avoid longer ($n > 5$) segments of amino acid residues with P_β-potentials larger than 1. If necessary, and when the sequence permits, insert a ß-breaking residue in a central position of the concerned segment; proceed to the synthesis of these segments by the stepwise approach and finally establish the total sequence by segment condensation. Such a procedure which gives due consideration to the conformational transitions and the dependent physicochemical properties of peptides should help to overcome two main obstacles in peptide synthesis, namely the low solubility of segments and the reduced reactivity of the terminal amino groups with increasing chain length.

Acknowledgement

We thank the Deutsche Forschungsgemeinschaft for the financial support of this work through the Sonderforschungsbereich 41, "Chemie und Physik der Makromoleküle".

References

1. Mutter, M., Bayer, E.: in "The Peptides: Analysis, Synthesis and Biology", Vol. II, Gross, E., Meienhofer, J.: Eds., Academic Press, New York, 1980, p. 286

2. Mutter, M.: Macromolecules, 10, 1413 (1977).

3. Mutter, M., Mutter, H., Uhmann, R., Bayer, E.: Biopolymers, 15, 917 (1976).

4. Rahman, S.A., Anzinger, H., Mutter, M.: Biopolymers, 19, 173 (1980).

5. von Dreele, P.H., Lotan, N., Ananthanarayanan, V.S., Andreatta, R.H., Poland, D., Scheraga, H.A.: Macromolecules 4, 408 (1971).

6. Ribeiro, A., Goodman, M., Naider, F.: J. Am. Chem. Soc., 100, 3903 (1978).

7. Toniolo, C., Bonora, G.M., Mutter, M., Pillai, V.N.R.: Makromol. Chem., 182, 1997 (1981).

8. Toniolo, C., Bonora, G.M., Mutter, M., Pillai, V.N.R.:

Makromol. Chem., 182, 2007 (1981).

9. Tanaka, S., Scheraga, H.A.: Macromolecules, 9, 142 (1976).

10. Chou, P.Y. Fasman, G.D.: Ann. Rev. Biochem., 47, 251 (1978).

SYNTHETIC STUDIES OF NEUROTOXIN II FROM VENOM OF CENTRAL ASIAN COBRA *Naja naja oxiana*

Vladislav Deigin, Victor Ulyashin, Inessa Mikhaleva, Vadim Ivanov

Shemyakin Institute of Bioorganic Chemistry, USSR Academy of Sciences, Moscow, USSR

At the previous FRG-USSR Symposium and at the 15th European Peptide Symposium we reported some results of our investigation aimed at total synthesis of neurotoxin from the venom of Central Asian cobra *Naja naja oxiana* (Fig. 1) (1).

Synthesis of neurotoxin II is a component part of the structure-functional investigation of polypeptide neurotoxins and the development of methodology for synthesis of large peptides and proteins.

NT-II is a 61-membered polypeptide with 4 disulfide bridges.

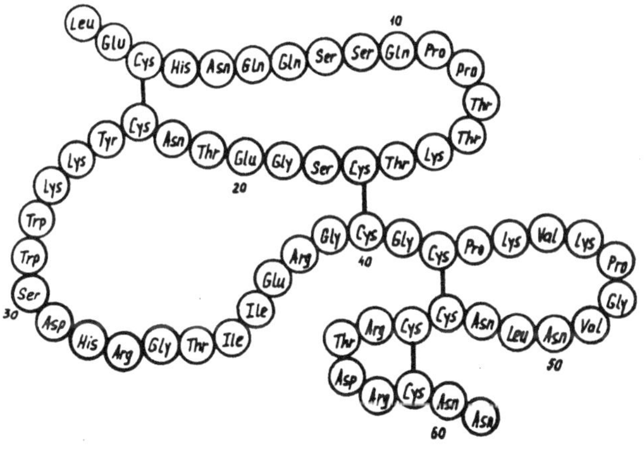

Fig. 1. Amino acid sequence of neurotoxin II from the venom of Central Asian cobra *Naja naja oxiana*.

Chemistry of Peptides and Proteins, Vol. 1

The synthesis was elaborated in such a manner so that we could employ the strategy of maximal protection of the trifunctional residues with the help of hydrophobic groupings of the benzyl type as was described in an earlier paper for the case of α-bungarotoxin (2). In view of the fact that we encountered some difficulties in synthesizing α-bungarotoxin, for instance, side reactions involving the liberating cysteine SH group due to HF treatment, we utilized the acetamidomethyl protection instead of p-methoxybenzyl. Such a combination of the protecting groups allows the purification of the resulting product following HF deblocking to be carried out without any risk of oxidation of the cysteine residues. The glutamine carboxamide groups and tryptophan indole remained unprotected. The following protections were used for the remaining amino acids: benzhydryl for asparagine, benzyl esters for aspartic and glutamic acids, benzyl ethers for serine and threonine, 2,6-dichlorobenzyl for tyrosine and the tosyl groups for the arginine guanidine group. Tos, Boc and Dnp groups were used to protect the histidine imidazole. The α-amino group was protected either by Boc- or Z-group. Stepwise synthesis was the method used to obtain all the short fragments corresponding to the total amino acid sequence of the NT-II molecule (Fig. 2), namely 1-3, 4-8, 9-12, 13-15, 16-19, 20-26, 27-33, 34-47, 48-51 and 51-61. Each fragment was synthesized in more than one way that allowed us to optimize the yield and facilitate the purification of the final product.

The present paper is devoted to further studies on condensation of the fragments, as well as on methods of their purification. Several attempts have been made to purify the protected peptides by gel-filtration on soft strongly swelling sorbents, as for example, enzacryl K-2 (3), Sephadex LH-20 and LH-60 (4) or Bio-Beads Sxl (5). However, under these conditions it appeared difficult to realize the separation of peptides as they differ in molecular mass by less than twofold. Besides, difficulties connected with their purification increased with a rise in the molecular mass of the peptides.

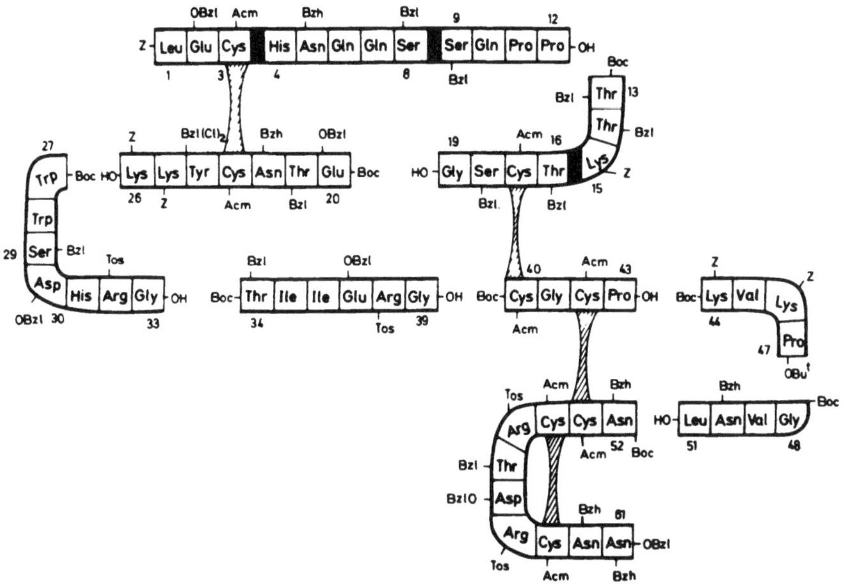

Fig. 2. Molecule fragment scheme for neurotoxin II.

Moreover, we encountered other difficulties in the process of purification of the intermediate protected NT-II fragments, especially when using the conventional gel-filtration methods (1).

Analytical and preparative size-exclusion HPLC methods were employed in the case of the protected hydrophobic peptides in organic solvents. But in order to optimize these procedures we had to investigate various types of gels - soft, semirigid and rigid.

Among the investigated soft gels, Sephadex LH-20 is in virtue of its properties the most suitable for separation and analysis of synthetic protected peptides with a molecular mass below 2000 dalton (Fig. 3). The mechanical properties of Sephadex LH-20 favour a rise in the efficiency of the column and linear flow rate.

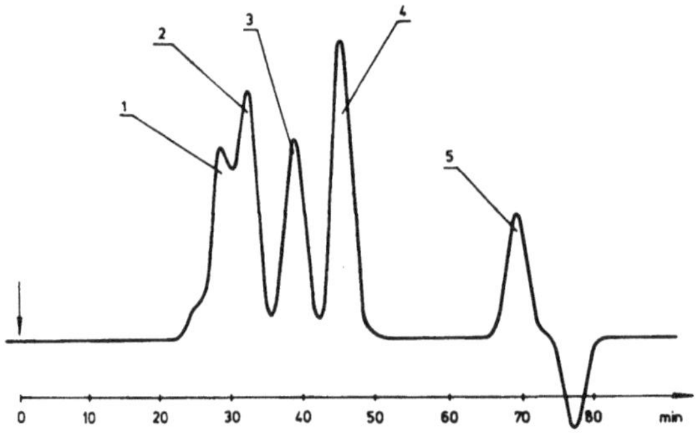

Fig. 3. Chromatogram of the protected peptides and naphthalene mixture:

		M.M.
1.	Aoc-56-61-OBzl	1963
2.	Aoc-58-61-OBzl	1265
3.	Boc-60-61-OBzl	768
4.	Boc-Asn(Bzh)-OBzl	488
5.	Naphtalene	130

8 x 600 mm column, Sephadex LH-20 gel, eluent DMF . Flow rate 0.4 ml/min, refractive index detector.

With the help of narrow fractions of sorbent particles (40 - 50 μ) we managed to obtain an efficient column operation (about 8000 t.p. per meter). A more detailed investigation of the soft and semirigid gels was carried out in (6).

We used silica gels of different types to separate and analyze peptides with a molecular mass above 2000 dalton.

The application of rigid sorbents in size-exclusion chromatography has opened the way for new opportunities for performing separation and analysis of peptides. Thus, for example, the accessibility of silica gel with small-size particles makes it possible to obtain an extremely high efficient column (about 100000 t.p. per meter). Additionally, the mechanical properties allow operating at high flow rates. One can select silica gels for working in any molecular mass range. By modifying the silica gel surface one can obtain a sorbent suitable for the separation of substances of any class, including labile

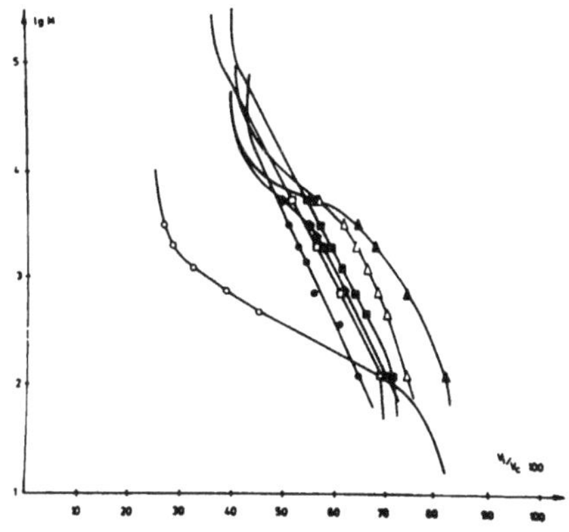

Fig. 4. Calibration curves for the studied sorbents with synthetic protected peptides in DMF.

1. Sephadex LH-20
2. Silica gel "L"
3. Silasorb 600
4. Spherical silica gel
5. Si-40
6. PSM-60S
7. Zorbax Sil

Synthetic peptides indicated in Fig. 4 caption as well as a Boc-52-61-OBzl fragment (M.M.=2575) and naphtalene were used for calibration. Void volume was determined using high molecular impurities.

biopolymers. Therefore, silica gels can be applied for analysis and separation in preparative quantities of practically any substance. With the help of different modifications it is also possible to prepare sorbents on which the separation proceeds according to a purely size-exclusion mechanism.

We investigated the selectivity of different silica gels and the behaviour of peptides on them, with the aim to choose an optimal form for size-exclusion chromatography of protected peptides in organic solvents (Fig. 4).

As is seen in Fig. 4 Zorbax Sil and PSM-60S have practically a linear calibration curve but their selectivity value is relatively low. Silica gel Si-40(Merck) also has the same linear calibration curve, although its selectivity is apparently less and the operating range narrower, most likely due to the larger

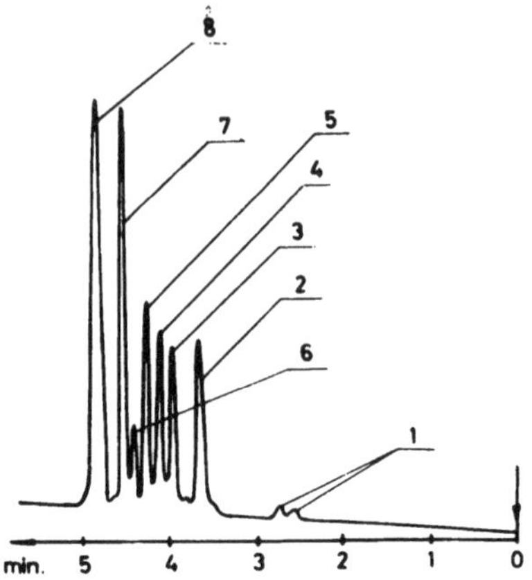

Fig. 5. Chromatogram of artificial peptide mixture:

		M.M.
1.	High molecular weight impurities . .	-
2.	Boc-34-61-OBzl	5417
3.	Boc-48-61-OBzl	3125
4.	Aoc-55-61-OBzl	1962
5.	Aoc-58-61-OBzl	1265
6.	Boc-60-61-OBzl	768
7.	Boc-Asn(Bzh)-OBzl	488
8.	DMSO	78

8 x 250 mm column, Silasorb 600 sorbent, eluent DMF. Flow rate 2 ml/min, refractive index detector.

volume of the micropores.

The silica gel prepared according to Unger's methods (7) appeared to have the best selectivity, especially in the high molecular weight region. The calibration curve of this silica gel has a non-linear shape.

Besides, the nature of the silica gel surface also plays an essential role in achieving a suitable separation of the protected peptides. For instance, only totally protected peptides were eluted from the column with a non-modified surface by the pure size-exclusion mechanism (Fig. 5). Peptides containing free imidazole, guanidino or amino groups sorbed on a gel

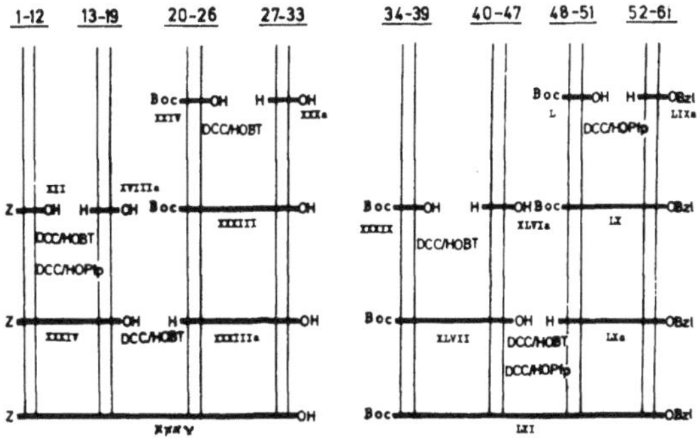

Fig. 6. Preparation scheme for fragments 1-33 and 34-61.

matrix and eluted a little later then the protected peptides of the same molecular mass.

In the course of synthesis and condensation of the fragments we observed the formation of high molecular weight impurities

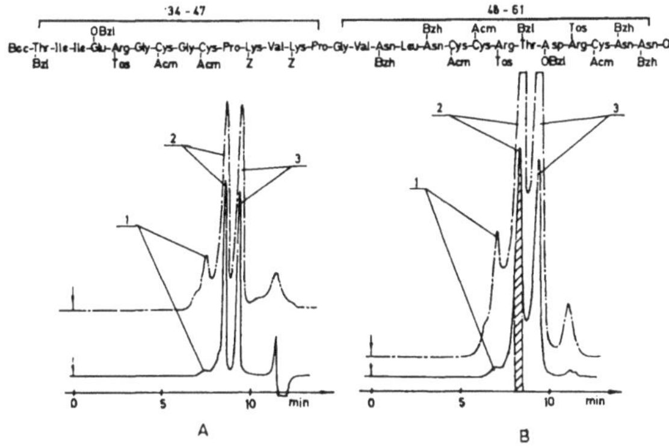

Fig. 7. Analytical (A) and preparative (B) chromatography of synthetic peptide of 34-61 sequence on a silica gel "L" column, 25 x 330 mm in DMF. Flow rate 10 ml/min.

1. High molecular impurities
2. Boc-34-61-OBzl
3. Initial fragments

UV-detector,respect.refractive index detector. Loading under preparative conditions 50 mg per 20 ml.

of peptide nature whose amount increased with reaction time. Not always is it possible to successfully remove these polymer products by standard methods, especially on large peptides. According to our data these polymers are not products of peptide association, for their retention times do not change with rechromatography.

The fragments were condensed mostly by DCC/HOBT method (Fig.6). Fragments 1-33 were obtained by the condensation of two blocks, that of 1-19 and 20-33. The C-terminal part of the molecule sequence was obtained by the same DCC/HOBT, as well as by DCC in the presence of pentafluorophenol. In both cases the yield varied within the range of 25-30%. The reaction product was separated by size-exclusion chromatography on a silica gel column. An example of the analytical (A) and preparative (B) separation of the condensation reaction products of fragments 34-37 and 48-61 is given in Fig. 7.

The individuality of all the resulting fragments was also controlled chromatographically and by determining the N-terminal amino acids and was confirmed by amino acid analysis data. Thus, two large molecule fragments were obtained, those of 1-33 and 34-61 (Fig. 2).A search for optimal conditions for condensation of these blocks into a uniform polypeptide chain is still in progress.

References

1. Mikhaleva, I.I., Deigin, V.I., Ulyashin, V.V., Ivanov, V.T.: "Peptides 1978", Proceedings of the 15-th European Peptide Symposium, Poland, 391-395 , 1979

2. Ivanov, V.T., Mikhaleva, I.I., Volpina, O.M., Myagkova, M.A., Deigin, V.I.: "Peptides 1976" (ed. A.Loffet), Proceedings of the 14-th European Peptide Symposium, Belgium 219-229, 1976

3. Galpin, I.J., Handa, B.K., Kenner, G.W., Moore, S., Ramage, R.: J. Chromatogr. 123, 237-242 (1976).

4. Danho, W., Föhles, T.: Hoppe-Seyler's Z. Physiol. Chem. 36, 839-847 (1980).

5. Volpina, O.M., Deshko, T.N., Mikhaleva, I.I., Ivanov, V.T.: Bioorg. Khim. 6, 1155-1162 (1980).

6. Deigin, V.I., Ulyashin, V.V., Nefedev, P.P., Zhmakina, T.P., Belen'ky, B.G., Ivanov, V.T.: Bioorg. Khim. (USSR) 8, in press

7. Unger, K., Check-Kalb, J., Straube, B.: J. Polym. Col. Sci. 253, 658 (1974).

INVESTIGATION ON THE SYNTHESIS OF SALMON CALCITONIN II FRAGMENTS

Gennadij Vlasov, Valentina Glushenkova, Valerij Lashkov, Natalija Kozhevnikova, Lidija Nadezhdina, Leonid Krasnikov, Irina Ditkovskaja, Oskana Glinskaja and Tatjana Komogorova
Institute of Macromolecular Compounds of the Academy of Sciences of the USSR, Leningrad, USSR

Introduction

Synthesis of the salmon calcitonin II fragments was investigated using a classic approach. A number of by-reactions in the course of the synthesis was investigated.

Results

Calcitonins are 32-membered polypeptide hormones with a disulfide bond between the first and the seventh amino acid (1). They regulate the calcium and phosphorus levels in the blood (2). Salmon calcitonin is more active than other hormones except eal calcitonin (3).

In order to develop a salmon calcitonin synthesis the investigation of the synthesis of its fragments was carried out in solution by using a minimum amount of blocking groups. The fragments 1-5, 6-10, 11-15, 16-21, 22-24, 25-28 and 29-32 were synthesized according to schemes 1-7.

The synthesis of fragment 1-5 was carried out by the use of the NPS blocking group for the amino functions and DCC and PFP methods for coupling (scheme 1). The synthesis of fragment 6-10 was carried out with NPS (Leu^8, Val^9) and Boc ($AcmCys^7$, Thr^6) blocking groups and DCC as the coupling rea-

Chemistry of Peptides and Proteins, Vol. 1

gent (scheme 2).

Scheme 1

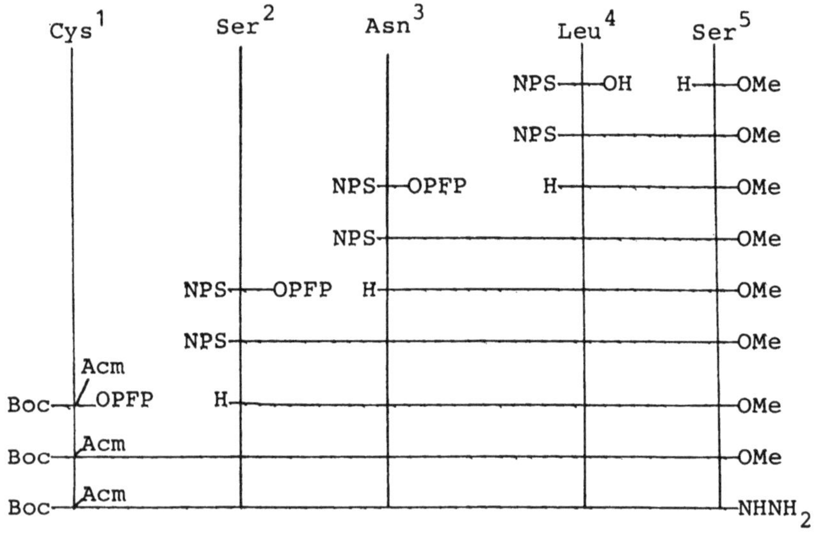

Scheme 2

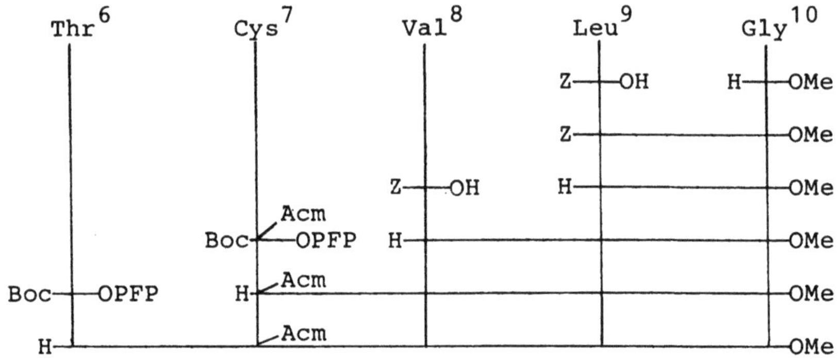

The pentapeptide 11-15 was synthesized from two fragments: 11-13 and 14-15. The carbobenzoxy blocking group was used for the protection of the α-amino function and the tert-butyloxycarbonyl group for blocking the ε-amino function of lysine[11]. Condensation of fragments was carried out by Rudinger's modification of the azide method (scheme 3). The hexapeptide 16-21 was synthesized from two fragments (16-17 and 18-21) by the DCC/HOSu coupling method. The dipeptide 16-17 was prepared by the reaction of the N-hydroxysuccinimide ester of Z-leucine with histidine. The tetrapeptide 18-21 was synthesized according to the scheme 4. The carbobenzoxy and tert-butyloxycarbonyl blocking groups were used for lysine [18]. The condensation of the acylamino acids and peptide fragments was carried out with the PFP, NP ester and DCC/HOSu method.

Scheme 3

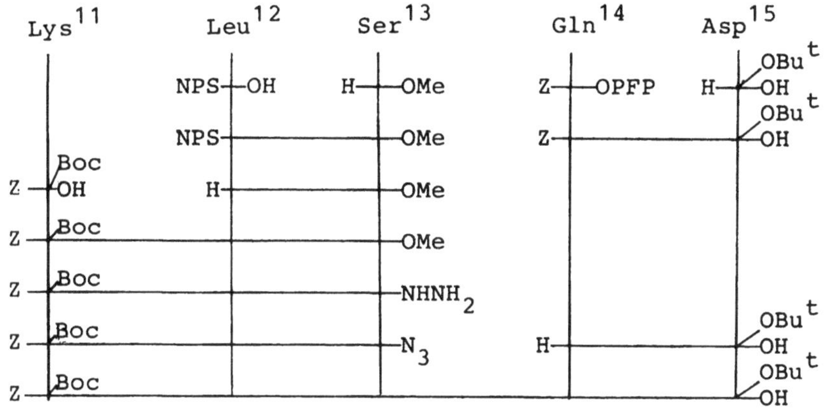

The synthesis of the fragment 22-24 (scheme 5) was carried out without blocking the guanidino group of arginine[24] with the help of the activated ester method. The tetrapeptide 25-28 was prepared according to scheme 6 from the azide of the tripeptide 25-27 and glycine. The tetrapeptide 29-32 was synthesized according to scheme 7 from two dipeptides by the DCC/PFP method.

Crystallization, gel chromatography and adsorption chromatography were used for the preparation of the calcitonin fragments in pure form. Many by-reactions were investigated. Dipeptide 31-32 can be prepared only with the use of prolin-

Scheme 4.

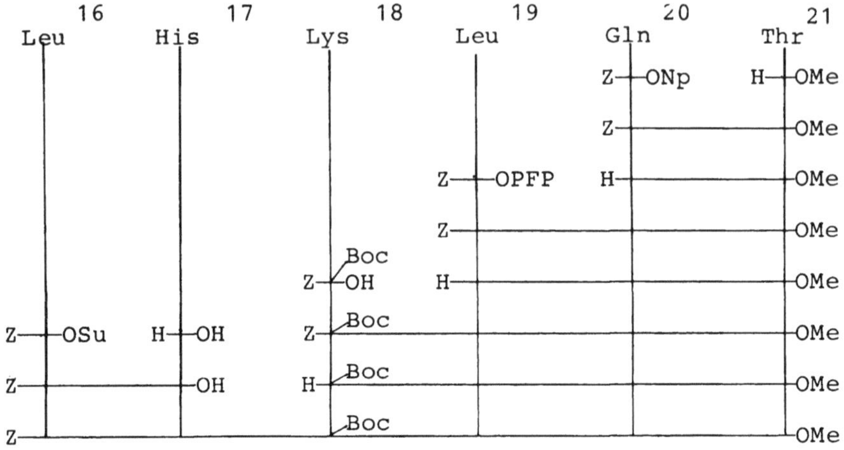

Scheme 5.

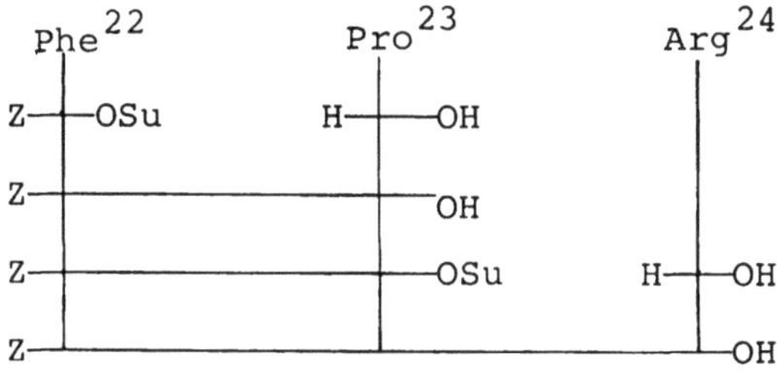

amide but not from the methyl ester. All attempts to transform the carbobenzoxy dipeptide ester into the amide were unsuccessful. The reaction of acylation of the methyl ester of threonine[25] or the methyl ester of threonyl-asparaginyl-threonine (25-27) by acyl-(Z or Boc)nitroarginine with the help of DCC yield only N-acyl-urea (70-80 %).

Only the 2.5-diketo-piperazine of valine and proline was found during the catalytic hydrogenation of carbobenzoxy-valyl-prolinamide in the absence of strong acid. However, in the presence of p-toluenesulfonic acid hydrogenation yields the dipeptide amide.

Scheme 6.

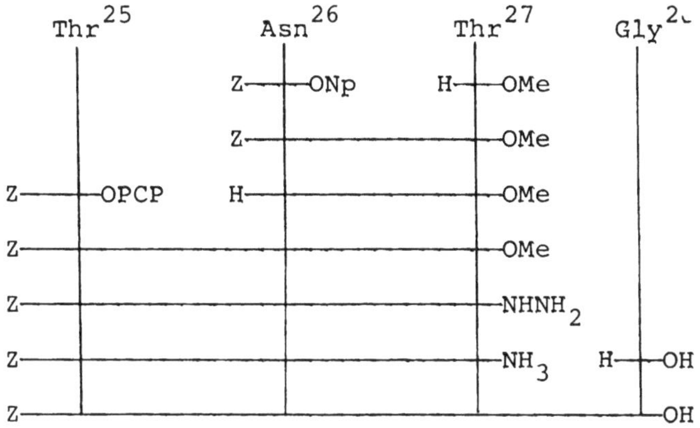

Scheme 7.

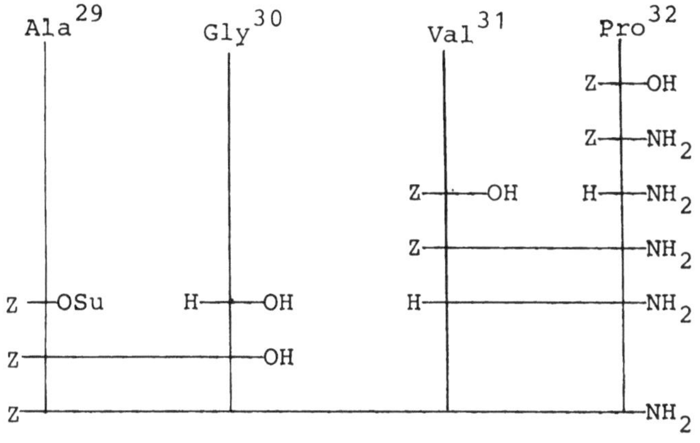

Attempts to synthesize more complex peptide fragments were made and the decapeptide 1-10, pentadecapeptide 1-15, octapeptide 25-32 and heptadecapeptide 16-32 were prepared.

Optimal conditions of coupling bulky fragments to the peptide of the salmon calcitonin sequence are under investigation.

References

1. Niall, H.D., Keutmann, H.T., Copp, D.H., Potts, Y.T.:Proc. Natl.Acad.Sci.64, 771 (1969).

2. Riniker, B., Reher, R., Maier, R., Byfield, P.G.H., Gudmundsson, T.V., MacIntyre, J.: Nature 220, 984 (1968).

3. Keutmann, H.T., Passons, J.A., Potts, J.T.: J.Biol.Chem. 245, 1491 (1970).

TOTAL SYNTHESIS OF SOMATOSTATIN WITHOUT HYDROXYL GROUP PROTECTION OF HYDROXYL AMINO ACID RESIDUES

Yu.P.Shvachkin, S.K.Girin, A.P.Smirnova, A.A.Shishkina, N.M. Ermak

Institute of Experimental Endocrinology and Hormone Chemistry, Academy of Medical Sciences of the USSR, Moscow

Introduction

The hormone somatostatin is a tetradecapeptide of structure (I) (all asymmetric amino acids have L-configuration) (1).

```
   1              5                   10                14
H-Ala-Gly-Cys-Lys-Asn-Phe-Phe-Trp-Lys-Thr-Phe-Thr-Ser-Cys-OH
          |_____________________________________________|
```

(I)

At present this hormone is of interest for peptide chemists as well as for clinicians. In spite of a number of published papers concerning the synthesis of somatostatin in solution and on polymeric carriers the problem of preparative methods for somatostatin remains nevertheless an actual one.

Results

In pursuit of a research study on somatostatin derivatives we have performed a total chemical synthesis of this hormone according to a new scheme whose specific features are:

1) The possibility to use amino acid derivatives (Thr^{10}, Thr^{12} and Ser^{13}) without hydroxyl group protection;

2) the application of the tetrahydropyranyl protection for masking mercapto groups in the Cys^{3} and Cys^{17} residues at intermediate stages of the synthesis.

Chemistry of Peptides and Proteins, Vol. 1

Preparation of the protected linear tetradecapeptide was performed by a fragment condensation starting from the previously synthesized 1-7 and 8-14 fragments. From the point of view of the tactics of the peptide synthesis this variant which makes allowance for "a docking" of the mentioned heptapeptide fragments is the most preferable one since in this case properties of the final desirable product differ maximally from those of the initial materials. This would facilitate the separation of the conceived tetradecapeptide from the reaction mixture. All condensation reactions were carried out under conditions which reduced racemization of the individual amino acid residues to a minimum. The key stages of the fragment docking were performed by the azide method or by the method of activated (pentafluorphenyl) esters.

The tert-butyloxycarbonyl group or the o-nitrophenylsulfenyl group were used to protect the α-amino groups. The carboxy groups were protected either by esterification (benzyl and methyl ester) or by salt formation. The benzyloxycarbonyl group was used for protection of the ε-amino group in the Lys^{4} and Lys^{9} residues. The use of derivatives of all hydroxy amino acids - constituents of the somatostatin molecule (Thr^{10}, Thr^{12} and Ser^{13}) - with the hydroxyl groups unprotected made it possible to diminish the number of experimental stages, to reduce the laboriousness of the procedure and, consequently, to increase the overall efficiency of the synthetis which is particularly important for the hormone production on a preparative scale.

To prevent undesirable reactions of O-acylation of hydroxy amino acid residues the 8-14 fragment was synthesized according to the 6 + 1 scheme rather than to that of 5 + 2. This allowed to reduce the number of the condensation stages from seven to three with participation of the hydroxyl-unprotected Ser^{13} residue.

The results of kinetic studies previously obtained in our laboratory have demonstrated that the above mentioned amino acids with unprotected hydroxyl groups were reacted under conditions which suppressed the undesirable reaction of O-acylation, namely: Under a minimum excess of base (2), in dimethylformamide (3) and as pentafluorophenyl esters (4). Phenylalanine methyl ester, tert-butyloxycarbonylphenylalanine, N-tert-butyloxycarbonylasparagine p-nitrophenyl ester, N-tert-butyloxycarbonyl-N-benzyloxycarbonyllysine p-nitrophenyl ester, S-tetrahydropyranylcysteine methyl ester, glycine benzyl ester and tert-butyloxycarbonylalanine were used as starting materials to synthesize the 1-7 fragment.

To make the 8-14 fragment S-tetrahydropyranilcysteine, N-tert-butyloxycarbonylserine pentalfuorophenyl ester, threonine methyl ester, tert-butyloxycarbonylphenylalanine pentafluorophenyl ester, N-tert-butyloxycarbonylthreonine pentafluorophenyl ester, N^{α}-tert-butyloxycarbonyl-N^{ε}-benzyloxycarbonyllisine pentafluorophenyl ester, N-tert-butyloxycarbonyltrypto-phan pentafluorophenyl ester, N-(o-nitrophenylsulfenyl)tryptophan pentafluorphenyl ester and N-(o-nitrophenylsulfenyl)-serine pentafluorophenyl ester were taken.

The azide condensation of the 1-7 and 8-14 fragments resulted in a partially protected tetradecapeptide which was treated with hydrogen bromide, dissolved in trifluoroacetic acid. This treatment and the chosen combination of the protecting groups provided a total removal of all protecting groups in the course of a single working step. Anisole and mercaptoethanol were used as protectors.

The formation of the disulfide bridge between the Cys^{3} and Cys^{14} residues was achieved at the final stage of the synthesis by oxidizing the deblocked tetradecapeptide with atmospheric oxygen at pH 6.9.

Somatostatin (I) was isolated from the reaction mixture by the partition chromatography method on a 2 x 100 cm Sephadex G-25 column and the system n-butanol-acetic acid-water (4:1:5).

The synthetic somatostatin was obtained as an analytically pure substance and its physical data and biological activity did not differ from those of the natural hormone.

References

1. Brazeau, P., Vale, W., Burgus, R., Ling, N., Butcher, M., Rivier, J., Guillemin, R.: Science 179, 77-79 (1973).
2. Girin, S., Shvachkin, Yu.: J.Gen.Chem. 48, 902-908 (1978).
3. Girin, S., Shvachkin, Yu.: J.Gen.Chem. 48, 1887-1891 (1978).
4. Girin, S., Shvachkin, Yu.: J.Gen.Chem. 48, 451-457 (1978).

TOTAL SYNTHESIS OF SOMATOSTATIN-28

Luis Moroder, Manfred Gemeiner, Ernst Jaeger and Erich Wünsch
Max-Planck-Institute of Biochemistry, Department of Peptide Chemistry.
8033 Martinsried / Munich, FRG

Introduction

The presence of higher molecular weight forms of somatostatin has been recognized in various tissues of different species by means of immunological assays, thus strongly suggesting that somatostatin (1) in common with other peptide hormones, is synthetized as part of a larger percursor protein which is then processed intracellularly by a series of enzymatic cleavages. Several laboratories have attempted in the last years to isolate such putative prohormonal forms of somatostatin from different organs. Pradayrol et al. (2) succeeded to prepare somatostatin-active material from a "post-secretin" side fraction used for the preparation of vasoactive intestinal peptide and deriving from an extract of porcine upper small intestine tissue. Upon preliminary sequencing work the tetradecapeptide

1 ... 14
H-Ser-Ala-Asn-<u>Arg</u>-Asn-Pro-Ala-Met-Ala-Pro-Arg-Glu-<u>Ser</u>-Lys-

was proposed as the N-terminal extension of somatostatin for the newly isolated peptide factor, which was therefore named somatostatin-28. Chromatographic comparison of tryptic fragments with synthetic reference peptides prepared in our laboratory, and repetition of the sequence analysis led Pradayrol et al. (3) to finally propose for somatostatin-28 its primary structure as the octacosapeptide

1
H-Ser-Ala-Asn-<u>Ser</u>-Asn-Pro-Ala-Met-Ala-Pro-Arg-Glu-<u>Arg</u>-Lys-Ala-Gly-
Cys-Lys-Asn-Phe-Phe-Trp-Lys-Thr-Phe-Thr-Ser-Cys-OH (28; disulfide bridge Cys–Cys)

Chemistry of Peptides and Proteins, Vol. 1

with one intramolecular disulfide bridge. This primary structure was definitely confirmed by comparing the chromatographic, biological and immunological properties of the natural preparation with those of our synthetic somatostatin-28 (4).

Synthesis

A) General strategy

The S-tert-butylthio-group (5), an asymmetric disulfide type protection of cysteine thiol functions has successfully been applied in our synthesis of somatostatin-14 (6): it was found to be optimally suited for the preparation of cysteine-containing peptides according to the strategy of overall acid-labile side-chain protection in combination with the 2-nitrophenylsulfenyl group for the intermediate reversible N^{α}-blocking in the chain-elongation steps. In fact, this asymmetric disulfide represents one of the few thiol protecting groups stable under the conditions of HCl- or HBr-promoted cleavage of the sulfenamide bonds. Most of the other cysteine protecting groups are partially or quantitatively removed by the 2-nitrophenylsulfenyl halide formed during the sulfenamide-cleavage reaction (7,8).

These positive experiences prompted us to apply an identical protection scheme and to adapt it to fragment condensation procedure for the elaboration of an efficient synthetic route to somatostatin-28.

The design of the fragments in the C-terminal region relies upon the plan of synthesis for somatostatin-14 (6), whereas in the N-terminal region it was conditioned by the presence of a methionine residue in position 8 (scheme 1).

I) H-Lys(AdOC)-Asn-Phe-Phe-Trp-Lys(BOC)-Thr(tBu)-Phe-Thr(tBu)-Ser(tBu)-Cys(StBu)-OtBu	H-[18-28]-OtBu
II) NPS-Ala-Gly-Cys(StBu)-OH	NPS-[15-17]-OH
III) NPS-Met-Ala-Pro-Arg(HBr)-Glu(OtBu)-Arg(HBr)-Lys(BOC)-OH	NPS-[8-14]-OH
IV) BOC-Ser(tBu)-Ala-Asn-Ser(tBu)-Asn-Pro-Ala-OH	BOC-[1-7]-OH

Scheme 1. Suitably protected fragments for the synthesis of somatostatin-28.

B) Protected dihydrosomatostatin-28

In our first synthetic approach to somatostatin-28 fragments III* and IV* corresponding to sequence 8-14 and 1-7 of the preliminary erroneous primary structure (scheme 2 and 3), were obtained using the arginine guanido function to trap the C-terminal carboxyl group (zwitterion) in the stepwise performed acylation steps. To convert the two fragments into a form suitable for dicyclohexylcarbodiimide/N-hydroxysuccinimide (9) mediated condensation steps, a posteriori protonation of the guanido functions had to be performed. No difficulties were encountered in the protonation of the heptapeptide derivative BOC-Ser(tBu)-Ala-Asn-Arg-Asn-Pro-Ala-OH (fragment IV*), whereas even in aqueous/organic media addition of HBr in equivalent amounts to NPS-Met-Ala-Pro-Arg-Glu(OtBu)-Ser(tBu)-Lys(BOC)-OH (fragment III*) produced partial sulfenamide-cleavage.

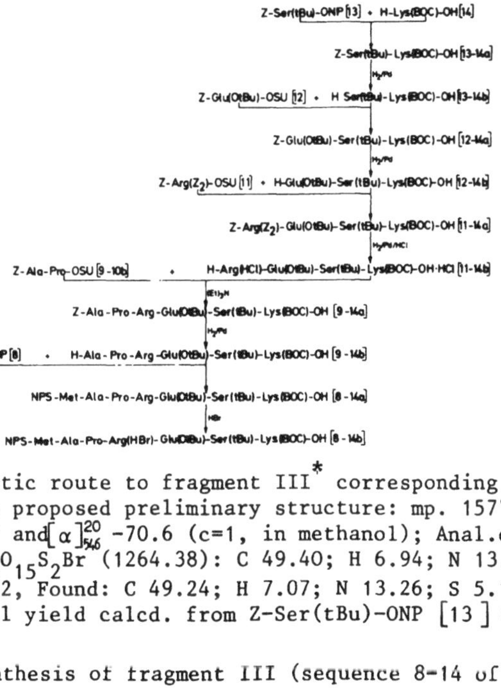

Scheme 2. Synthetic route to fragment III* corresponding to sequence 8-14 of the proposed preliminary structure: mp. 157° (dec.); $[\alpha]_D^{20}$ -63.1° and $[\alpha]_{546}^{20}$ -70.6 (c=1, in methanol); Anal.calcd. for $C_{52}H_{87}N_{12}O_{15}S_2Br$ (1264.38): C 49.40; H 6.94; N 13.29; S 5.07; Br 6.32, Found: C 49.24; H 7.07; N 13.26; S 5.13; Br 6.21. Overall yield calcd. from Z-Ser(tBu)-ONP [13] : 22%.

Thus for the synthesis of fragment III (sequence 8-14 of the revised structure), i.e. NPS-Met-Ala-Pro-Arg(HBr)-Glu(OtBu)-Arg(HBr)-Lys(BOC)-OH, the hexapeptide derivative H-Ala-Pro-Arg(HBr)-Glu(OtBu)-Arg(HBr)-Lys(BOC)-OH (scheme 4) was acylated directly in the zwitterion-form with the two arginyl residues already protonated by hydrogen bromide; nevertheless the

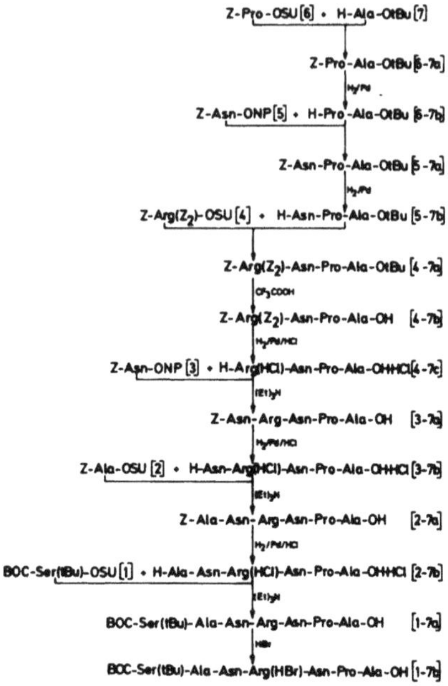

Scheme 3. Synthetic route to fragment IV* corresponding to sequence 1-7 of the proposed preliminary structure: mp. 170-172°C; $[\alpha]_D^{20}$ -68,0° and $[\alpha]_{546}^{20}$ -81.6° (c=1, in methanol); Anal. calcd. for $C_{37}H_{65}N_{12}O_{13}Br\cdot 1/2H_2O$ (974.9): C 45.58; H 6.82; N 17.29; Br 8.20. Found: C 45.37; H 6.85; N 16.98; Br 8.28. Overall yield calcd. from Z-Pro-OSU[7]: 23%.

reaction proceeds smoothly leading in good yield to the desired product. The synthesis of fragment IV, i.e. BOC-Ser(tBu)-Ala-Asn-Ser(tBu)-Asn-Pro-Ala-OH, corresponding to sequence 1-7 of the revised structure, was significantly facilitated by the absence of an arginyl residue. In fact, by stepwise elongation via active esters the heptapeptide derivative was obtained over the 11 steps in a yield of 43.5% at a high degree of purity (scheme 5).

Fragment II., i.e. NPS-Ala-Gly-Cys(StBu)-OH (sequence 15-17) was obtained by acylation of H-Cys(StBu)-OH with NPS-Ala-Gly-OSU, analogously to the N-terminal fragment, BOC-Ala-Gly-Cys(StBu)-OH, of our somatostatin-14 synthesis (6). Similarly the undecapeptide derivative corresponding to sequence 18-28 (fragment I) represents an intermediate segment in the somatostatin-14 synthesis (6).

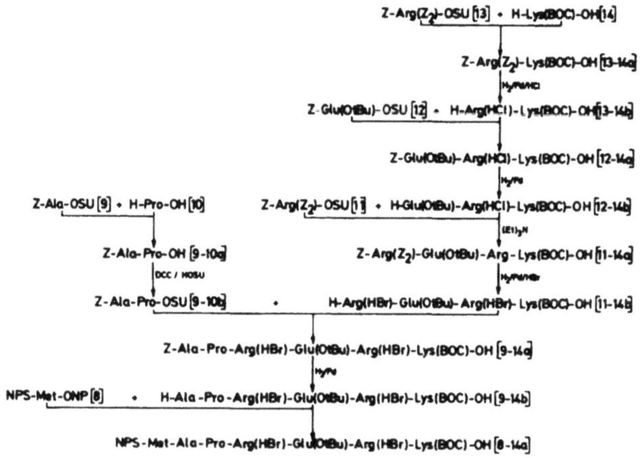

Scheme 4. Synthetic route to fragment III corresponding to sequence 8-14 of somatostatin-28; for analytical data see Ref. 14; overall yield calcd. from Z-Arg(Z_2)-OSU [13]: 29%.

Scheme 5. Synthetic route to fragment IV corresponding to sequence 1-7 of somatostatin-28; for analytical data see Ref. 14; overall yield calcd. from Z-Pro-OSU [7] : 43.5%.

Subsequently the fragments were assembled in sequence order via the N-hydroxysuccinimide/dicyclohexylcarbodiimide (9) method as shown in scheme 6. The intermediate N^{α}-deprotection by exposure to HCl or HBr was performed in the presence of large excesses of 2-methylindole to prevent sulfenylation of the tryptophan residue (10) as well as to markedly lower the proton activity and thus to avoid partial hydrolysis of acid-labile side-chain protection (11).

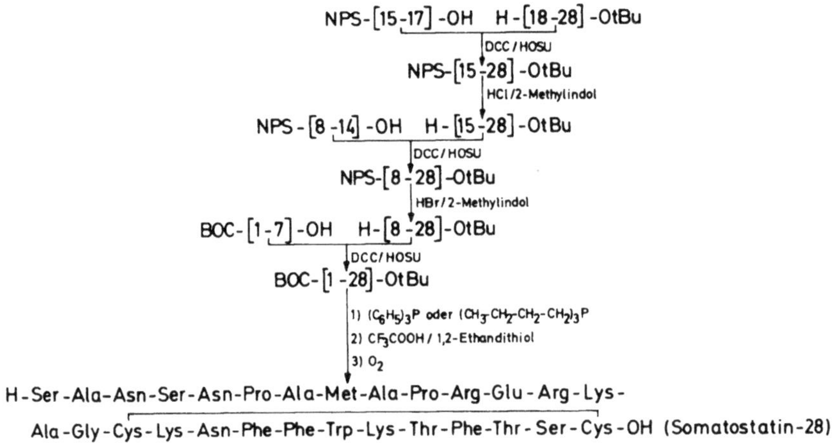

Scheme 6. Synthetic route to somatostatin-28.

No pronounced solubility problems were encountered during the building up steps of the protected dihydrosomatostatin-28: the protection scheme applied by interchanging tert-butanol derived protecting groups with the adamantyloxycarbonyl residue, and the fragment design based on alternating hydrophobic with hydrophylic segments, produced good solubility along the assembly steps. Protonated arginines may lead to conformation dependent effects similar to those observed on model peptides by Mutter (12).

C) Somatostatin-28

The reductive cleavage of the S-tert-butylthio groups was performed in conditions determined as optimal in model studies (13,14), i.e. in trifluoroethanol using an excess of tributylphosphine. The reaction was followed both by thin layer chromatography and NMR-measurements: the quantitative removal of the cysteine protecting groups is monitored by disappearance of the signal at 1.27ppm (13,14). The resulting partially protected dihydrosomatostatin-28 was fully deblocked by exposure to trifluoroacetic acid in the presence of a large excess of 1,2-ethanedithiol to prevent as much as possible tryptophan- and cysteine-tert-butylation. The tryptophan residue in somatostatin-14 and -28 under these conditions was found to be less prone to latter side reaction than that in the gastrins (15), since no tert-butylated tryptophyl-side products could be detected so far in the pooled side fractions collected during purification of crude somatostatins. Even if the reaction rates of electrophylic substitution at the indole function markedly increase from tryptophan to tryptophan derivatives and peptides (10), additional strong topochemical effects may lead to unpredictable results.

On the other hand under the experimental conditions used in the acid deprotection step no cysteine tert-butylation was observed, since in the ^{1}H-NMR-spectrum of crude dihydrosomatostatin-28 the typical signal of S-tert-butyl-cysteine is absent.

The conversion of dihydrosomatostatin-28 into its cyclic form was again performed by air oxidation at room temperature in high dilution. The use of chemical oxidants, especially of iodine, was avoided to prevent at priori both tryptophan oxidation or sulfenylation. Sieber et al. (16) have recently confirmed that the latter reaction occurs in the case of somatostatin-14.

The oxidation rate as shown in fig. 1 is slow, levelling off after 4 to 5 days. The residual thiol concentration has to be attributed to linear polymeric forms. A significant increase of latter polymers on standing has not been observed. The polymeric forms are easily separated and may be recycled as in the case of somatostatin-14, by reduction with phosphines in trifluoroethanol/ammonium acetate buffer.

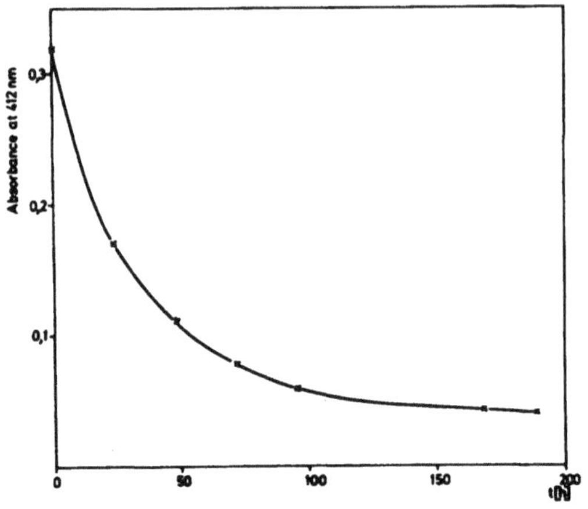

Fig. 1. Oxidation rate of crude dihydrosomatostatin-28 by exposure to air oxygen as monitored by Ellman's reagent: 5.10^{-5}M dihydrosomatostatin-28 in 0.1M ammonium acetate (pH 6.8) in the dark at room temperature.

Purification of the crude oxidation product by gel filtration on Biogel P6 followed by ion exchange chromatography on Biogel CM-2 led to somatostatin-28 contaminated in batch dependent extents only by small amounts of its methionine-S-oxide-derivatives. The high tendency to oxidation is structure dependent and observed even in natural preparations.

This synthetic somatostatin-28 has been compared with the natural product in different chromatographic tests, in the biological activity and immunological properties. In all assays performed so far the two preparations behaved identically (4).

References

1. Brazeau, P., Vale, W., Burgus, R., Ling, N., Butcher, M., Rivier, J., Guillemin, R.: Science 179, 77-79 (1973).
2. Pradayrol, L., Chayvialle, J.A., Carlquist, M., Mutt, V.: Biochem. Biophys. Res. Commun. 85, 701-708 (1978).
3. Pradayrol, L., Jörnvall, H., Mutt, V., Ribet, A.: FEBS Lett. 109, 55-58 (1980).
4. Wünsch, E., Moroder, L., Gemeiner, M., Jaeger, E., Ribet, A., Pradayrol, L., Vaysse, N.: Z.Naturforsch. 35b, 911-919 (1980).
5. Wünsch, E., Spangenberg, R.: Peptides 1969 (Scoffone, E. ed.) North-Holland, Amsterdam, 30-34 (1971).
6. Moroder, L., Gemeiner, M., Goehring, W., Jaeger, E., Thamm, P., Wünsch, E.: Biopolymers 20, 17-37 (1981).
7. Moroder, L., Marchiori, F., Borin, G., Scoffone, E.: Biopolymers 12, 493-505 (1973).
8. Chimiak, A., Pastuszak, J.: Peptides 1978 (Siemion, J.Z. and Kupryszewski, G. eds.), Wrocław University Press, Varsavia, Poland, 101-104, (1979).
9. Wünsch, E., Drees, F.: Chem.Ber. 99, 110-120 (1966).
10. Wünsch, E., Fontana, A., Drees, F.: Z.Naturforsch. 22b, 607-609 (1967)
11. Wünsch, E., Wendlberger, G.: Chem.Ber. 101, 341-345 (1968).
12. Mutter, M., Pillai, V.N.R., Anzinger, H., Toniolo, C., Bayer, E.: Peptides 1980 (Brunfeldt, K. ed.) Scriptor, Kopenhagen, 660-665 (1981).
13. Moroder, L., Gemeiner, M., Goehring, W., Jaeger, E., Wünsch, E.: Peptides 1980 (Brunfeldt, K., ed.) Scriptor, Kopenhagen, 121-126 (1981).
14. Moroder, L., Gemeiner, M., Göhring, W., Jaeger, E., Musiol, J., Scharf, R., Stocker, H., Wünsch, E.: Hoppe Seyler's Z.Physiol. Chem. 362, 697-716 (1981).
15. Jaeger, E., Thamm, P., Schmidt, I., Knof, S., Moroder, L., Wünsch, E.: Hoppe Seyler's Z.Physiol.Chem. 359, 155-164 (1978).
16. Sieber, P., Kamber, B., Riniker, B., Rittel, W.: Helv.Chim.Acta 63, 2359-2363 (1980).

SYNTHESIS, STRUCTURE AND MEMBRANE PROPERTIES OF GRAMICIDIN A DIMER ANALOGS

Larissa Fonina, Alla Demina, Sergei Sychev, Alexander Irkhin, Vadim Ivanov

Shemyakin Institute of Bioorganic Chemistry, USSR Academy of Sciences, Moscow, USSR

Jan Hlaváček

Institute of Organic Chemistry and Biochemistry, Prague, ČSSR

Introduction

The spatial structure of gramicidin A in solution and in a membrane is attracting the attention of researches already for a number of years. Gramicidin A is known to form four dimers in solution that are in equilibrium with each other and with the monomer (1). In nonpolar media equilibrium is established very slowly so that each of the dimeric species can be isolated in the individual state and its spectral characteristics recorded.

Urry et al. (2, 3) and Ramachandran and Chandrasekaran (4, 5) independently proposed a helical structure for gramicidin A, that differed from earlier proposed helices and received the name of a π_{LD}-helix. Two molecules of the antibiotic head-to-head associated by means of hydrogen bonds form a so-called $\overrightarrow{\pi_{LD}}\overleftarrow{\pi_{LD}}$-helix. Urry et al. (2, 3) considered several such helices differing in the number of residues per turn. Of these the $\overrightarrow{\pi_{LD}^{6.3}}\overleftarrow{\pi_{LD}^{6.3}}$-helix (with six intermolecular hydrogen bonds and 6.3 residues per turn) had a length of ~30 Å, approximating the width of the hydrocarbon part of the lipid bilayer. Veatch et al. (6) proposed a different dimer model wherein the peptide chain of the antibiotic forms a double helix stabilized by 28-30 intermolecular hydrogen bonds. The chain orienta-

Chemistry of Peptides and Proteins, Vol. 1

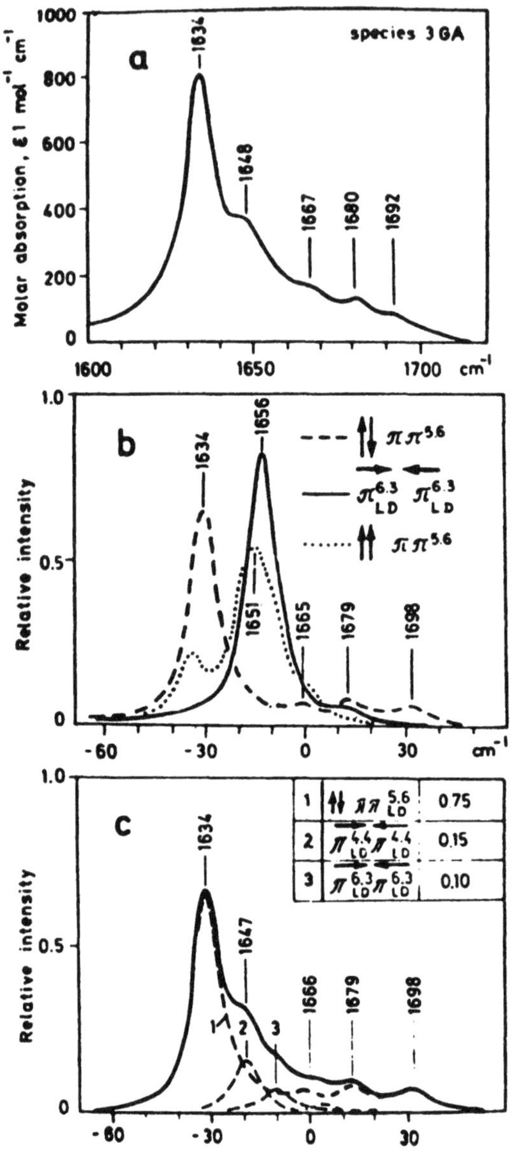

Fig. 1. Resonance frequency shift Δν (cm^{-1}). a) Observed IR spectrum of the gramicidin A species 3; b) computed IR spectra of selected helical structures; c) imitation of the experimental spectrum by superposition of weighted amounts of theoretical spectra.

tion in this model could be either parallel ($\uparrow\uparrow\pi\pi$) or antiparallel ($\uparrow\downarrow\pi\pi$), the diameter of the axial cavity and the length of the dimer being approximately the same as those for the $\pi_{LD}\pi_{LD}$-helix. However, no convincing physicochemical data had been produced to substantiate any of the proposed types of helices.

In order to fill in this gap we investigated gramicidin A and a large series of its shortened analogs by CD, fluorescence and IR spectroscopy techniques (11-13). It was shown that 1634-1637 cm^{-1} frequences should be attributed to antiparallel double $\uparrow\downarrow\pi_{LD}^{5.6}$ or $\uparrow\downarrow\pi_{LD}^{7.2}$ helices, 1645-1649 cm^{-1} bands might be due to $\pi_{LD}^{4.4}\pi_{LD}^{4.4}$ dimers while bands at ca 1656 cm^{-1} are associated with $\pi_{LD}^{6.3}\pi_{LD}^{6.3}$ dimers. Based largely on a comparison between theoretical and experimental IR spectra we arrived at the conclusion that none of the gramicidin A I-IV species is conformationally pure, but rather all are in an equilibrium mixture of the predominant $\uparrow\downarrow\pi\pi_{LD}$-helix with smaller amounts of $\pi_{LD}\pi_{LD}$-dimers. This is illustrated in Fig. 1 which shows some theoretical spectra, the experimental spectrum of species 3 of gramicidin A and optimal synthesis of its profile from the spectra of the individual species.

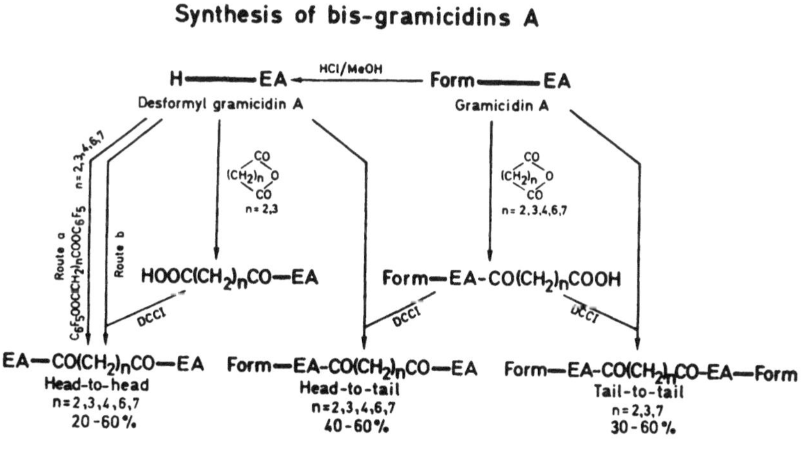

The highly complex conformational equilibrium of gramicidin A that had been thus revealed proved a difficult barrier to its further physicochemical study. Thus it precluded interpretation of the NMR spectra of the helices. Analysis of the structure-membrane activity relationship of gramicidin A also turned out to be practically impossible. In order to simplify the conformational equilibrium so as to be able to surmount these difficulties, we undertook the synthesis and study of a series of new analogs wherein the head-to-head, head-to-tail and tail-to-tail disposition of the terminal groups was fixed by inserting covalent bridges (Scheme).

Results

To obtain dimers of head-to-head type the interaction of desformylgramicidin with various activated derivatives of dicarboxylic acids was investigated. The attempts to use dicarboxylic acid chlorides in the condensation with desformylgramicidin resulted in a low yield of desired dimer analogs. With N-hydroxysuccinimide esters of dicarboxylic acids the reaction was too slow while carbodiimide condensation gave rise to considerable amounts of hardly separable side products, especially with succinic and glutaric acids. Satisfactory results were obtained only with dipentafluorophenyl esters, the final yields ranging from 20 to 60 %. A two stage synthesis appeared to be the most convenient way to bis-gramicidins (succinyl, glutatryl derivatives). At the first stage amino group of desformylgramicidin was acylated with 80 % yield to the corresponding cyclic anhydride. The resultant N-gramicidincarboxylic acids were then coupled with another molecule of desformylgramicidin with the aid of N,N'-dicyclohexyl carbodiimide. Head-to-tail and tail-to-tail derivatives were also obtained by the two stage method as shown in the Scheme.
The analogs obtained were separated from the ionogenic impuri-

Table. 1. Last stage yields and chromatographic mobilities of bis-gramicidins.

N	Junction type	n	Yield	R_f	
				1[a]	2[b]
1	Head-to-head	2	20[c]	0.39	
2		3	45[c]	0.56	
3		4	50[d]	0.69	
4		6	60[d]	0.75	
5		7	60[d]	0.78	
6	Head-to-tail	1	50		0.30
7		2	58		0.32
8		3	60		0.34
9		4	55		0.36
10		6	50		0.47
11		7	40		0.50
12	Tail-to-tail	2	30		0.51
13		3	50		0.49
14		7	50		0.44

[a]Chloroform - methanol - water, 65:20:2
[b]Chloroform - methanol - acetic acid - water, 65:80:1:1
[c]Route a (see the Scheme)
[d]Route b

ties by ion-exchange chromatography and purified by preparative thin-layer chromatography on silica gel. The analogs are colourless amorphous compounds, readily soluble in methanol, dimethylformamide, chloroform and slightly soluble in nonpolar media. Homogeneity of the compound obtained was controlled by reverse phase HPLC on Separon STC 18 column in methanol (90%) - 0.005 M $(NH_4)_2SO_4$ (10 %). Some characteristics of the compounds obtained are listed in Table 1.
Analysis of amide I IR spectral region was used to monitor the conformational parameters of bis-gramicidins; intermolecular association was also followed by CD measurements. According to

Table 2. Conformations of bis-gramicidins A, as deduced from 264 their IR spectra in dioxan.

Compound No	Concent.	Conformations and their molar fractions			
		$\pi\pi^{5.6}_{LD}$	$\pi^{4.4}_{LD}\pi^{4.4}_{LD}$	$\pi^{6.3}_{LD}\pi^{6.3}_{LD}$	unordered
1,3,5	10^{-3}-10^{-5}	0.25	0.15	0.05	0.55
	10^{-3} [a]	0.15	0.30	0.05	0.50
6,7,10,11	10^{-3}	0.60	0.30	0.10	-
10	10^{-4}	0.15	0.25	0.15	0.45
11	$3\cdot10^{-5}$	0.35	0.40	-	0.25
12,13	10^{-3}	0.65	0.25	0.10	-
	$3\cdot10^{-5}$	0.35	0.40	-	0.25

[a] in 2-chloroethanol

the earlier work (9) 1634-1637 cm^{-1} frequences should be attributed to antiparallel double $\uparrow\downarrow\pi\pi^{5.6}_{LD}$ or $\pi\pi^{7.2}_{LD}$ helices, 1645-1649 cm^{-1} bands should be due to $\pi^{4.4}_{LD}\pi^{4.4}_{LD}$ dimers while bands at ca 1656 cm^{-1} are associated with $\pi^{6.3}_{LD}\pi^{6.3}_{LD}$ dimers. Analogs 1, 3, 5 at 10^{-3} M concentration in dioxan produced $\uparrow\downarrow\pi^{5.6}_{LD}$ and $\pi^{4.4}_{LD}\pi^{4.4}_{LD}$ helices in addition to considerable fraction of unordered forms (Table 2). Since antiparallel double helices cannot be formed within one molecule of the head-to-head dimer their presence points to considerable intermolecular aggregation. Stability constants are quite high since a 50-fold dilution practically does not affect the spectra. In 2-chloroethanol there are more $\pi_{LD}\pi_{LD}$ and less $\uparrow\downarrow\pi\pi_{LD}$ conformers, presumably due to less aggregation.

IR spectra of the head-to-tail and tail-to-tail analogs in dioxan at 10^{-3} M are very close to each other. All of them show a major peak at 1634 cm^{-1} and a less intensive 1648 cm^{-1} band; there is no indication of the random coil band. On passing to 10^{-4} M and 10^{-5} M concentrations there occurs a gradual spectral shift, the equlibration time being 5-6 days. In the resultant spectra the 1648 cm^{-1} band is more intensive than the

Table 3. Channel parameters of lipid bilayers modified by gramicidin A and its bis-derivatives,

Compound	Single channel experiments[a]			Noise experiments[b]	
	Min. active concentration M	Channel conduct. pS	Mean channel life-time, sec	Channel conduct. pS	Mean channel life-time, sec
Gr	10^{-13}	30.2	0.23	18	0.2
1	10^{-16}	24.2	12.6	-	-
2	$5\cdot10^{-16}$	27.6	8.2	-	-
3	10^{-16}	24.6	10.1	-	-
5	10^{-11}	24	~9	-	-
7	10^{-13}	24.3	9.7	-	-
		30.3	0.2	13	0.5
12	10^{-13}	24.8	7.5	-	-
		29.4	~0.2	14	1.42

[a]Membrane were formed from glycerolmonooleate - cholesterol (2:1); 0.5 M KCl, 21°C, 100 mV

[b]Membrane were formed from dioleyllecithin - cholesterol (9:1); 1 M KCl, 21°C

1634 cm^{-1} band. Therefore, in the head-to-tail and tail-to-tail series the $\uparrow\downarrow\pi\pi_{LD}$ conformations are also attributed to intermolecular associates while $\overrightarrow{\pi_{LD}}\overrightarrow{\pi_{LD}}$ and $\overleftarrow{\pi_{LD}}\overrightarrow{\pi_{LD}}$ helices are predominant in the monomeric state. In summary, a dominating conformation of all monomeric bis-gramicidins are the single-stranded helices - $\overrightarrow{\pi_{LD}}\overleftarrow{\pi_{LD}}$, $\overrightarrow{\pi_{LD}}\overrightarrow{\pi_{LD}}$ or $\overleftarrow{\pi_{LD}}\overrightarrow{\pi_{LD}}$, depending on the position of the covalent bridge. Double helices are found only in the aggregated state.

The membrane properties of bis-gramicidins were studied with lipid bilayers. Three series of experiments were carried out:

i. Measurements of discrete fluctuations in membrane conductance in the presence of minimal concentration of the peptide (i.e. at single channel conditions).

ii. Registration and analysis of spontaneous fluctuations in

membrane potential under constant current (noise analysis). Much higher (10^{-6} - 10^{-8} M versus 10^{-11} - 10^{-15} M in i.) concentrations of the peptide are required for this work.
iii. Measurements of biionic potentials which develop at the membrane separating different electrolyte solutions.
All bis-gramicidins demonstrated their channel-forming capacity, some of them at extremely low concentrations (see Table 3). The ionic selectivity sequence determined from biionic potentials is the same as with non-modified gramicidin A:

$$H^+ > NH_4^+ > Cs^+ > Rb^+ > K^+ > Na^+ > Li^+.$$

Of considerable interest are the mean life-time and conductance values of these channels. As seen in Table 3 all bis-gramicidins, regardless of their bridge position and spatial structure form channels with similar conductance but with the mean life-time by 1-2 orders of magnitude longer than with the channels formed by gramicidin A. In addition to the long-living channels the head-to-tail and tail-to-tail analogs also form short-living channels. It are these channels which are responsible for the observed conductance in the noise experiments.
Bearing in mind the aggregation tendency of these analogs revealed by spectral measurements in solution, and their high concentrations at noise experiments we assume that the short-living channels are formed by bis-gramicidin associates (possibly, by $\uparrow\downarrow\pi\pi$ double helices). The head-to-head derivatives do not seem to form short-living channels at the great variety of membrane modifications.
So, the covalent crosslinking applied to gramicidin A in this work failed to produce a conformationally homogenous species. Nonetheless the data obtained clearly indicate that the channel-forming capacity in the gramicidin series should not be considered as a prerogative of a single molecular structure. In particular, a whole family of $\overrightarrow{\pi}_{LD}\overleftarrow{\pi}_{LD}$, $\overrightarrow{\pi}_{LD}\overrightarrow{\pi}_{LD}$ and $\overleftarrow{\pi}_{LD}\overrightarrow{\pi}_{LD}$ helices are efficient channel formers. As for the double helices, their channel forming capacity still requires further study.

References

1. Veatch, W.R., Blout, E.R.: Biochemistry 13, 5257-5264 (1974).

2. Urry, D.W.: Proc. Nat. Acad. Sci. USA 68, 672-676 (1971).

3. Urry, D.W., Goodall, M.C., Glickson, J.D., Mayers, D.F.: Proc. Nat. Acad. Sci. USA 68, 1907-1911 (1971).

4. Ramachandran, G.N., Chandrasekaran, R.: "Progress in Peptide Research" (S. Lande, ed.), Gordon and Breach, New York, p. 195 (1972).

5. Ramachandran, G.N., Chandrasekaran, R.: Ind. J. Biochem. Biophys. 9, 1-11 (1972).

6. Veatch, W.R., Fossel, E.T., Blout, E.R.: Biochemistry 13, 5249-5256 (1974).

7. Shepel, E.N., Jordanov, St., Ryabova, I.D., Miroshnikov, A.I., Ivanov, V.T., Ovchinnikov, Yu.A.: Bioorg. Khim. 2, 581-593 (1976).

8. Ovchinnikov, Yu.A., Ivanov, V.T.: "Biochemistry of Membrane Transport" FEBS-Symp. No. 42 (G. Semenza, E. Carafoli, eds) Springer-Verlag, Berlin, pp. 123-146, 1977

9. Sychev, S.V., Nevskaya, N.A., Jordanov, St., Shepel, E.N., Miroshnikov, A.I., Ivanov, V.T.: Bioorg. Chem. 9, 121-151 (1980).

CYCLIC ANALOGUES OF BRADYKININ AND KALLIDIN

Gunars Chipens, Felikss Mutulis, Natalia Mishlyakova
Division of Peptide and Protein Bioregulators, Institute of Organic Synthesis, Latvian SSR Academy of Sciences, 226006, Riga, USSR

Introduction

Total semi-empirical space structure calculations for bradykinin have shown that energetically preferable are the bent, quasi-cyclic structures (1) characterized by ionic bonding between the guanidyl group of arginine in position 1 within the common fragment (2) and the carboxyl group of arginine in position 9. Experimental evidence in favour of the existence of bent structures in aqueous solution was obtained by Ovchinnikov, Ivanov and co-workers investigating labelled analogues of bradykinin by EPR and fluorescence spectroscopy (3,4). However, further analysis of NMR spectra of bradykinin (5) and titration data (6) have shown that there is no ionic bonding between the terminal arginine residues of the bradykinin molecule in aqueous solution.
In our view, this fact can be easily explained by the existence of a hydration layer, screening the electromagnetic fields of ions. However, quasi-cyclic structures are possibly formed in apolar biophase during hormone-receptor interaction after the dehydratation of ionogenic groups. This assumption is supported by the fact that dehydratation of the contacting surfaces is the first step during substrate-enzyme interaction(7).

Chemistry of Peptides and Proteins, Vol. 1

Table.

Cyclokinins

Compound	Formula	Depressor action*	Miotropic action**
cyclo-Bradykinin (I)	$ArgPro_2GlyPheSerProPheArg$ (cyclic)	250/500(>120)	0.52/7.86
cyclo-[(ε-Lysine^1,glycine^6)-bradykinin] (II)	$LysPro_2GlyPheGlyProPheArg$ (cyclic)	5/50(>120)	inactive
α-Arginyl-cyclo-[(ε-lysine^1, glycine^6)-bradykinin] (III)	$ArgLysPro_2GlyPheGlyProPheArg$ (cyclic)	5/50(>120)	inactive
cyclo-ε-Kallidin (IV)	$LysArgPro_2GlyPheSerProPheArg$ (cyclic)	5/50(>120)	1.00/7.56
cyclo-[(ω-Aminododecanoyl-bradykinin] (V)	$ArgPro_2GlyPheSerProPheArg$ — $CO(CH_2)_{11}NH$ (cyclic)	50/250(35)	0.81/6.41
cyclo-[(ω-Aminododecanoyl-ω--aminododecanoyl)-bradykinin] (VI)	$ArgPro_2GlyPheSerProPheArg$ — $/CO(CH_2)_{11}NH/_2$ (cyclic)	inactive	inactive

* tested on anaesthetized rats, threshold dose mkg/kg / effective dose mkg/kg (duration of effect in minutes)

** tested on rat uterus, α / pD_2

Results

To check this hypothesis, we have synthesized in solution a number of bradykinin cycloanalogues, widely using the method of pentafluorophenyl esters (see Table). In order to prevent racemization at the cyclization stage, the linear precursors had glycine, proline or ω-aminododecanoic acid residues at the C-terminus.

Comparison of CD spectra of bradykinin and its cyclic analogues in aqueous solution demonstrates that the curves derived for analogue I (a compound, maximally simulating the quasi-cyclic conformation) and the parent hormone virtually coincide (Fig.1), thus indicating that bradykinin in aqueous solution exists in a bent, quasi-cyclic-like conformation.

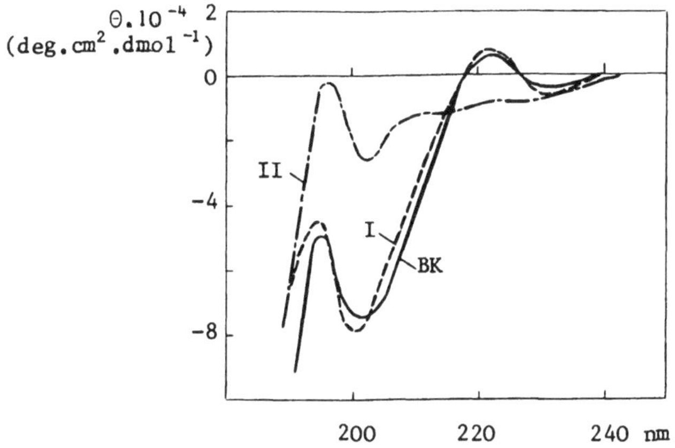

Fig. 1. Circular dichroism spectra of bradykinin (BK) and cyclopeptides I and II in water.

The study of biological activity conducted on the synthesized compounds revealed that cyclization gives analogues with a prolonged and selective action (see Table, Fig.2). Such effects have been observed for the first time among bradykinin analogues and deserve further attention, since it opens possibilities of obtaining selectively acting drugs. The lowest threshold doses are posessed by substances with structures closely resembling the quasi-cyclic conformation of bradykinin (analogues II-IV). The absence of miotropic action in analogues II and III can be due to the lack of guanidyl group in the proper site of their molecules.

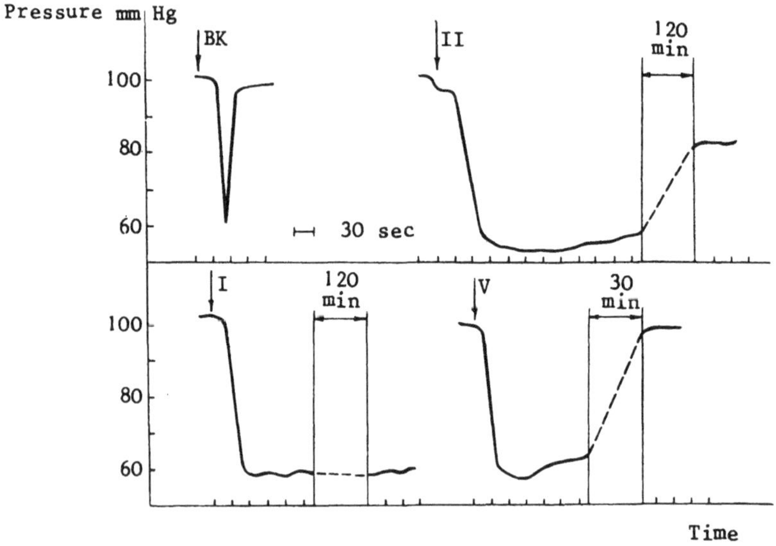

Fig.2. Effect of bradykinin (BK) and cyclopeptides I, II and V on the arterial pressure of anaesthetized rats following their administration at doses 50, 500, 50 and 250 mkg/kg, respectively.

The obtained experimental evidence supports our hypothesis postulating the fixation of bradykinin structure in receptor biophase, the common fragment ArgProPro being involved in this process. As the common fragments have been discovered in the structure of more than 50 natural peptides, it is possible that the formation of compact quasi-cyclic structures is a universal principle of their biological action.

References

1. Galaktionov, S.G., Sherman, S.A., Shenderovich, M.D., Nikiforovich, G.V., Leonova, V.I.: Bioorgan. Khimia 3, 1190-1197 (1977)(in Russian).
2. Chipens, G.I., Krikis, A.J., Polevaya, L.K.: Biophysical and Biochemical Information Transfer in Recognition, Plenum Press, 23-48 (1979).

3. Filatova, M.P., Reissmann, Z., Reutova, T.O., Ivanov, V.T., Grigoryan, G.L., Shapiro, A.M., Rozantsev, E.G.: Bioorgan. Khimia 3, 1181-1189 (1977) (in Russian).

4. Efremov, E.S., Filatova, M.P., Reutova, T.O., Stepanova, L.N., Reissmann, Z., Ivanov, V.T.: Bioorgan. Khimia 3, 1169-1180 (1977) (in Russian).

5. London, R.E., Stewart, J.M., Cann, J.R., Matwiyoff, N.A.: Biochemistry 17, 2270-2283 (1978).

6. Paiva, A.C.M., Juliano, L.: Peptides, A.Halster Press Book, J.Wiley & Sons, New York, 337-339, 1977

7. Huber, R., Bode, W.: Acc. Chem. Res. 11, 114-122 (1978).

INHIBITION OF RNA POLYMERASE BY ANALOGS OF AMANINAMIDE

Theodor Wieland, Christian Birr, Alain E. Vaisius
Max-Planck-Institut für medizinische Forschung
Abteilung Naturstoff-Chemie
D-6900 Heidelberg, Germany

Giancarlo Zanotti
Centro per le Studio della Chimica del Farmaco del CNR
Università di Roma
Roma, Italia

Introduction

The amatoxins are potent inhibitors of the eukaryotic DNA-dependent RNA polymerase B (or II). Previously it has been shown that the side chain of dihydroxy-L-isoleucine, as in α-amanitin, or at least with one γ-hydroxyl group, as in γ-amanitin, is responsible for a high affinity of the toxin to the enzyme. Shortening of this side chain by one C atom (γ-hydroxyvaline) led to a reduction of the affinity (and toxicity) to one tenth of the original value (1). We have now further modified side chain no. 3, and additional features by synthesis of various analogs of amaninamide (I).

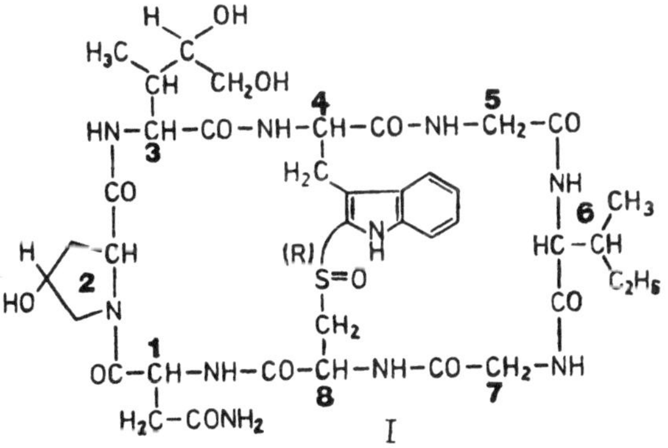

I

Chemistry of Peptides and Proteins, Vol. 1

Amaninamide is a minor toxin of the white mushroom Amanita virosa (2). It differs from the main amatoxins by the lack of the 6'-hydroxyl group at the indole nucleus, which does not contribute to the toxicity. Since this also is true for the sulfoxide oxygen of I the S-deoxo compounds were considered in this study.

Syntheses

The syntheses were performed using the thioether cyclization by the Savige-Fontana reaction as published in a previous paper (3). As formulated with Ile^3-S-deoxoamaninamide (Ia) the linear heptapeptide II synthesized via mixed anhydrides was treated with water free trifluoroacetic acid. Hereby, after loss of all protecting groups, internal thioether formation occurred yielding the monocyclic secocompound III which, via mixed anhydride or with DCCI, formed the desired analog Ia.

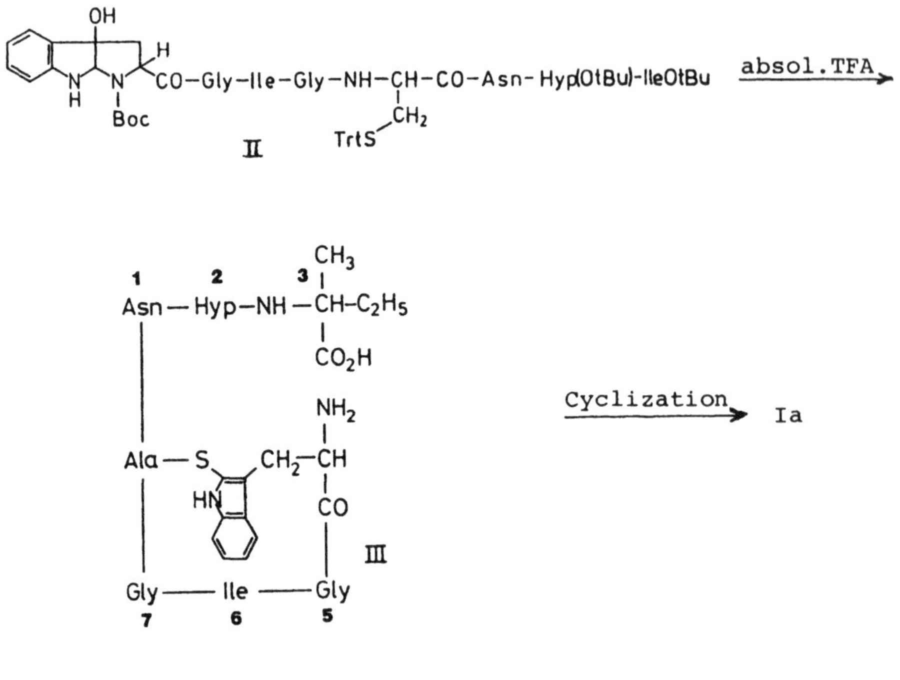

Results

The inhibitory effects or the analogs I(a-h) have been measured at RNA polymerase B from calf thymus, or in one case, from embryos of Drosophila melanogaster. The inhibition constants K_i (50 percent inhibition at K_i molarity) are shown in table 1. For comparison: K_i of amaninamide (I) is ca. 0.05×10^{-7}.

Table 1. Inhibition constants at RNA polymerase B from calf thymus of several synthetic analogs of S-deoxoamaninamide

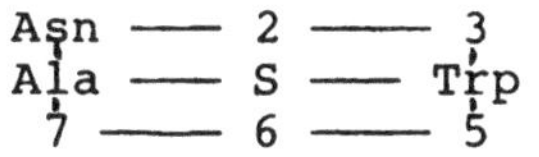

No.	Position no. 2	3	5	6	7	$K_i \times 10^7$
Ia	Hyp	Ile	Gly	Ile	Gly	0.8
b	-	Val	-	-	-	3.2
c	-	Leu	-	-	-	33
d	-	Ala	-	-	-	68
e	-	Ile	Ala	-	-	200
f	-	-	Gly	Ala	-	560
g	-	-	-	Ile	Ala	2.8
h	Pro	-	-	-	Gly	22

Discussion

For efficient binding to RNA polymerase B side chain no.3 must consist of at least three branched C atoms (Ia, Ib). Yet the maximal K_i's of table 1 are about 20 fold greater than those of the natural toxic α-, β- or γ-amanitin or of amaninamide (I).

This is a consequence of the lacking of the γ-standing OH group. If an OH is missing in position 2 (1 h) the affinity is also reduced to 5% as compared with Ia. The greatest diminution of the inhibitory capacity (greatest K_i) follows by replacement of Ile^6 by Ala. It appears that a hydrophobic interaction of this side chain plays a decisive role for binding of amatoxins to their receptor enzyme.

References

1. Wieland, Th., Fahrmeir, A.: Liebigs Ann. Chem. 736, 95-103 (1970).
2. Buku, A., Wieland, Th., Bodenmüller, H., Faulstich, H.: Experientia 36, 33-34 (1980).
3. Zanotti, G., Birr, Ch., Wieland, Th.: Int. J. Peptide Protein Res. 13, 204-216 (1978).

STRUCTURE-ACTIVITY RELATIONSHIP OF ACTIN-BINDING PEPTIDES

Heinz Faulstich
Max-Planck-Institut für Medizinische Forschung, Abt.Naturstoffchemie, D-6900 Heidelberg, Germany

Introduction

Among the toxic cyclopeptides of the green death cap Amanita phalloides, the amatoxins cause the fatal cases of mushroom poisoning. Poisoning by the second group of toxic peptides, the phallotoxins, has so far been afforded only in laboratory animals, and, only by intravenous administration. For example, if 2 mg phalloidin per kg body weight are injected into the mouse, (or 0,35 mg/kg into the rat), the animals die within 2-5 hours by a severe hemorrhagia of the liver.

Under the electron microscope, the structural lesions caused by phalloidin become evident: the liver cells are invaginated from the sinusoidal space and contain numerous vacuoles filled with erythrocytes and fibrin clots (1). A corresponding picture is found when isolated hepatocytes are treated with phalloidin; however, in this case the cells, normally round, are covered with protrusions and blebs (2). Both lesions have probably a common cause, namely, that the toxin induces a labilisation of liver cell plasma membranes. With phalloidin, the membranes of the liver cells become susceptible to even small pressure gradients. For example, hepatocytes in situ are invaginated by the small pressure of the portal vein, while after isolation, the small intracellular pressure caused by osmotic forces induces the development of protrusions. Concomitantly with the lesions of the membrane, the continuous web underlying the plasma membrane is disrupted.

Chemistry of Peptides and Proteins, Vol. 1

This web contains ordered actin filaments, which recently have been visualized by a fluorescent labeled phallotoxin (3). After treatment with phalloidin the actin is found accumulated in clusters. We assume that only the intact web provides elasticity to the cell membrane.

There is convincing evidence that all toxic effects of phalloidin are caused by the interaction of the toxin with the actin of the liver cells. Mechanistically, the toxin either induces a thorough polymerisation of liver actin as shown for muscle actin (4), or, the complexation of phalloidin with actin disturbs the interaction of actin with other structural proteins involved in the architecture of the submembrane web of hepatocytes. In the first case phalloidin would extremely decrease the pool of monomeric actin; in the second case, phalloidin would induce the breakdown of the submembrane structure described above. Anyway, the toxicity of phalloidin is closely related, or even identical, to its affinity for actin.

Method

The affinity to actin was assayed by equilibrium dialysis experiments. A radioactively labeled phallotoxin, [^{3}H]-demethylphalloin, bound to filamentous muscle actin, was substituted by increasing concentrations of the phallotoxin to be tested. The unit of measure was that concentration necessary for the substitution of 50 % of the labeled toxin. The lower the concentration of the toxin needed, the higher its affinity for actin. In a typical experiment, 5 x 10^{-7}M muscle actin in KCl/Tris (0.1 M, 1 mM; pH 7.4) was incubated with 5 x 10^{-7}M [^{3}H]-demethylphalloin to form the 1 : 1 complex. This solution was equilibrated with 10^{-5}M to 5 x 10^{-8}M of the phallotoxin to be tested for 16 hours. For example, the concentration of phalloidin, which substitutes 50 %

of the labeled toxin was determined to be 4.6×10^{-7}M. Among all phallotoxins the affinity of phalloidin to actin is the highest. Therefore this value is taken = 100 %.

Results and discussion

In the past 8 years a host of phallotoxin derivatives has been synthesized, either partially, starting from native phallotoxins, or totally (5, 6). The work was done to gain information on the relationship of structure and biological activity. First results in this field were obtained by relating the structural modifications of these phallotoxins to their toxicity in the white mouse (7). However, in this case also pharmacokinetic parameters may be involved. Therefore, data, representative for solely the peptide-protein interaction, were collected by the equilibrium dialysis experiments. In the following, several outstanding derivatives of phallotoxins will be discussed under this aspect.

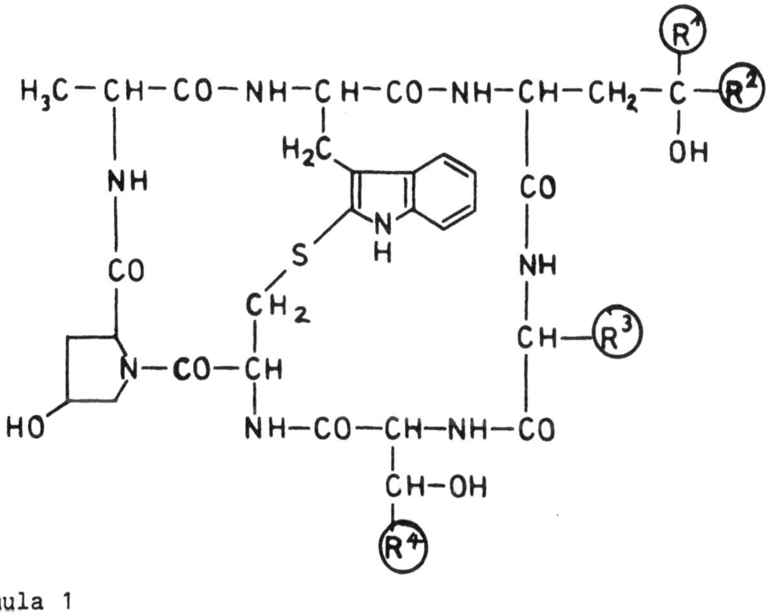

Formula 1

In order to discuss the chemical modifications we have to understand the structure of the naturally occurring phallotoxins (formula 1):

Table 1

	R^1	R^2	R^3	R^4	affinity to muscle actin [%]
Phalloidin	CH_3	CH_2OH	CH_3	CH_3	100
Phalloin	CH_3	CH_3	CH_3	CH_3	74
Phallisin	CH_2OH	CH_2OH	CH_2OH	CH_3	55
Phallacidin	CH_3	CH_2OH	$CH(CH_3)_2$	COOH	91

As shown in table 1, there are only small differences in the affinity to actin among the natural phallotoxins. Obviously, the side chain of the hydroxylated leucine is, to some extent, involved in binding, but seems not essential for activity. The tightest interaction with the protein was found for the compounds containing dihydroxylated leucine, i.e. phalloidin and phallacidin.

In contrast, one of the other hydroxy groups of the molecule must be involved directly in the interaction with the protein. This is shown by the fact that triacetylated phalloidin, acetylated at the D-threonine, the allo-hydroxyproline, and in the δ-position of leucine, is no longer bound by actin (affinity < 1 %). The hydroxyl group essential for binding was identified to be that in allo-position of proline. This was concluded from the l-hydroxyproline analog of norphalloin (8) and from a natural precursor peptide, the pro-phalloin (9), which, both, have no affinity to actin (< 1 %).

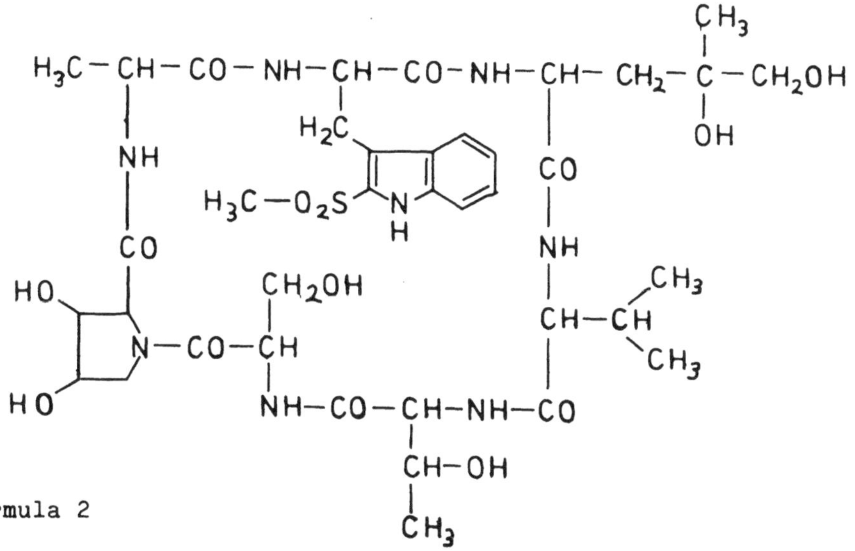

Formula 2

From previous ORD studies we know that all monocyclic derivatives of phalloidin are non-toxic, caused by a changed conformation (7). The present binding studies confirm these results. For example, secophalloidin, a product of partial hydrolysis of phalloidin at the amino terminus of the alanine unit in the 18-membered ring, has an affinity to actin of only 3 %; similarily, the monocyclic peptide obtained by reductive cleavage of the sulfur bridge, dethiophalloidin, is not bound at all. Its affinity is < 1 %. However, monocyclic peptides can be bound to actin, provided there exist additional facilities for the interaction with the protein, i.e. additional hydroxyl groups.

This became evident after the elucidation of the structure of the virotoxins, which are peptides with phalloidin-like structure and activity, which were isolated from the mushroom Amanita virosa (10). These peptides are monocyclic, but contain dihydroxyproline (11), and D-serine (formula 2), two additional hydroxyl groups which may achieve a better interaction with actin. Actually, the affinity to actin is even higher than that for phalloidin. For viroidin, a main component of this toxin group we determined an affinity value of 145 %.

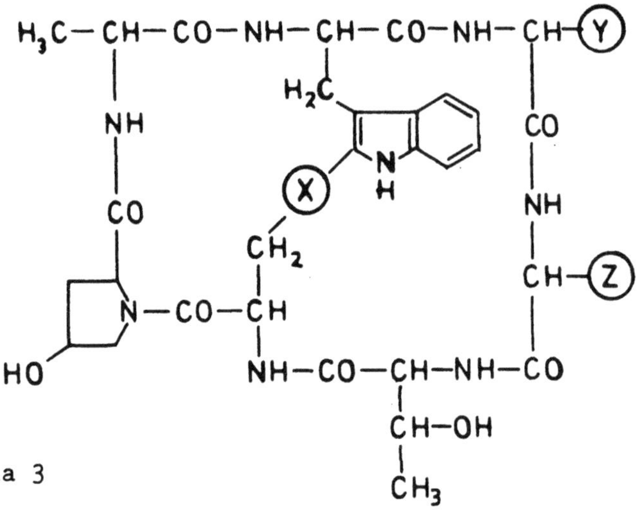

Formula 3

According to the natural variability of the leucine side chain, which, as we know, does not seriously affect the activity, we prepared phallotoxins chemically modified in pos. 7; (formula 3, table 2). As expected, all derivatives are more or less active. One of them, the dithiolaneamino-phalloidin (12), is even as active as phalloidin itself with an affinity value of 102 %. By reaction of this compound with fluoresceinisothiocyanate (FITC) we obtained the fluorescent labeled phallotoxin mentioned above. Dithiolaneamino-fluoresceinylphalloidin has only 15 % of the affinity of phalloidin to actin, which is, however, sufficient for excellent staining of actin-containing structures (13, 14).

In position 1 of the phallotoxins, the affinity to actin requires a short and lipophilic side chain (table 3). By synthesis or semisynthesis (15), several analogs were pre-

Table 2

	X	Y	Z	affinity to muscle actin [%]
Phalloidin	S	$CH_2-C(OH)(CH_3)-CH_2OH$	CH_3	100
Ketophalloidin	S	$CH-C(=O)-CH_3$	CH_3	34
Norphalloin	S	$CH_2-CH_2-CH_3$	CH_3	64
Leu[7]-phalloidin	S	$CH_2-CH(CH_2)(CH_3)$	CH_3	40
Dithiolano-ph.	S	$CH_2-C(S\,S)-Ch_3$ (S–S ring)	CH_3	64
Dithiolanamino-ph.	S	$CH_2-C(S\,S)-CH_3$ (S–S ring) $-CH_2-NH_2$	CH_3	102
Dithiolanamino-fluoreceinylo-phalloidin	S	$CH_2-C(S\,S)-CH_3$ (S–S ring) $-CH_2-NH-C(=S)-NH-Fl.$	CH_3	15

pared, showing that the maximum interaction is realized with the methyl group, while the isopropyl group reduces the interaction by 1/3 and the isobutyl group by about 2/3. Practically no interaction was found for the benzyl group, which is probably too big and rigid. Nor was their any activity for the D-configurated methyl group or for the glycine derivative.

Table 3

	X	Y	Z	affinity to muscle actin [%]
Phalloidin	S	$CH_2-\overset{CH_2OH}{\underset{OH}{C}}-CH_3$	CH_3	100
Gly[1]-phalloidin	S	"	H	2
D-Ala[1]-phalloidin	S	"	CH_3(D-)	2
Val[1]-phalloidin	S	"	$CH(CH_3)_2$	62
Leu[1]-phalloidin	S	"	$CH_2CH(CH_3)_2$	28
Phe[1]-phalloidin	S	"	CH_2-	1

Unexpectedly, the biological activity was found to be strongly dependent on the nature of the sulfur bridge (table 4).
By oxidation two sulfoxides were obtained (16). One of them, the (R)-configurated one, had a reduced affinity (10 %), while the (S)-configurated one showed no affinity at all (<1 %). Further oxidation of both forms yielded the sulfone, which had an affinity of 38 %. These data confirmed the results obtained previously with the in-vivo toxicity. In these experiments only the sulfone and the (R)sulfoxide exhibited toxicity in the white mouse, while the (S)-configurated sulfoxide was non-toxic. A possible explanation is that the strong dipole of the sulfoxide disturbs the interaction of the peptide and the protein, and, in the S-configuration, even completely prevents the formation of the toxin complex.

Table 4

	X	Y	Z	affinity to muscle actin [%]
Phalloidin	S	$CH_2-C(CH_3)(OH)-CH_2OH$	CH_3	100
Phalloidinsulfoxide(R)	SO	"	"	10
Phalloidinsulfoxide(S)	SO	"	"	1
Phalloidinsulfone	SO_2	"	"	38

References

1. Miller, F., Wieland, O.: Virchow's Arch. Pathol. Anat. Physiol. 343, 83 (1967).
2. Weiss, E., Sterz, I., Frimmer, M., Kroker, R.: Beitr. Phathol. 150, 345 (1973).
3. Jahn, W., Faulstich, H., Deboben, A., Wieland, Th.: Z. Naturforsch. 35 c, 467 (1980).
4. Faulstich, H., Schäfer, A.J., Weckauf, M.: Hoppe-Seyler's Z. Physiol. Chem. 358, 181 (1977).
5. Wieland, Th., Faulstich, H.: Crit. Rev. Biochem. 5, 185 (1978).
6. Wieland, Th., Faulstich, H.: in Frontiers in Bioorganic Chemistry and Molecular Biology, Ovchinnikov, Yu. A., Kolosov, M.N., eds., Elsevier, Amsterdam, 97, 1979
7. Faulstich, H., Wieland, Th.: Europ. J. Biochem. 22, 70 (1971).
8. Faulstich, H., Nebelin, E., Wieland, Th.: Liebigs Ann. Chem. 50 (1973).
9. Munekata, E., Faulstich, H., Wieland, Th.: Liebigs Ann. Chem. 776 (1978).
10. Faulstich, H., Buku, A., Bodenmüller, H., Wieland, Th.: Biochem. 19, 3334 (1980).
11. Buku, A., Faulstich, H., Wieland, Th., Dabrowski, J.: Proc. Natl. Acad. Sci. USA 77, 2370 (1980).

12. Wieland, Th., Deboben, A., Faulstich, H.: Liebigs Ann. Chem. 416 (1980).

13. Wulf, E., Deboben, A., Bautz, F.A., Faulstich, H., Wieland, Th.: Proc. Natl. Acad. Sci. USA 76, 4498 (1979).

14. Jahn, W., Faulstich, H., Deboben, A., Wieland, Th.: Z. Naturforsch. 35 c, 467 (1980).

15. Munekata, E., Faulstich, H., Wieland, Th.: J. Amer. Chem. Soc. 99, 6151 (1977).

16. Faulstich, H., Wieland, Th., Jochum, C.: Liebigs Ann. Chem. 713, 186 (1968).

SYNTHESIS, SPECTRAL AND BIOLOGICAL PROPERTIES OF DSIP AND ITS ANALOGS

Inessa Mikhaleva, Asmic Sargsyan, Tamara Balashova, Vadim Ivanov

Shemyakin Institute of Bioorganic Chemistry, USSR Academy of Sciences, Moscow, USSR

Introduction

Over the past three years we have been involved (1, 2) in the structure-function study of δ-sleep inducing peptide (DSIP) which was isolated by Monnier, Swiss physiologist, from dialysates of cerebral venous blood obtained from rabbits in which sleep was induced by electrical stimulation of the thalamus (3). According to data obtained by Monnier this peptide is capable of enhancing the slow-wave (delta) and spindle patterns in the electroencephalogram after intracerebroventricular or intravenous injection.

Results

With the aim of studying the relationship between the structure and function of this endogenous peptide we have synthesized DSIP and 11 of its analogs (4). The series of analogs (Fig. 1) includes four Tyr-containing peptides, 2-5 , analogs with amidated β- or γ-carboxyls of Asp and Glu, 6, 7 , analogs with D-Ala residue in one of the positions 3, 4, 5 (analogs 9, 11, 12), an analog with δ-aminovaleric acid instead of the Gly-Gly moiety, 8 , and a compound with Gly-Pro segment at the N-terminus , 10.
All of these peptides were synthesized by solution methods

Chemistry of Peptides and Proteins, Vol. 1

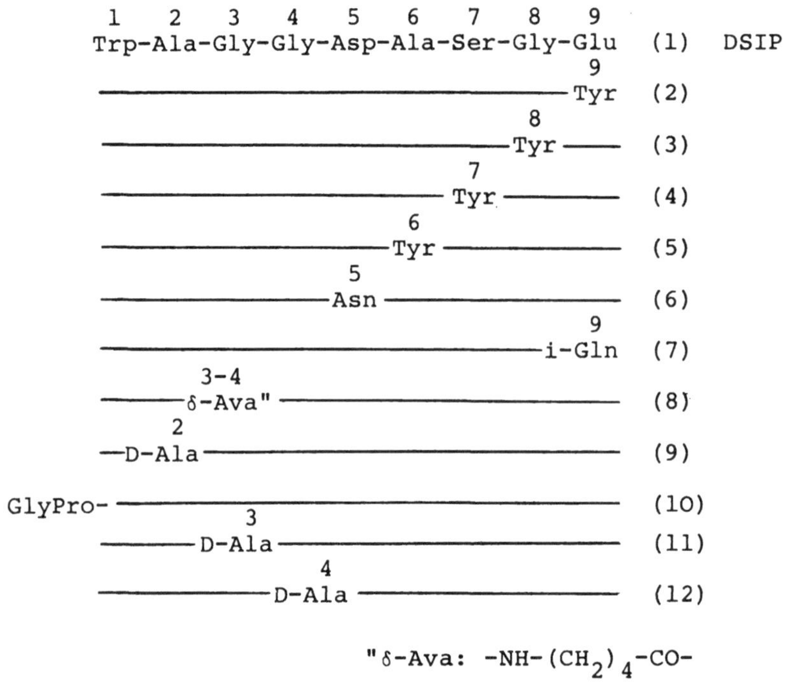

Fig. 1. DSIP and its analogs.

using a mixed anhydride condensation of two segments (1-4 and 5-9), prepared stepwise. We utilized benzyl-type protective groups for side chains and Boc or Z for N^{α}-function. The amino acid derivatives were coupled by activated ester (-ONSu, -ONp) or DCC/HOBT procedure. The protected final products were purified by high performance size exclusion chromatography. After catalytic hydrogenation the free peptides were isolated by ion-exchange chromatography on DEAE-Sephadex A-25 column. Purity of the final products was checked by thin-layer and column chromatography, electrophoresis, amino acid analysis and was in addition confirmed by NMR spectra (300 MHz instrument).

The synthesis of these peptides did not appear to be so complicated as were numerous experiments performed by our colleagues, biologists, in order to clarify the biological effects of DSIP.

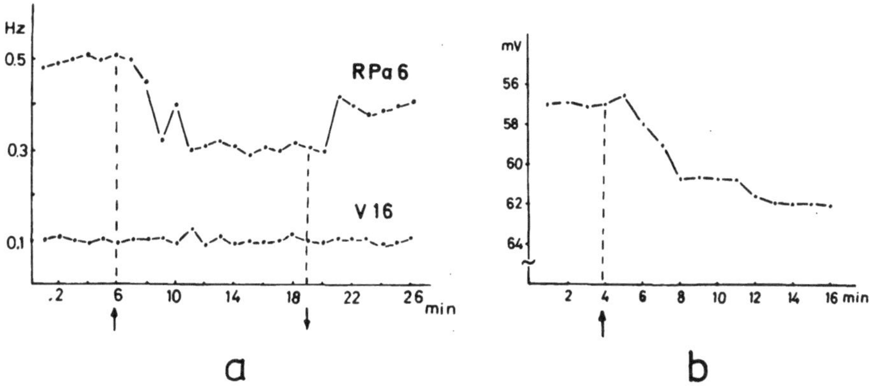

Fig. 2. Effect of 10^{-8} M DSIP at the frequency of spontaneous firing of RPa6 and V-16 neurons (a) and the resting potential of RPal neurone (b). Arrows indicate beginning and the end of DSIP application.

Although there is a great interest in this peptide its real physiological role in sleep remains unclear today: in some studies this activity was found, 5-8 , in others it failed to appear, 9-11 . Yet, much work has to be done before DSIP can be labelled functionally. It seems likely that sleep involves a complex interaction of many factors and transmitters, therefore the elucidation of the function of DSIP as a single regulatory substance is a very complicated problem.
Consequently we had to study more reliable *in vitro* test-systems, one of them being the electrophysiological study with identified neurons of the isolated nervous system from snail *Helix Lucorum L.* It was found (4) that DSIP in very small concentration (5 nM) inhibited the endogenous electrical rhythm by 30-40% in a number of neurons; it also gave rise to a 1-4 mV increase in the resting potential (Fig. 2). As to analogs, compounds 8 and 2-5 have similar effects but at 10-50 higher concentrations whereas others are not active at all 6, 9-12 . Analogue 7 showed an excitatory effect with a dose of $5 \cdot 10^{-6}$ M.
Another *in vitro* system tested on DSIP was the DSIP action on the enzymatic system responsible for the breakdown of acetyl-

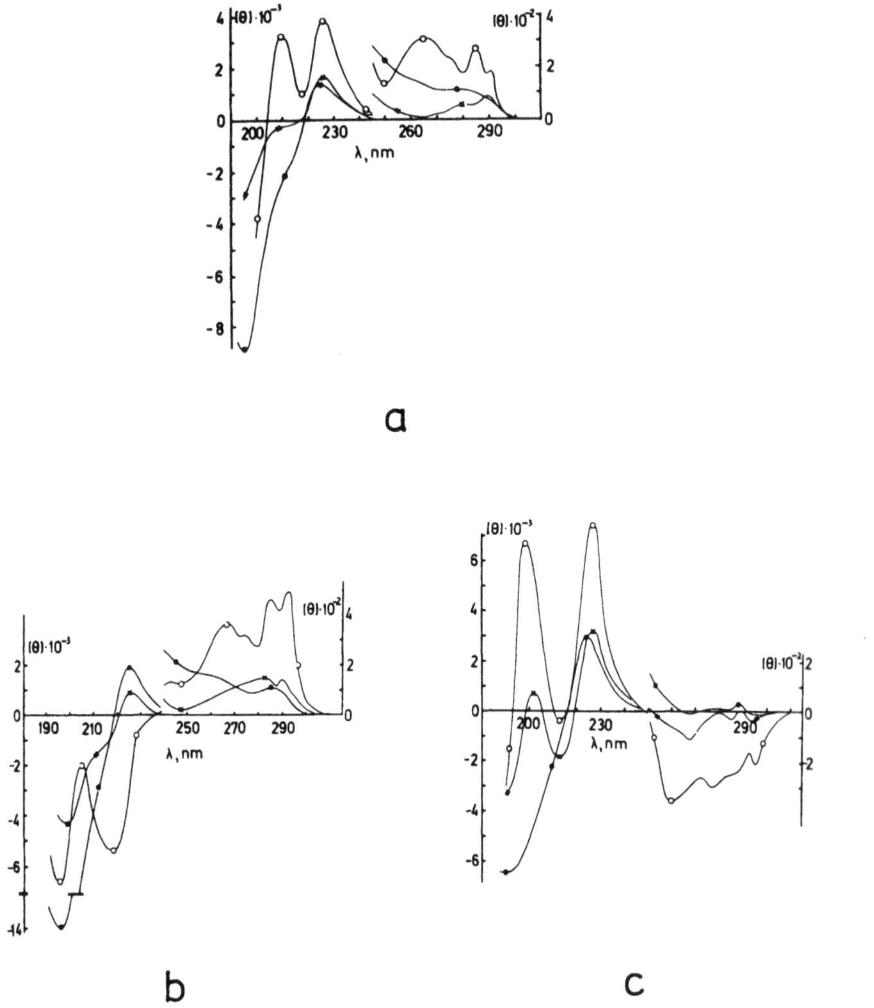

Fig. 3. CD curves of DSIP (a) and its analogs 8 (b) and 11 (c).
-o- Water, 20°; -x- ethanol-water (19:1), 20°; -●- ethanol-water (19:1), -85°.

choline, serotonine and catecholamines. DSIP in a concentration of 10^{-5} -10^{-6}M proved to be efficient in a 2-4 fold activation of monoaminooxidase A and has no effect on monoaminooxidase B and acetylcholine-esterase (12). None of the analogs affected the activity of the above enzymes.

So, DSIP acts strongly and selectively on the serotoninergic system which, according to general considerations, is involved in the regulation and maintenance of sleep, mainly NREM sleep.

Summing up the results of the biological data obtained for DSIP in both tests (*in vitro*) it may be said that the biological effects of DSIP are highly sensitive to structural modifications.

In order to find out the reasons that give rise to a structural specificity of DSIP, we carried out a conformational study of the compounds by spectral methods.

CD spectra of DSIP at room temperature shown in Fig 3 (a) are typical of an unordered conformation while on cooling, a spectral shift occurs due to some specific predominant conformation(s).

As seen from Fig. 3 (b, c) the analogs in water have similar CD curves but all of them differ considerably from the CD curve of DSIP on cooling. Some of the analogs differ in the backbone conformation (8, for inst.), others in the chromophore orientation, 11 . Taking into account the biological properties of the analogs it might be assumed that the conformation of DSIP which is realized in ethanol on cooling is responsible for their biological activity.

Further information about the spatial structure of DSIP and its analogues was obtained from Raman Laser spectra (2) shown in Fig. 4. The DSIP spectrum (a) reveals amide III bands at 1242-1256 cm^{-1} from the unordered parts of the peptide and a high frequency at 1281-1291 cm^{-1} due to presence of β-turn(s) (13). Formally a possible assignment of the latter bands to the α-helical conformation seems unbelievable if we bear in mind, for instance, the CD results. High frequency bands are present in spectra of Tyr^7- and Tyr^8-analogs (Fig. 4, c, d) and are absent in spectra of Tyr^6- and Tyr^9-analogs (Fig. 4,

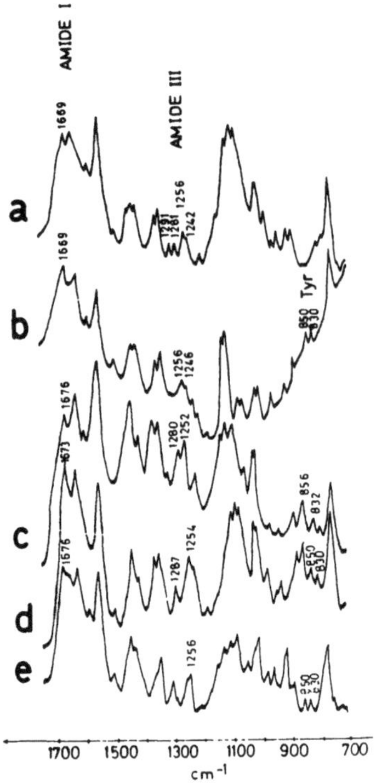

Fig. 4. Raman spectra of DSIP (a) and its analogues 5 (b), 4 (c), 3 (d) and 2 (e) in water, pH 6.5.

b, c), indicating that β-turn might be located at the 6-9 site. Some additional evidence of the folded structure(s) of DSIP analogs may be obtained from analysis of the tyrosine band intensity ratio 850/830. This doublet is the sensor of the Tyr microenvironment. For Tyr^6- and Tyr^9-derivatives the intensity ratios are 1.2 and 1.1 respectively that is they are typical of a normal environment, while for Tyr^7- and Tyr^8-analogues these ratios are 2.4 and 2.6 respectively and are indicative of a positively charged group, i.e. the protonated amino terminus in the vicinity of the tyrosin hydroxyl, as shown schematically in Fig. 5.

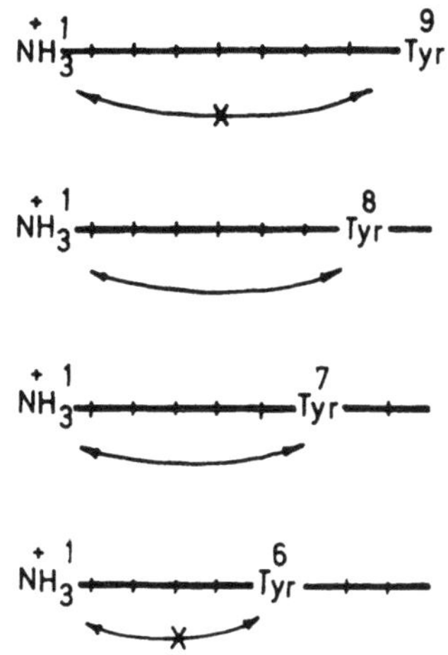

Fig. 5. Schematical representation of the proximity of N-terminus and tyrosin hydroxyl in analogues 2-4.

More detailed information was obtained from NMR spectra of DSIP and related compounds. NMR parameters of NH signals of DSIP chemical shifts, their temperature dependencies ($\Delta\delta/\Delta T=(5\text{-}6)\cdot 10^{-3}$ ppm/degree), are characteristic of an unordered conformation containing neither intramolecular hydrogen bonds nor salt bridges. These data are in good agreement with the stability of $C^{\alpha}H$, $C^{\beta}H_2$ and aromatic proton signals of Trp residue to acid titration (20°, D_2O) (Fig. 6). The Ala-2 and Ser-7 NH protons are sensitive to aspartic acid β-COOH titration (5°, H_2O), the indole NH signal is slightly shifted near pK of glutamic acid γ-COOH (Fig. 6b). Titration of the NH_2 terminal group is accompanied by changes in all the $C^{\alpha}H$ chemical shifts and NH signals of Gly-3 and Gly-4.
Thus, NMR data revealed that DSIP exists in water as a folded structure(s) with extended ionogenic groups, the central position of β-COOH inside. NMR study of Tyr^7-DSIP and Asn^5-DSIP showed that all the general features of DSIP structure remained

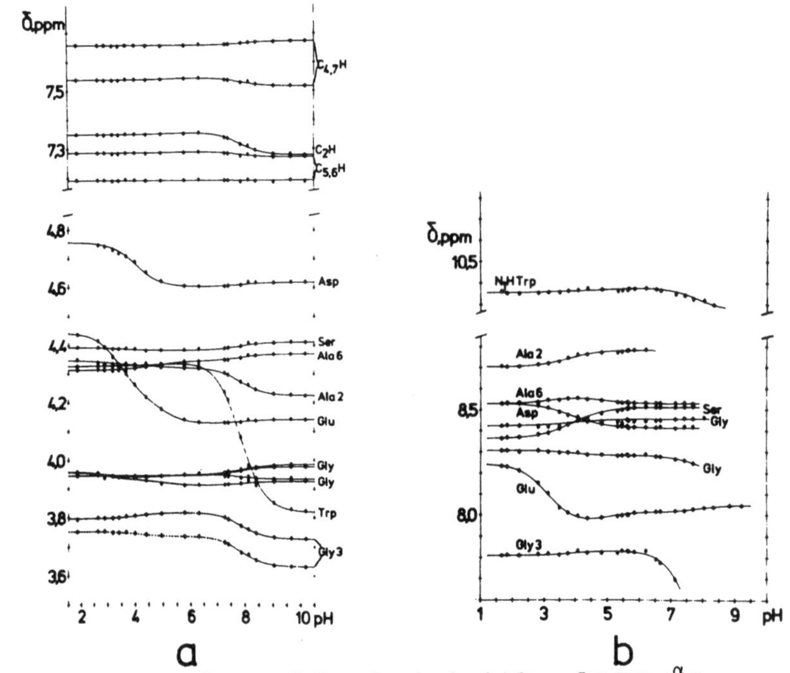

Fig. 6. pH Dependence of the chemical shifts of DSIP $C^{\alpha}H$, aromatic proton signals at 20° C in D_2O (a) and NH proton signals at 5° in H_2O (b); 300 MHz.

in these analogs. This structure is schematically represented below, its detailization being our present work.

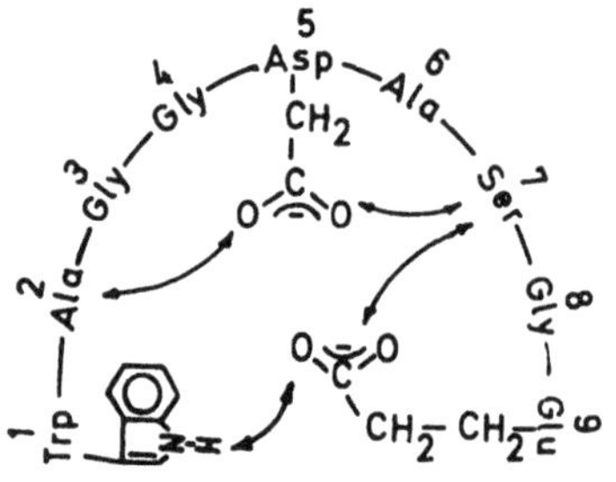

References

1. Mikhaleva, I.I., Sargsyan, A.S., Sumskaya, L.V., Balashova, T.A., Deshko, T.N., Efremov, E.S., Ivanov, V.T.: "Peptides. Structure and Biological Function" (E.Gross and J.Meienhofer, eds), Pierce Chemical Company, Rockford pp. 901-904, 1979
2. Ivanov, V.T., Mikhaleva, I.I., Sargsyan, A.S., Balashova, T.A., Efremov, E.S., Deshko, T.N., Nabiev, I.R.: Peptides 1980, in press.
3. Schoenenberger, C. A., Monnier, M.: Proc. Natl. Acad. Sci. USA 74, 1282-1286 (1977).
4. Sargsyan, A.S., Sumskaya, L.V., Aleksandrova, I.Yu., Besrukov, M.V., Mikhaleva, I.I., Ivanov, V.T., Balaban, P.M.: Bioorg. Khim. (Russian) 7, 1125-1149 (1981).
5. Schoenenberger, G.A., Maier, P.F., Tobler, H.J., Wilson, K., Monnier, M.: Pflügers Arch. 376, 119-129 (1978).
6. Kafi, S., Monnier, M., Gallard, J.-M.: Neuroscience Letters 13, 169-172 (1979).
7. Karmanova, I.G., Maksimuk, V.F., Voronov, I.B., Bogoslovsky, M.M., Demin, N.N., Rubinskaya, N.P., Albertin, S.V.: J. Evol. Biochem. Physiol. (Russian) 15, 583-589 (1979).
8. Polc, P., Scheeberger, J., Haffely, W.: Neuroscience Letters 9, 33-36 (1978).
9. Krueger, J.M., Bacsik, J., Garcia-Arraras : Am. J. Physiol. 238, E116-E123 (1980).
10. Tobler, I., Borbely, A.: Waking and Sleeping 4, 139-153 (1980).
11. Kovalson, V.M., Tsybulskii, V.L.: Higher Nerve Action (Russian) 5, 1064-1066.
12. Ashmarin, I.P., Dovedova, E.L.: Bull. Acad. Sci. USSR 255, 1501-1503 (1980).
13. Deslauriers, R., Leach, S.J., Maxfield, F.R., Minasian, E., McQuie, J.R., Meinwald, Y.C., Nemethy, G., Tottle, M.S., Rae, I.D., Scheraga, H.A., Stimson, E.R., Van Nispen, J.W.: Proc. Natl. Acad. Sci. USA 76, 2512-2514 (1979).

POLYOXYETHYLENEOXIDE(POE) PEPTIDES, MODELS FOR RHODOPSIN AND MYOGLOBIN

Ernst Bayer
Institut für Organische Chemie der Universität Tübingen, Auf der Morgenstelle 18, D-7400 Tübingen, FRG

The properties of POE peptides may be much more similar to biologically active proteins and enzymes than those of molecules of low molecular weight. The synthetic polymer support may, under certain circumstances, protect the active site.

Essential parts of a protein not belonging to the active site of the biomacromolecule may be replaced by a synthetic polymer chain and already small polymer-bound peptide units may show natural protein-like hemoglobin reactions. POE peptides serving as models for myoglobin and hemoglobin were synthesized with the liquid-phase method.

A similar steric arrangement, compared to the natural protein, is achieved by the amino acid sequence linking the synthetic polymer and the heme moiety. The synthetic polymer causes additional protection of the active site. Following this concept allows to synthesize for the first time hemopolymers which have reversible oxygenation properties under physiological conditions. The oxygen partial pressures (0.2-3 Torr) are of similar values as those found for the myoglobin.

As models for rhodopsin the following POE peptides were synthesized: POE-Ala-Lys-Glu, POE-Ala-Lys-Ala-Glu, POE-Ala-Lys-Ala-Asp and POE-Ala-Lys-Trp. In all peptides 11-cis as well as 11-trans-retinal was reacted with the ε-amino group of lysine and the spectroscopic properties, the kinetics and quantum yield of the cis-trans isomerizations were studied. Similar to the photoisomerization of rhodopsin the cis-trans iso-

Chemistry of Peptides and Proteins, Vol. 1

merization of the model peptides takes place in an excited singlet state in 100 nanoseconds. The quantum yields of the synthetic models vary between 0.2 and 0.7 and are dependent on the peptide sequence. The values are similar to those found for rhodopsin. These POE peptides are the first known synthetic Schiff's bases reacting after the same mechanism as rhodopsin.

CHEMICAL MUTATION: REPLACEMENT OF REACTIVE SITE RESIDUE P_1=LYS IN BOVINE KUNITZ INHIBITOR BY OTHER AMINO ACIDS AND CONCOMITANT SPECIFICITY CHANGE

Harald Tschesche and Herbert R. Wenzel
Lehrstuhl für Biochemie, Fakultät für Chemie, Universität Bielefeld
D-4800 Bielefeld 1, FRG

Introduction

Protein proteinase inhibitors form strong 1:1-stoichiometric complexes with their target enzymes. The family of serine proteinases e.g. contains such important enzymes as trypsin, acrosin, plasmin, chymotrypsin, cathepsin G, elastase, kallikrein, thrombin etc. They all have similar threedimensional structures, related from the process of divergent evolution and differ in substrate specificity due to differing geometrics of their substrate specificity pocket in the binding site of the enzymes. Proteinase inhibitors evolve as antagonists of one family of enzymes. The rigid structure of the inhibitors is preserved as a means of providing the rigid and proper surface geometry for the almost perfect fit to the enzyme substrate binding site. The interaction between enzyme and inhibitor resembles the classical picture of a lock and key interaction. The primary specificity of protein proteinase inhibitors is thought to be mainly determined by the geometry and polarity of the residue P_1 [1,2] in the reactive site peptide bond P_1-P_1' in the inhibitor according to the notation of Schechter and Berger [3]:

$$P_3 - P_2 - P_1 - P_1' - P_2' - P_3'$$

A straigtforward approach to test this hypothesis would be to replace P_1 by a single other amino acid and thus to accomplish "mutations" at the late protein stage.

In our experiments toward this aim we used as a model compound the Kunitz inhibitor from bovine organs (TKI), a powerful inhibitor of trypsin, plas-

Chemistry of Peptides and Proteins, Vol. 1

min and kallikrein. The single polypeptide chain of TKI consists of 58 amino acid residues which are cross-linked by three disulfide bridges. The inhibitor is folded in such a way that a pear-shaped molecule results with residue P_1=Lys^{15} located in an exposed position at the top of the pear (Fig. 1) [4].

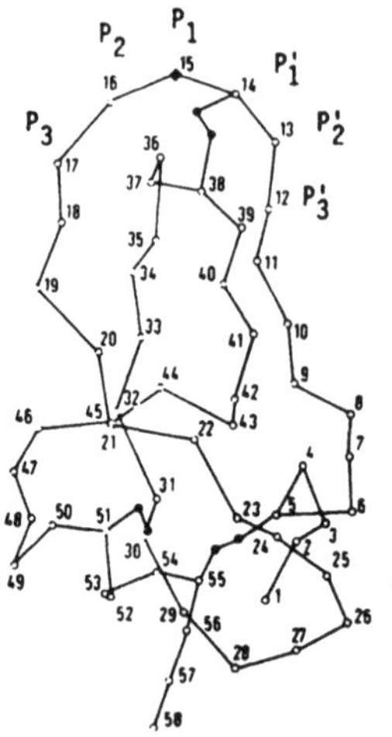

Fig. 1. Projection of the three-dimensional structural model of the trypsin-kallikrein inhibitor (Kunitz) [4]. The positions of the α-carbon atoms of the peptide chain are indicated. ● = sulfur atoms of the disulfide bridges, ◆ = Lys^{15}.

Results and Discussion

Replacement of residue P_1 at first requires a split of the entire polypeptide chain at the P_1-P_1' reactive site peptide bond to give modified inhibitor (TKI*, Lys^{15} -Ala^{16} peptide bond hydrolyzed) [5] and second removal of P_1 by carboxypeptidase B to yield modified des-Lys^{15} -inhibitor. Reinsertion of P_1 was possible by a reversed carboxypeptidase A (or B) catalyzed reaction by which the inactive modified des-Lys ^{15}-inhibitor is resynthesized to an inhibitory active Phe-, Trp- (or Arg)- modified inhibitor which can be trapped from the equilibrium by chymotrypsin (or trypsin) to yield the enzyme inhibitor complex. The complex may be isolated and subjected to kinetic controlled dissociation [6] to yield enzyme and resynthesized single chain (virgin) inhibitor with a new amino acid residue in P_1 position. It was possible to replace Lys^{15} by Arg [6] and ^{13}C-Lys [7, 8] using carboxypeptidase B and by Phe and Trp [6] using carboxypeptidase A. The essential steps of this "enzymatic mutation" [9] are shown in Fig. 2 (left side, introduction of amino acids X). Insertion of Arg yielded a product with inhibitor properties almost indistinguishable from those of native TKI but somewhat better inhibition of pancreatic kallikrein. Introduction of the aromatic amino acids led to strong chymotrypsin inhibitors with only weak anti-tryptic activity.

Several attempts to perform similar insertions with various aliphatic amino acids have been unsuccessful due to the lack of a proper enzyme system for resynthesis.

Recently we have developed a chemical method which overcomes these difficulties and allows insertion of almost any amino acid [10, 11].

This is achieved by coupling an amino acid methyl ester to the Cys^{14} carboxyl group of inactive modified des-Lys^{15} inhibitor [6] with water-soluble carbodiimides yielding new active inhibitors. These active inhibitors are Y^{15} methyl ester modified inhibitors, which react almost as readily with the cognate enzymes as does the virgin inhibitor with e.g. trypsin. The reaction of the methyl ester derivative to form the stable complex is much more rapid than with modified inhibitor. This may be explained by the fact that the necessary acylenzyme intermediate which has to be passed in order to resynthesize the peptide bond P_1-P_1' is more easily attained from the methyl ester by transacylation than from the free carboxylate form of the modified inhibitor, Fig. 2.

Fig. 2 shows the reactions employed in the course of this "chemical mutation" (right side, introduction of amino acids Y).

The table represents a nonquantitative evalution of the inhibitory properties of the various semisynthetic Kunitz inhibitors having different residues P_1 at the reactive site.

Insertion of Met yields a potent chymotrypsin inhibitor, which also weakly inhibits trypsin and the elastases from porcine pancreas and human leucocytes. A strong inhibitor of both elastases and chymotrypsin is obtained by insertion of Leu: The Leu^{15}-inhibitor still inhibits trypsin weakly.

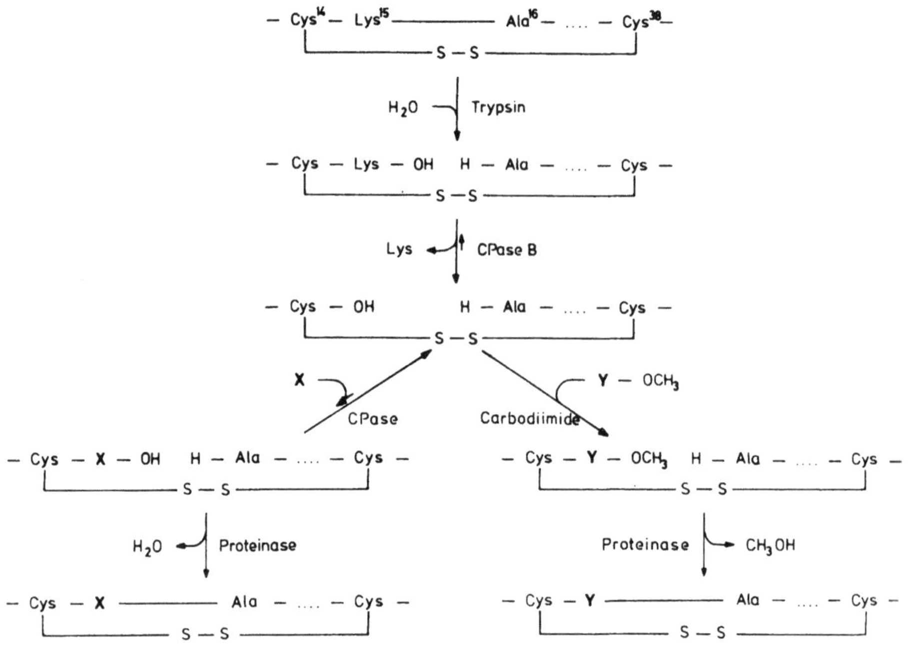

Fig. 2. Schematic representation of the reactions involved in the "enzymatic mutation" (left) [6, 7] and the "chemical mutation" [10, 11].

The Val^{15}-inhibitor seems to be rather specific. It exhibits strong inhibition of human leucocyte elastase and weak inhibition of trypsin, chymotrypsin and porcine pancreas elastase.

The latter four enzymes are only weakly inhibited by the Ala^{15}-inhibitor. Gly^{15}-inhibitor, without anti-elastase effect, is an efficient chymotrypsin inhibitor and, rather surprisingly, an excellent trypsin inhibitor.

Subtilisin is inhibited neither by the native (virgin) inhibitor (TKI) nor by its semisynthetic derivatives.

Amino acid[15]	Trypsin (bovine pancreas)	Chymotrypsin (bovine pancreas)	Elastase (porcine pancreas)	Elastase (human leucocytes)	Subtilisin (bacillus subtilis)
Lys	+++	++	-	+	-
Gly	+++	++	-	-	-
Ala	+	+	+	+	-
Val	+	+	+	+++	-
Leu	+	+++	+++	+++	-
Met	+	+++	+	+	-

Table: Inhibitory properties of five semisynthetic inhibitors with Gly, Ala, Val, Leu, Met in the P_1 position compared to that of the native Lys^{15} inhibitor (Kunitz) from bovine organs against trypsin, chymotrypsin, and elastases with differing specificity [11].
Symbols -, +, ++, +++ denote no, weak, strong and very strong inhibition.

These results indicate that it is possible to a certain degree to intentionally change the specificity of an inhibitor by a semisynthetic single amino acid replacement. An educated guess on the amino acid to be inserted in order to obtain an inhibitor for a certain enzyme can be made, if the specificity of the target proteinase is known. Numerous additional experiments are being carried out in order to draw general conclusions as to the impact of the P_1 residue on the primary specificity.
During the insertion synthesis coupling of the amino acid methyl esters also takes place to some extent at the five carboxyl groups already present in native TKI. These groups are, however, remote from the reactive site, and their modification does not generally affect the complex formation with the proteinases. Attempts to reversibly protect these five groups are in progress.
It is obvious that our replacement procedure is not limited to L-amino acids, uncommon compounds such as D-amino acids and various amino acid analogues can be used as well. Moreover, experiments are currently under way in our laboratory to extend our method to other proteinase inhibitors.

References

1. Laskowski, M., Jr., Sealock, R. W. in The Enzymes (Boyer, P. D., ed.) Vol. III, Academic Press, New York (1971) pp 375-483.
2. Tschesche, H.: Angew. Chem. Internat. Ed. 13, 10-28 (1974).
3. Schechter, J., Berger, A.: Biochem. Biophys. Res. Commun. 27, 157-162 (1967).
4. Huber, R., Kukla, D., Rühlmann, A., Epp, O., Formanek, H.: Naturwissenschaften 57, 389-392 (1970).
5. Jering, H., Tschesche, H.: Eur. J. Biochem. 61, 443-452 (1976).
6. Jering, H. Tschesche, H.: Eur. J. Biochem. 61, 453-463 (1976).
7. Richarz, R., Tschesche, H., and Wüthrich, K.: Biochemistry 19, 5711-5715 (1980).
8. Richarz, R., Tschesche, H., and Wüthrich, K.: Eur. J. Biochem. 102, 563-571 (1979).
9. Sealock, R. W., Laskowski, M., Jr.: Biochemistry 8, 3703-3710 (1969).
10. Wenzel, H. R., Tschesche, H.: Hoppe-Seyler's Z. Physiol. Chem. 361, 345 (1980).
11. Wenzel, H. R., Tschesche, H.: Angew. Chem. 93, 292-293; Angew. Chem. Internat. Ed. 20, 295-296 (1981).

Acknowledgement: This work was supported by: Deutsche Forschungsgemeinschaft, Bonn, Ministerium für Wissenschaft und Forschung, Nordrhein-Westfalen, Düsseldorf and Stiftung Volkswagenwerk, Hannover.

SEMISYNTHETIC MODIFICATION OF THE N-TERMINUS OF THE INSULIN A-CHAIN

Peter Trindler and Dietrich Brandenburg

Deutsches Wollforschungsinstitut
D-5100 Aachen, Federal Republic of Germany

Introduction

The invariant A1-glycine (Fig. 1) plays an important role in the stabilization of the three-dimensional structure of insulin and its biological activity, but it is as yet impossible to interpret the observations obtained with presently known analogues satisfactorily. As part of our structure-function studies[1] we are aiming at "atom-by-atom" alterations of the residue in position A1. In this communication we report on the replacement of glycine by several amino, hydroxy, and unsubstituted carboxylic acids[2].

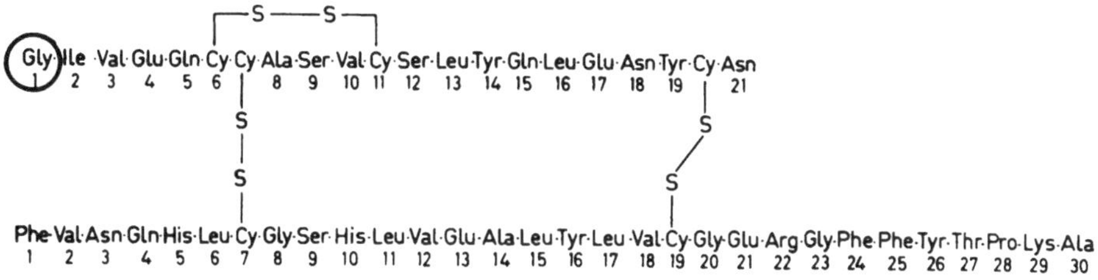

Figure 1. Primary structure of bovine insulin[3].

Abbreviations:
Abu = aminobutyryl, HO-Ac = hydroxyacetyl,
Msc = methylsulphonylethyloxycarbonyl,
Ptc = phenylthiocarbamoyl

Chemistry of Peptides and Proteins, Vol. 1

Strategic considerations

Insulin analogues with a modified N-terminus of the A-chain are accessible via the following four routes:

Total synthesis of	a) the whole analogue
	b) the A-chain
Partial synthesis with	c) the A-chain
	d) the whole insulin molecule.

a) Obviously the most demanding route is total synthesis. Märki et al.[4] have recently obtained [Pro^{A2}]insulin and several other analogues by joining A(1-13) to the asymmetrical disulphide A(14-21)/B(1-30).

b) Several A-chain analogues have been obtained by synthesis in solution, e.g. des-Gly^1-A-chain[5], des-amino[6], and [Ala^1]A-chain[7]. Synthetic efforts are thus reduced to the assembly of one chain, but the combination with B-chain of natural origin remains as a major yield-limiting problem.

c) Semisynthesis appears to be the method of choice, particularly for terminal modifications. However, the two-chain molecule presents difficulties. There are no specific enzymatic methods for removal of A1, and Edman degradation is highly selective for B1[1]. Des-Gly-A-chain has been obtained by SS-cleavage of des-Gly, des-Phe-insulin, or Edman degradation of A-chain, and has been elongated with different amino acids, and then combined with natural B-chain[1]. Alternatively, it has been combined with B1,B29-$(Boc)_2$-B-chain, and the protected intermediate acylated subsequently [1]. The work necessary for obtaining the chain analogues was thus markedly economized, but the problem of chain combination was not.

d) Due to the high reactivity of the A-chain amino group towards acylating agents, the direct preparation of B1,B29 diprotected derivatives is not possible. A successful replacement of A1-glycine in intact insulin thus necessitates differential protection, and acid-stable blocking groups at the

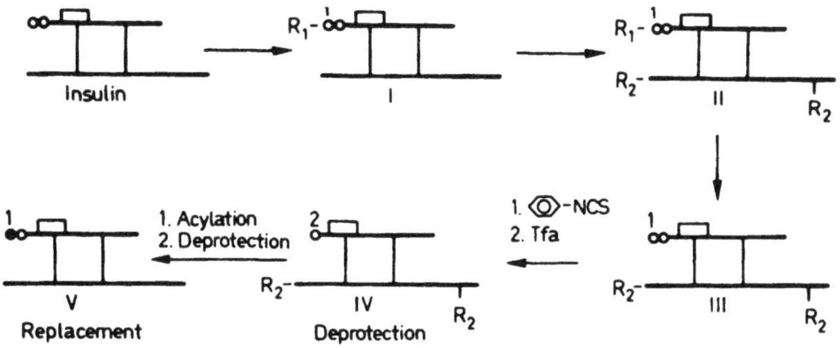

Figure 2. Scheme for the replacement of A1-glycine .

B-chain. Geiger et al.[8] started with A1,B29-$(Boc)_2$-insulin, introduced the Msc-group at B1, and liberated the amino groups A1 and B29. Preferential reaction with phenylisothiocyanate at A1 was followed by blocking the remainder of free B29-amino groups with the Msc-moiety. Crude intermediate IV (Fig.2) was acylated with Boc-amino acid active esters (e.g. of D-Ala). Side products with Ptc-groups at B29 were removed in the course of subsequent purifications. Saunders and Offord[9] followed the scheme outlined in Fig.2 with R_1 being Boc, and using Z-methionine as an acid and alkali-stable protecting group (R_2).

The tactics tested in our work were based on Boc protection at A1 (R_1) and the Msc residue as R_2, featuring the isolation of intermediate IV in homogeneous form.

Protection and degradation of insulin (bovine)

The reaction and purification steps for the preparation of the key intermediate are summarized in Fig. 3. A1-Boc-insulin (I) was obtained by adapting and optimizing a procedure developed by Friesen[1] for porcine insulin. Complete acylation with Msc-ONSu was effected with 15 equivalents in DMSO/triethylamine . Treatment with trifluoroacetic acid resulted in a product which contained ca 20% of apparently non-cleaved material. Reaction with 40 eq. phenylisothiocyanate in 90% pyridine was quantitative. The derivative IV (net charge + 4 at pH 3) was separated from all side-products, $(Msc)_3$-desamido-insulin, "$(Msc)_2$-Boc-insulin" and a possible comtamination with Ptc derivative from incomplete Edman degradation by one final SP-Sephadex chromatography at pH 3.

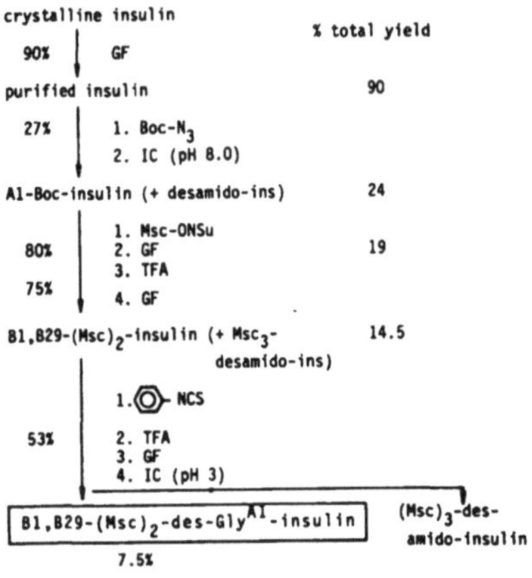

Figure 3.

Preparation of the key intermediate.

GF = Gel filtration Sephadex G-50

IC (pH 8)= DEAE cellulose, urea/Tris-HCl, NaCl gradient

IC (pH 3)= SP-Sephadex, isopropanol/ acetic acid 1.5 M 40 : 60 (v/v), NaCl gradient

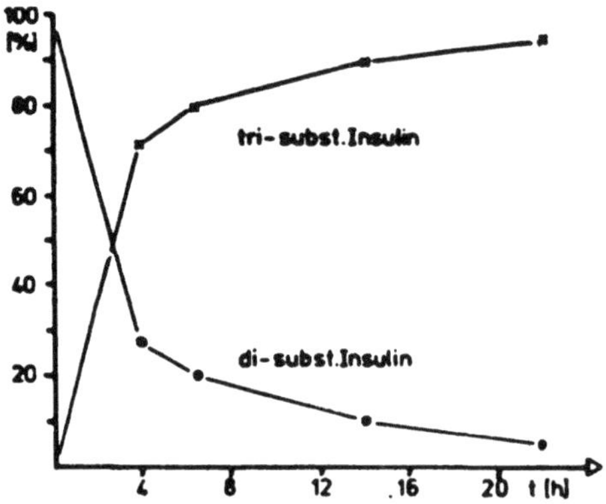

Figure 4. Time course of the acylation of B1,B29-$(Msc)_2$-des-Gly^{A1}-insulin (4 µmole/ml) with Msc-γ-Abu-ONp (3 eq.) in DMSO containing 20 eq. N-methylmorpholine at 22°C. Aliquots were precipitated with ether and subjected to cellulose acetate electrophoresis (pH 2). Strips were stained with Ponceau-S and evaluated by densitometry (Friesen et al., in ref. 1).

Syntheses

The homogeneous intermediate IV obtained by selective Edman degradation of insulin was acylated with 3 groups of p-nitrophenyl esters (see Table 1): a) of carboxylic acids[11], b) of O-protected glycolic acid[12], and c) of N^{α}-Msc-amino acids. The latter had been prepared according to Tesser and Balvert-Geers[13]. The optimal conditions for each coupling were previously determined by analytical time-course studies (Fig.4).

Preparative acylations of 70-80 mg IV were effected with 3-9 equivalents of nitrophenyl esters in DMSO in the presence of 15-20 equivalents of N-methylmorpholine (Msc derivatives) or imidazole, and carried through to 60-100% completeness. Reaction times varied between 3 and 24 h at room temperature and took a maximum of 48 h in the case of Msc-γ-Abu-ONp (98% acylation). Reactions were stopped by acidification with acetic acid, followed by dialysis and gel filtration (Sephadex G-50, 10% acetic acid). Unreacted Msc-des-Gly-insulin was,

1. Msc-D-Pro-ONp + $(Msc)_2$-des-Gly^{A1}-Insulin (quant.)
   9 eq., 3h, 23°C
2. Deprotection (NaOH/methanol/dioxane)
3. DEAE-Sephadex (pH 8.1, 7 M urea, 5°C)

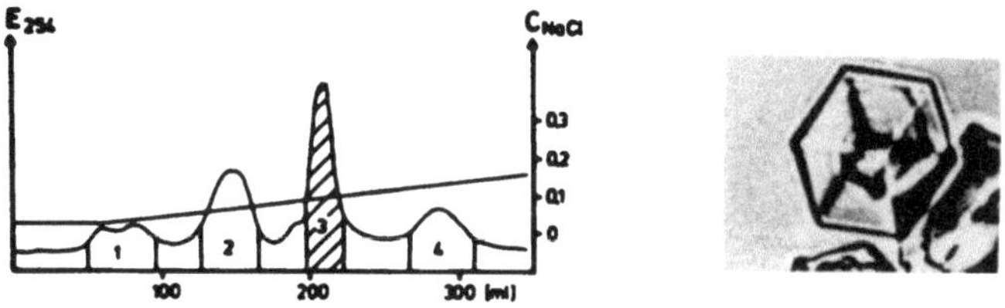

Yield : 19 %
on Insulin: 1.4 %

Figure 5. Preparation of [D-Pro^{A1}]insulin, and rhombohedral 2-zinc crystals of the analogue.

whenever necessary, removed by chromatography on SP-Sephadex at pH 3[10]. Msc groups were removed with NaOH in dioxane/methanol[14,10]. Subsequent gel filtration[14,10] gave homogeneous [t-But-OAc] insulin. The analogues with β-Ala, γ-Abu and D-Pro were subjected to a final purification by DEAE-cellulose chromatography at pH 8 (see Fig. 5), while the acetyl- amd propyl-insulins were chromatographed on SP-Sephadex at pH 3. The removal of the t-butyl group with trifluoroacetic acid gave rise to the formation of small amounts of desamido insulin which was removed by DEAE-cellulose chromatography.

All compounds were characterized by cellulose acetate electrophoresis at pH 2 and 8.6, end group analysis with dansyl chloride, amino acid analyses, UV spectra, and oxidative sulphitolysis followed by paper electrophoresis of the resulting chains. The analogues and yields are listed in Table 1.

Table 1. Insulin analogues with replacements in position A1.

Analogue	Yield		
	mg	a	b
$[Ac^{A1}]$[1,6]	21.5	33 %	2.5 %
$[Prop^{A1}]$	23	33	2.5
$[Bu^{t}\text{-}OAc^{A1}]$	45	60	4.6
$[HO\text{-}Ac^{A1}]$	11	37	1.7
$[\beta\text{-}Ala^{A1}]$	11	13	1
$[\gamma\text{-}Abu^{A1}]$	14	18.7	1.4
$[D\text{-}Pro^{A1}]$	15	18.7	1.4

a based on B1,B29-$(Msc)_2$-des-Gly^{A1}-insulin, b based on insulin

Discussion and conclusions

The semisynthesis of the analogues proceeded without major difficulties. They appear to be homogeneous within the limits of the analytical procedures which were applied. If O-acylation at tyrosines had occurred, it was reversed during the removal of the Msc-groups[10]. Self-crosslinking[15], at least between A- and B-chains, could not be detected by sulphitolysis and electrophoresis of the purified analogues. The absolute yields of ca 10 - 20 mg allowed for a variety of biological and physico-chemical investigations, the results of which will be reported elsewhere. The full biopotency of [D-Pro]insulin can be regarded as additional indication of homogeneity, since, both, des-Gly-insulin and the L-Pro analogue are only 1-2% active[1].

Saunders and Offord[9] reported yields of ca. 35% for resynthesized insulin and the [D-Lys^{A1}]analogue. No purification step was included in the sequence of reactions, and the homogeneity of the final products was about 85%. In the procedure developed by Geiger (see ref. 8) which led to several analogues in yields of about 7%, the intermediate IV is obtained by preferential, but not exclusive reaction of A1-glycine with phenylisothiocyanate. ε-Ptc contaminants are not removed before the end of the semisynthesis.
The route described in this paper aimed at the preparation of a homogeneous intermediate (IV) from which all analogues could be prepared. The absence of small amounts of impurities was crucial, since the biological potency of analogues was to be expected in a wide range (ca 1 to 100%).

It could be realized by purification at different stages during the semisynthesis, but at the expense of yields, which were 1 - 2.5%, based on insulin. The purification of intermediates is a safe-guard against the accumulation of small amounts of side products.

Acknowledgements

We are grateful to the Deutsche Forschungsgemeinschaft for support (Za 5/30) and wish to thank Dr. J. Föhles for amino acid analyses and Mr. K. Freude for excellent experimental help.

References

1. Brandenburg, D.: Relations Between the Structure and Biological Activity of Insulin (Review). Medicinal Chemistry Advances, ed.F.G. de las Heras and S. Vega, Pergamon Press, Oxford and New York, 487-502, 1981
2. Trindler, P.: Thesis, Technische Hochschule Aachen, 1980
3. Ryle, A.P., Sanger, F., Smith, L.F. and Kitai, R.; Biochem. J. 60, 541 (1955).
4. Märki, F., De Gasparo, M., Eisler, K., Kamber, B., Riniker, B., Rittel, W. and Sieber, F.; Hoppe-Seyler's Z. Physiol. Chem. 360, 1619 (1979).
5. Berndt, H., Gattner, H.-G. and Zahn, H.; in ref. 1.
6. Katsoyannis, P.G. and Zalut, C.; Biochemistry, 11, 3065, (1972).
7. Cosmatos, A., Cheng, K., Okada, Y. and Katsoyannis, P.G.: J. Biol. Chem., 253, 6586 (1978).
8. Geiger, R., Geisen, K., Regitz, G., Summ, H.-D. and Langner, D.: Hoppe-Seyler's Z. Physiol. Chem., 361, 563 (1980).
9. Saunders, D.J. and Offord, R.; Hoppe-Seyler's Z. Physiol. Chem., 358, 1469 (1977).
10. Weimann, H.-J.: Thesis, Technische Hochschule Aachen, 1977
11. Zahn, H. and Schade, F.: Chem. Ber. 96, 1747 (1963).
12. Wasielewski, C. :C.A. 67, 44074[9] (1967).
13. Tesser, G.I. and Balvert-Geers, I.C.: Int. J. Peptide Protein Res., 7, 295 (1975).
14. Geiger, R., Obermeier, R. and Tesser, G.I.: Chem. Ber. 108, 2758 (1975).
15. Brandenburg, D.: Hoppe-Seyler's Z. Physiol. Chem., 352, 8 (1971).

ENZYMATIC-CHEMICAL TRANSFORMATION OF PORCINE INSULIN INTO HUMAN INSULIN

E.N.Voluyskaja, S.P. Krasnoschokova, V.V.Knjazeva, M.N.Rjabtsev, S.M.Funtova, T.I.Zujanova, A.I.Ivanova, V.P.Fedotov, Yu. P.Shvachkin

Institute of Experimental Endocrinology and Hormone Chemistry, Academy of Medical Sciences of the USSR, Moscow

Introduction

In connection with investigation of the modes of the transformation of animal insulins into human insulin(I) the method of enzymatically-chemical conversion of porcine insulin(II) into human insulin(I) has been elaborated with the hybride B-chain(III) being used as an intermediate compund.

Results

The process of transformation of porcine insulin(II) into human insulin(I) involves six stages.

The first stage consists in oxidative sulfolysis of porcine insulin(II) and the separation of resulting tetra-S-sulfonate A-chain(IV) and bis-S-sulfonate B-chain(V). The second step is the treatment of the compound(V) with trypsin followed by isolation of bis-S-sulfonate des(octapeptide$^{23-30}$)-B-chain porcine insulin(VI). The third stage is the interaction of the compound(VI) with t-butyloxycarbonyl azide with the formation of bis-S-sulfonate Boc-des(octapeptide$^{23-30}$)-B-chain porcine insulin(VII). The fourth step includes condensation of substance (VIII) with the synthetic octapeptide(VIII) corresponding to the fragment B^{23-30} human insulin. As a result

Chemistry of Peptides and Proteins, Vol. 1

of this reaction the bis-S-sulfonate of the partially protected B-chain of human insulin(IX) is formed. The fifth stage consists in the treatment of compound(IX) with trifluoroacetic acid producing the removal of all protecting groups and isolation of the hybride B-chain of bis-sulfonate(III) which is structurally identical to the human insulin B-chain. The sixth step is the combination of the porcine insulin A-chain (IV) with the hybride B-chain(III).

The first operation was performed using sodium sulfite and sodium tetrathionate. The next stage was accomplished by the use of trypsin which was earlier treated with tosylaminophenylethylchloromethyl ketone (TPCK). The reaction of enzymatic hydrolysis of the porcine insulin B-chain bis-S-sulfonate(V) was carried out at pH 8.6 and 30°C. The isolation and purification of the bis-S-sulfonate of the porcine insulin des(octapeptide$^{23-30}$)-B-chain(VI) was accomplished by means of gel filtration on a Sephadex G-25 column, equilibrated by a 0.05 M solution of ammonium bicarbonate. The third stage was realized using fresh prepared tert-butyloxycarbonyl azide; the reaction of the latter with compound(VI) being performed in the presence of triethylamine in the mixture of dimethylformamide and dimethylsulfoxide (1:1). The partially protected octapeptide(VIII) was synthesized in accordance with the scheme which involves the use of acid-labile protecting groups only and combination of the solid phase and liquid phase methods for the formation of peptide bonds.

$$\overset{23}{\text{R-Gly}}\text{-}\overset{24}{\text{Phe}}\text{-}\overset{25}{\text{Phe}}\text{-}\overset{26}{\text{Tyr}}(\text{Bu}^t)\text{-}\overset{27}{\text{Thr}}\text{-}\overset{28}{\text{Pro}}\text{-}\overset{29}{\text{Lys}}(\text{Boc})\text{-}\overset{30}{\text{Thr}}(\text{Bu}^t)\text{-OBu}^t$$

VIII. R = H;   VIIIa. R = Z

The solid phase method was used for preparing the fragment B^{23-25}, and the fragments B^{26-37} and B^{28-30} were synthesized in solution. The condensation of the fragments was carried out according to the scheme (3+2)+3). Boc-L-Phe-OH, Z-Gly-OH, Z-L-Tyr(Bu^t)-OH, H-L-Thr-OMe, Z-L-Pro-OH, Z-L-Lys(Boc)-OH and H-L-Thr(Bu^t)-OBu^t served as starting materials. For the solid phase synthesis of the fragment B^{23-25} a chloromethylated copolymer of styrene with 2 % divinylbenzene was used as support. The synthetic octapeptide was initially produced in the form of (VIIIa), and then was transformed into compound (VIII) by catalytic hydrogenolysis. The interaction of the synthetic octapeptide(VIII) with compound(VII) was performed under conditions which provide the formation of bis-S-sulfonate of the partially protected B-chain of human insulin (IX). Compound(IX) was then treated with trifluoroacetic acid in the presence of anisole. After purification by means of ion-exchange chromatography on SP-Sephadex C-25 and gel filtration on Sephadex G-25 the bis-S-sulfonate of the hybride B-chain (III) was isolated in analytically pure form. Enzymatic hydrolysis of the compound proved the bis-S-sulfonate structure(III). The final sixth stage, consisting in the combination of the reduced A-chain of porcine insulin(IV) with the hybride B-chain(III), was accomplished at pH 10.6. After purification by means of ion-exchange chromatography on CM-Sephadex C-25, precipitation of the complex with picric acid and transformation of the picrate into the hydrochloride the crystalline human insulin(I) was obtained in analytically pure form. As judged by the mouse convulsion assay method the biological activity of the insulin obtained is 100 % (in comparison with the activity of international standard).

TRYPSIN CATALYZED PEPTIDE SYNTHESIS: MODIFICATION OF THE B-CHAIN C-TERMINAL REGION OF INSULIN

Hans-Gregor Gattner, Waleed Danho, Reinhard Knorr, Helmut Zahn
Deutsches Wollforschungsinstitut
D-5100 Aachen, FRG

Introduction

Proteases are well-known as enzymes which can split peptide bonds in proteins or polypeptides. The reverse reaction, i.e. peptide or amide bond formation, can also be catalyzed by the same enzymes under special conditions. In semisynthesis the enzyme mediated peptide synthesis is - if possible - the method of choice because blocking and later unblocking of other functional groups can be greatly reduced or eliminated. A further advantage of enzymatic peptide synthesis is that, as a result of the stereospecificity of the enzyme, racemization is avoided. Figure 1 shows the equilibria involved in enzymatic peptide synthesis for the example of trypsin.

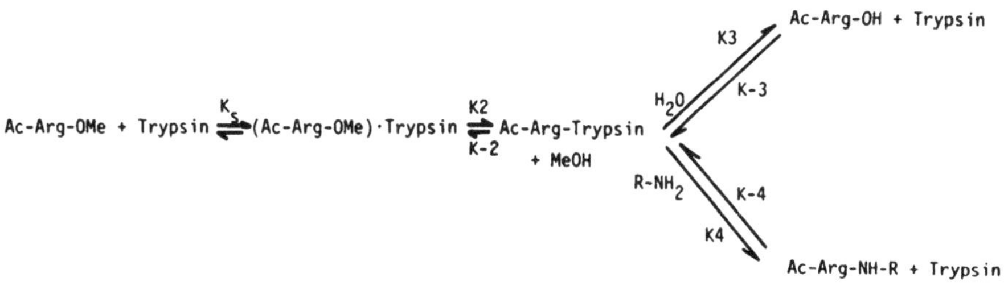

Figure 1: Scheme for the enzyme catalyzed peptide synthesis with trypsin.

Chemistry of Peptides and Proteins, Vol. 1

The scheme illustrates that in the presence of a nucleophile ($R-NH_2$) other than water, the acylenzyme intermediate (Ac-Arg-Trypsin) will be subject to aminolysis in addition to hydrolysis. The ratio of aminolysis to hydrolysis depends on K_4/K_3 as well as on the concentration of the nucleophile $R-NH_2$ and on the concentration of organic solvent which diminishes not only the concentration of water but also improves the yield of coupling (1).

Biological activity and aggregation of insulin are known to be closely related to the C-terminal region of insulin. The B-chain sequence B23-26, Gly-Phe-Phe-Tyr, has been remained invariant during evolution and therefore the residue has been assumed to be part of the receptor binding region of the hormone (2).

Chemical modification of specific amino acid residues in this region should give more information about the role of these amino acids regarding the structure-activity, receptor binding and aggregation behaviour of the insulin molecule. Insulin has two peptide bonds susceptible to tryptic digestion and therefore, under appropriate conditions, available for enzyme mediated synthesis. Trypsin catalyzed couplings with the carboxyl components Boc_2-des-octapeptide(B23-30)-insulin and des-Ala(B30)insulin were performed with synthetic octapeptides modified in the range B24-B26 and amino acid esters respectively in 60% organic solvent at pH 6.5 (3).

Experimental and results

Des-Ala(B30)insulin (DAI) (porcine) was prepared by digestion of insulin with carboxypeptidase-A in 0,2 M NH_4HCO_3/NH_4OH, pH 8.4, at 25°C for 4 h (4).

Des-octapeptide(B23-30)insulin (DOI) was obtained by trypsinization of insulin at pH 9.5 (5). To prevent self condensation during the trypsin-catalyzed synthesis with the synthetic octapeptides the α-amino groups have to be blocked. DOI was reacted with a four-fold excess of di-tert.-butyl-dicarbonate ($(Boc)_2O$) in DMF/N-ethylmorpholine for 16 h. The peptides were synthesized by the conventional solution method using the strategy of fragment condensation and acid labile protecting groups. The ε-amino group of Lys(B29) has to be protected because in the free state it would be an additional substrate for trypsin. To shift the equilibrium more to synthesis the synthetic peptides and amino acid derivatives were used in excess (10-100-fold, molar). A typical coupling experiment on a preparative scale and purification procedure in shown in scheme 1 in more detail (6). No coupled amino and carboxyl components are separated and can be reused.

Boc_2-DOI 20 µmol

↓

H-Gly-<u>Leu</u>-Phe-Tyr-Thr-Pro-Lys(Boc)-Thr-OH 200 µmol

↓

dissolved in 0.05 M NH_4HCO_3/NH_4OH, lyophilized

↓

dissolved in 1.05 ml DMF + 0.70 ml 0.5 M Tris/HCL, pH 6.5

↓

added trypsin (10 mg), incubated at 37°C for 20 h

↓

acidified to pH 2-3, chromatographed on LH-20 in DMF/0.5 M acetic acid (1/1)

↓

deprotected with trifluoroacetic acid

↓

chromatographed on CM-cellulose at pH 4.5

↓

human insulin analogue (25% yield)

<u>Scheme 1:</u> Coupling procedure for insulin analogues.

Table 1 shows a series of peptides coupled with Boc_2DOI by trypsin catalysis. Amino acids changed against the natural sequence are underlined.

<u>Table 1:</u> Peptides coupled with Boc_2-DOI by trypsin catalyzed peptide bond formation.

Peptide	yield %
H-Gly-Phe-<u>Tyr</u>-Tyr-Thr-Pro-Lys(Boc)-Ala-OH	12
H-Gly-Phe-<u>Ala</u>-Tyr-Thr-Pro-Lys(Boc)-Ala-OH	15
H-Gly-<u>Leu</u>-Phe-Tyr-Thr-Pro-Lys(Boc)-Thr-OH	25
H-Gly-Phe-<u>Leu</u>-Tyr-Thr-Pro-Lys(Boc)-Thr-OH	10
H-Gly-Phe-Phe-Tyr-OMe	15
H-Gly-Phe-Phe-<u>Phe</u>-OMe	45

For coupling of des-Ala(B30)insulin with amino acid esters 100 mg DAI and a 70-fold molar excess of the ester were dissolved in 1.71 ml 60% DMF/pH 6.5 buffer and 10 mg trypsin were added. After 20 h the reaction was stopped by lowering the pH to 2-3. Trypsin and excess of amino acid ester were removed by gel filtration on Sephadex G-50 in 1 M acetic acid. The coupled analogue was separated from DAI by ion exchange chromatography on CM-cellulose at pH 4.5. The yield of trypsin catalyzed coupling was found to be dependent on the amino acid and the kind of the ester. In table 2 some results are summarized.

Table 2: Coupling yields of DAI with different amine components.

Amine component	yield %	Amine component	yield %
Thr-OMe	85	Glu-$(OMe)_2$	90
Thr-NH_2	80	Ala-OMe	65
Tyr-OMe	70	Ala-OBu^t	40
Trp-OMe	70	Ala-OBzl	45
p-NO_2-Phe-OMe	40	Ala-NH_2	50
3,(NH_2)-Tyr-OMe	35		

Studying the progress of coupling we found for DAI and Thr-OMe a maximum yield of 85% after 30 min were reached which decreased to 70% after 22 h. To our surprise there was no splitting of the Arg B22-Gly B23 peptide bond.
The effect of amine component to carboxyl component ratio on the yield is shown in figure 2. By the use of excess amine component the equilibrium of the system moved towards synthesis.

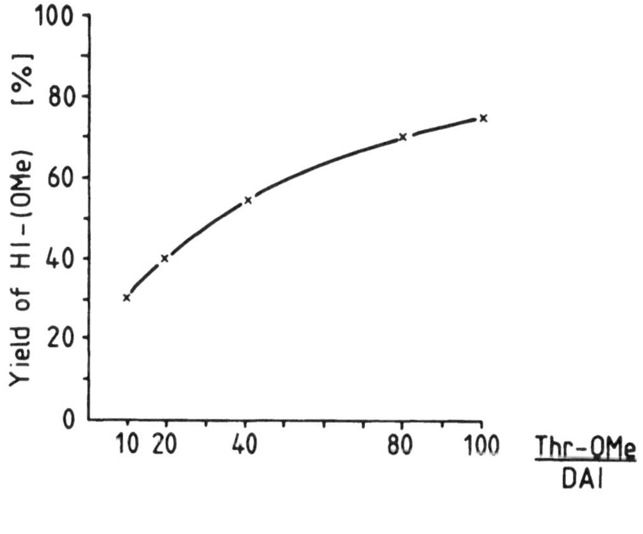

Fig. 2: Effect of Thr-OMe-concentration on the yield of Thr-OMe (B 30) insulin.

The equilibrium also depends on the concentration of water and the kind of organic solvent. Table 3 indicates that the addition of an organic solvent, such as DMF or DMSO, to the reaction medium greatly increases the synthesis of the peptide bond. The effect is much larger than that expected from the decrease in water concentration (1).

Table 3: Effect of organic solvent (60%) on the yield of trypsin catlayzed coupling: Boc_2DOI + Gly-OMe.

Solvent	% Boc_2DOI-Gly-OMe
Dioxane	20
i-Propanol	30
Glycerol	40
Dimethylformamide	55
Butanediol-1.4	60
Ethyleneglycol	60
Dimethylsulfoxide	75
4 M urea	25

The addition of organic solvents was also effective in increasing the solubility of reactants which governs the extend of peptide bond formation.

Conclusion

Following the procedure of Inouye et al. (3) for the enzyme assisted coupling of des(B23-30)insulin to human octapeptide (B23-30), we were able to synthesize new insulin analogues. This procedure opens a new perspective for the semisynthesis of analogues modified in the C-terminal region of the B-chain. It seems that the coupling yields are sequence dependent.

References

1. Homandberg, G.A., Mattis, J.A., Laskowski, Jr., M.: Biochemistry 17, 5220-5227 (1978).

2. Pullen, R.A., Lindsay, D.G., Wood, S.P., Tickle, I.J., Blundell, T.L., Wollmer, A., Krail, G., Brandenburg, D., Zahn, H., Gliemann, J., Gammeltoft, G.: Nature (London) 259, 369-373 (1976).

3. Inouye, K., Watanabe, K., Morihara, K., Tochino, Y., Kanaya, T., Emura, J., Sakakibara, S.: J. Am. Chem. Soc. 101, 751-752 (1979).

4. Schmitt, E.W., Gattner, H.-G.: Hoppe-Seyler's Z. Physiol. Chem. 359, 799-802 (1978).

5. Young, J.D., Carpenter, F.H.: J. Biol. Chem. 236, 743-748 (1961).

6. Gattner, G.-G., Danho, W., Behn, Chr., Zahn, H.: Hoppe-Seyler's Z. Physiol. Chem. 361, 1135-1138 (1980).

SEMISYNTHESIS WITH INSULIN-B-CHAIN

Erika E. Büllesbach, Ernst W. Schmitt, Hans-Gregor Gattner, Vinod K. Naithani and Josef Föhles

Deutsches Wollforschungsinstitut
D-5100 Aachen, Federal Republik of Germany

Introduction

In earlier work we had shown that mixed anhydrides of carbonic- and carboxylic acids are useful reagents for semisynthesis of completely protected carboxyl-components and native amino-components in aqueous-organic solvents[1-4]. Using this method, combined with enzymatic methods, we decided to prepare human proinsulin and proinsulin models with shortened C-peptides for reduction and reoxidation studies.

For the semisynthesis of human proinsulin (Fig.1) we used porcine A-chain (which is identical to human A-chain) as bis-disulphide[5]. For proinsulin models we used bovine bis-disulphide-A-chain. Initially Boc-Arg-OH was coupled by a trypsin-catalysed reaction[1,5]. The arginyl-A-chain was elongated by a mixed anhydride reaction according to Naithani et al.[1] with protected fragments of {Thr^{B30}}-human-C-peptide or modified {Thr^{B30}}-C-peptides.

Finally we attempted to couple the above products with the native insulin-B-chain sequence B_{1-29} (which is identical with the human proinsulin sequence 1-29). The amino component also contained free carboxyl groups, so the methods for activation were limited. In a first attempt, the N-protected B-chain was activated with isobutyl chloroformate and coupled with native insulin-A-chain. 5% product with an A : B-chain ratio of 1 : 1 was isolated. 70% of this products was a linear peptide and

Chemistry of Peptides and Proteins, Vol. 1

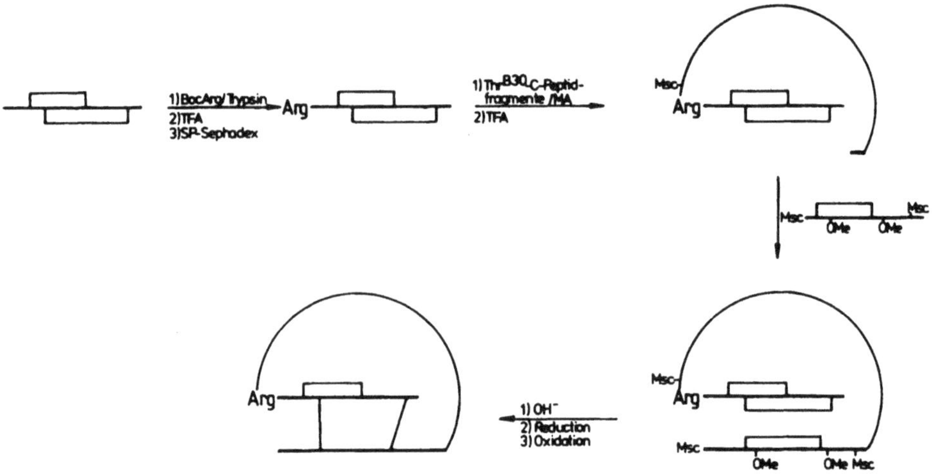

Figure 1. Pathway of the semisynthesis of human proinsulin or proinsulin models.

only 30% had reacted with the side chains of the two glutamic acids (B13 and B21) under chain branching[1].

Results and discussion

<u>Starting material</u>. For selective reaction at the C-terminal carboxyl group of the B-chain it is necessary to protect the glutamic acid side chains. One preparation of the partially protected B-chain is outlined in Fig. 2. Di(-sulfonate)-B-chain was transformed to sequence B_{1-29} by enzymatic digestion with carboxypeptidase A in 0.05M tris-buffer at pH 9.5 at 25°C for 6h (enzyme : substrate = 1 : 100). After reduction of the S-sulphonated groups with 2-mercaptoethanol the dithiol-B-chain was separated by desalting on Sephdex G-25 equilibrated in 30%

acetic acid[6] and - without isolation - oxidized with iodine in 50% acetic acid (concentration : 1.5 to 2 mg/ml) to the disulphide-B-chain. The monomeric des-Ala^{B30}-disulphide-B-chain was isolated by gel filtration on Sephadex G-50f in 10% acetic acid with a yield of 80% from the di(-S-sulphonate)-B-chain. The next step was the selective protection of the γ-carboxyl groups. BF_3-catalyzed esterfication in methanol yielded the tri-methyl ester (Glu$^{\gamma 13}$, Glu$^{\gamma 21}$, and Lys$^{\alpha 29}$). Under these conditions nearly 10% of N$\rightarrow$ O-acyl shiftwere observed. Saponification of the C-terminal methyl ester was specifically catalyzed by trypsin at pH 5.7. Both amino groups (N$^{\alpha 1}$Phe, and N$^{\varepsilon 29}$Lys) were acylated with methylsulphonylethyloxycarbonylsuccinimide. The N$^{\alpha 1}$,N$^{\varepsilon 29}$-bis(methylsulphonylethyloxycarbonyl)des-Ala^{B30}-B-chain-disulphide-O$^{\gamma 13}$O$^{\gamma 21}$-di-methyl ester (Msc$_2$DAB(SS)(OMe)$_2$) was purified via ion-exchange chromatography on SP Sephadex at pH 3 followed by CM cellulose at pH 4.8. The resulting B-chain derivative was transformed to the HCl salt by gel filtration on Sephadex G-25f in 10^{-4}M HCl. It had only alkaline labile protecting groups and a free α-carboxyl function.

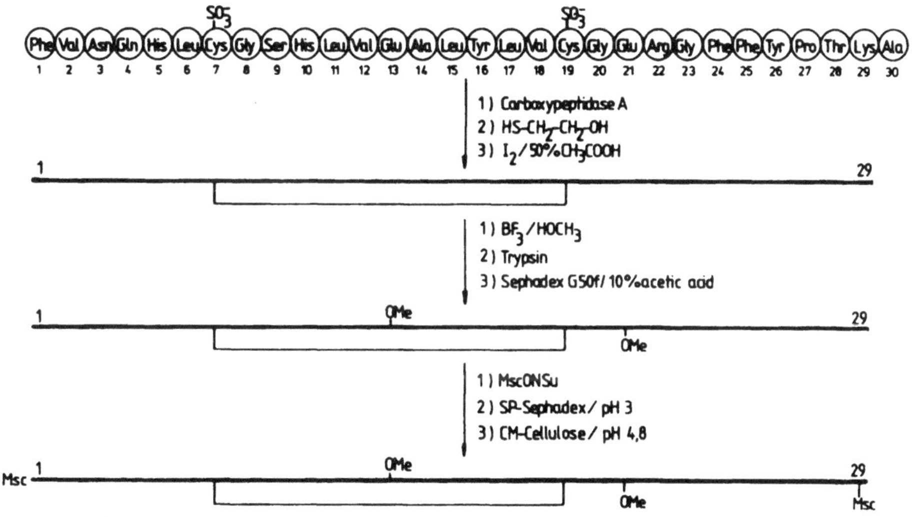

Figure 2. Preparation of Msc$_2$DAB(SS)(OMe)$_2$.

Activation and Coupling. $Msc_2DAB(SS)(OMe)_2$ was activated by the mixed anhydride method. To investigate the optimal conditions the reaction of the activated $Msc_2DAB(SS)(OMe)_2$ with a 20-fold excess of arginine was chosen. The content of arginine was determined by amino acid analysis after total hydrolysis. Standardized procedures were developed and used to attach insulin A-chain and N-terminal elongated insulin A-chain derivative to $Msc_2DAB(SS)(OMe)_2$.

Usually a solution of the B-chain derivative in DMF was converted to the N-methylmorpholinium salt. The mixed anhydride was formed with isobutyl chloroformate at -10°C. Acylations were performed in DMF/water at "pH 7.2-7.4" at room temperature. The reaction was terminated by addition of 1 M hydroxylamine, pH 7.5 in DMF/water. The crude product was passed over Sephadex G-25f in 10% acetic acid and lyophilized.

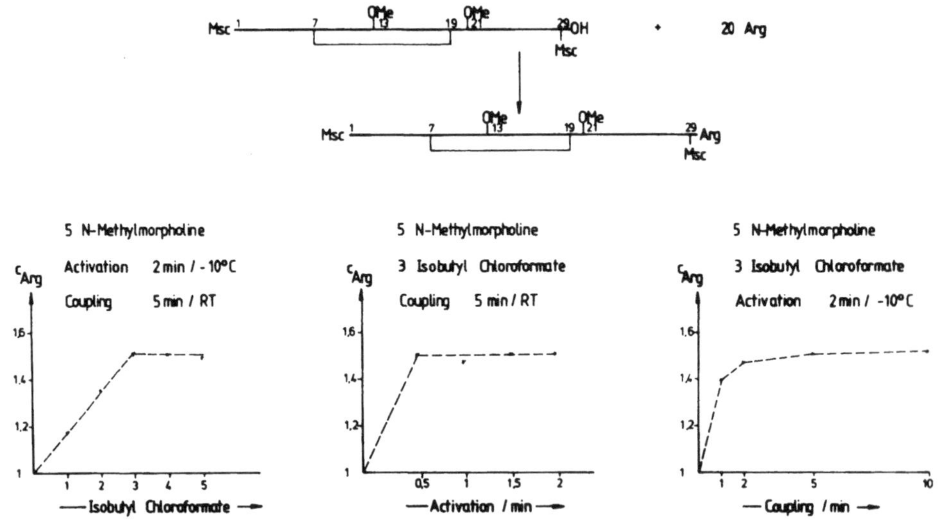

Figure 3. Standardization of the mixed anhydride coupling

Maximum yields of {Arg^{B30}}-B-chain derivative (51%) were achieved with the following activation conditions: 5 equivalents base, 3 equivalents isobutyl chloroformate, time for activation:30 sec and for coupling:2 min. For deprotonation of both histidine side chains 2 equivalents N-methylmorpholine were necessary (Fig. 3).

In no case were products with higher molecular weights found by gel chromatography on Sephadex G-50f. Inter-chain esterfication of the activated carboxyl group with the hydroxyl functions of the B-chain derivative was not observed. In the electropheresis at pH 8.6 the coupling mixtures showed the expected spots, but at pH 2.2 and 4.8 side products were observed with a lower mobility than the B-chain. These side reactions were almost completely reversed by treatment with hydroxylamine for 12 hours and by the alkaline deprotection conditions. It is known that the imidazole ring of histidine reacts with isobutyl chloroformate[7]. The N^{im}-isobutyloxycarbonyl group is removed in alkaline media[7].

Side reactions with the amino component were not detectable in the reaction of $Msc_2DAB(SS)(OMe)_2$ with a large excess of arginine. In an other experiment we coupled the B-chain derivative with native insulin A-chain and after the reaction we observed a large amount of N-acylated A-chain. Probably this side reaction was due to reaction with excess of isobutyl chloroformate.

Obviously milder conditions should be used for activating the B-chain derivative before coupling with equimolar or lower amounts of amino component. After activation with 4 equivalents base and 2 equivalents isobutyl chloroformate, subsequent reaction with 0.2 equivalents A-chain yielded 25.2% protected B_{1-29}-A_{1-21}, after isolation by ion-exchange chromatography on SP-Sephadex at pH 3. A similar reaction of Thr-Arg-A-chain gave 29% protected B_{1-29}-Thr-Arg-A_{1-21} after purification on CM-Cellulose at pH 4.8, and gel-filtration on Sephadex G-50f (Fig.4). These products were characterized by amino acid analysis, elec-

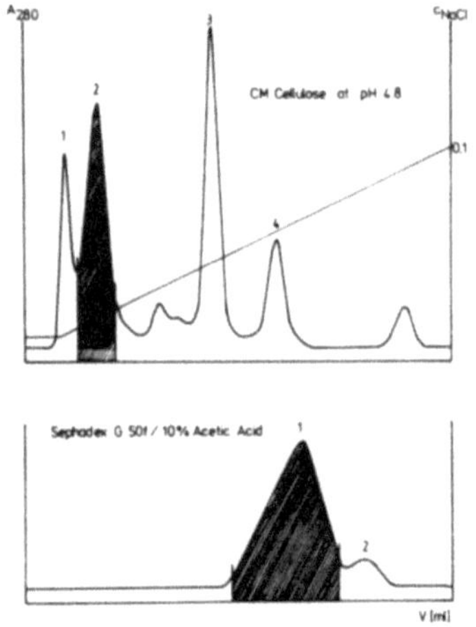

Figure 4. Purification of protected B_{1-29}-Thr-Arg-A_{1-21}

<u>above:</u> Ion-exchange chromatography on CM-Cellulose at pH 4.8. 1= N-blocked A-chain derivative, 2= protected B_{1-29}-Thr-Arg-A_{1-21} and Thr-Arg-A-chain, 3= $Msc_2DAB(SS)(OMe)_2$, 4= C-terminal blocked B-chain derivative;

<u>below:</u> Gel-filtration on Sephadex G-50f in 10% acetic acid (fraction 2). 1= protected B_{1-29}-Thr-Arg-A_{1-21}, 2= Thr-Arg-A-chain.

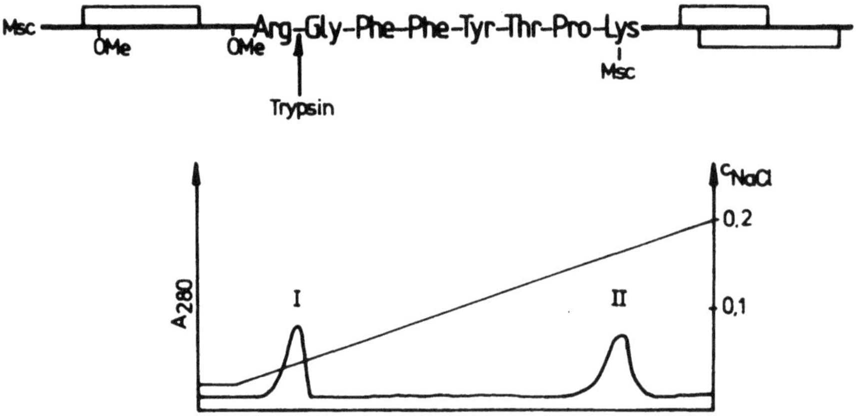

Figure 5. Tryptic digestion of protected B_{1-29}-A_{1-21} and separation of the fragments on SP-Sephadex in 10% acetic acid: I = B_{23-29}-A_{1-21}, II = B_{1-22}.

trophoresis, and sulphitolysis followed by electrophoresis. The protected derivative B_{1-29}-A_{1-21} was digested by trypsin. The expected two peptides, B_{1-22} and B_{23-29}-A_{1-21} (Fig.5), were separated on SP-Sephadex and characterized by electrophoresis and amino acid analysis (Table 1). These results proved the C-terminal elongation of the B-chain derivative.

Amino acid	B_{1-29}-A_{1-21}		B_{1-22}		B_{23-29}-A_{1-21}	
Asp	3.06	(3)	1.15	(1)	2.14	(2)
Thr	0.87	(1)	0	(0)	0.94	(1)
Ser	2.76	(3)	1.05	(1)	1.95	(2)
Glu	7.00	(7)	3.00	(3)	4.00	(4)
Pro	-	(1)	0.00	(0)	-	(1)
Gly	4.04	(4)	2.00	(2)	2.00	(2)
Ala	2.07	(2)	1.06	(1)	1.00	(1)
Val*	3.85	(5)	2.61	(3)	1.18	(2)
Cys_2	-	(3)	-	(1)	-	(2)
Ile*	0.21	(1)	0.00	(0)	0.18	(1)
Leu	5.77	(6)	3.97	(4)	1.95	(2)
Tyr	3.72	(4)	1.00	(1)	2.69	(3)
Phe*	2.66	(3)	0.62	(1)	1.97	(2)
Lys	1.02	(1)	0.00	(0)	0.98	(1)
His	1.91	(2)	1.82	(2)	0.00	(0)
Arg	0.99	(1)	0.95	(1)	0.00	(0)

Table 1. Amino acid composition of the bovine insulin derivative B_{1-29}-A_{1-21} and its tryptic fragments

* Low values due to incomplete hydrolysis under conditions used.

Conclusions

Partially protected $Msc_2DAB(SS)(OMe)_2$ with a free α-carboxyl group was prepared in six steps beginning with di(-S-sulphonate)-B-chain in an overall yield of 25%. It was activated to the mixed anhydride and elongated at the C-terminus with arginine, insulin A-chain and Thr-Arg-A-chain. The mixed anhydride method was also applicable to N-terminal elongation of native A-chain with completely protected peptides and C-terminal elongation of minimally protected B-chain. These studies show a way to prepare proinsulins with modifications in the C-peptide region.

Acknowledment

This work was supported by grants from the Deutsche Forschungsgemeinschaft. The skilful technical assistance of Mrs. G. Ortmann is gratefully appreciated.

References

1. Naithani, V. K., Gattner, H.-G., Büllesbach, E. E., Föhles, J., Zahn, H.: "Peptides Structure and Biological Function" Proc. 6th Amer. Peptide Symp.,(eds. E. Gross, J. Meienhofer), Pierce Chemical Co., Rockford, Ill., 571-576, 1979

2. Naithani, V. K., Büllesbach, E. E., Zahn, H.: Hoppe Seyler's Z.Physiol.Chem. 360, 1363-1366 (1979).

3. Büllesbach, E.E., Naithani, V.K.: Hoppe Seyler's Z.Physiol. Chem. 361, 723-734 (1980).

4. Naithani, V. K., Büllesbach, E. E., Zahn, H., Shield, J., Chance, R., Root, M. A.: "Insulin; Chemistry, Structure, and Function of Insulin and Related Hormones" Proc. 2nd Int. Insulin Symp., (eds. D. Brandenburg, A. Wollmer), Walter de Gruyter, Berlin 99-106, 1980

5. Gattner, H.-G., Danho, W., Naithani, V. K.:"Insulin; Chemistry, Structure, and Function of Insulin and Related Hormones" Proc. 2nd Int. Insulin Symp., (eds. D. Brandenburg, A. Wollmer), Walter de Gruyter, Berlin 117-123, 1980

6. Büllesbach, E. E., Danho, W., Helbig, H. J., Zahn, H.: Hoppe Seyler's Z. Physiol. Chem. 361, 865-873 (1980)

7. Gronvald, F. C., Johansen, N. L., Lundt, B. F.: "Peptides Structure ans Biological Function" Proc. 6th Amer. Peptide Symp., (eds. E. Gross, J. Meienhofer), Pierce Chemical Co., Rockford Ill., 571-576, 1979

INVESTIGATION OF THE INFLUENCE OF A1, B29 AMINO GROUP MODIFICATION ON THE STRUCTURE AND BIOLOGICAL ACTIVITY OF INSULIN

Gennadij Vlasov, Natalija Izvarina and Nina Illarionova
Institute of Macromolecular Compounds of the Academy of Sciences of the USSR, Leningrad, USSR

Introduction

Synthesis of A1,B29 azo-isobutyryl-isnulin, intramolecular crosslinked hormone, and its A1,B29 carbochain polymeric modifications was elaborated. Influence of the chemical modification on the insulin structure and biological activity was investigated.

Results

It is known that the chemical modification of insulin amino groups by various agents influences the structure and biological activity of the hormone (1,2). However, little information is available on the polymeric modification of insulin. The reason for this is the difficulty of the preparation of polymeric insulin conjugates with fixed sites of attachement of the polymer to insulin.

In order to study the influence of the modification of insulin amino groups by polymers on its structure and biological properties we developed a new method of the synthesis of covalent carbon chain polymeric conjugates with a fixed site of attachement of the polymer to hormone. In this work the preparation and properties of A1 (Glycine) and B29 (lysine) polymeric derivatives of insulin are described. The main idea of the method consists of the preparation of A1,B29 bis-

Chemistry of Peptides and Proteins, Vol. 1

imidoazo-isobutyryl-insulin, an intramolecularly crosslinked hormone, and the polymerization of various vinylic monomers with the crosslinked insulin as a polymeric initiator of polymerizytion. The A1,B29 crosslinked insulin was prepared by the reaction of insulin with bis-dimethylimidate of azo-bis-isobutyric acid, a derivative of a well-known radical initiator of polymerization of vinyl monomers (3). Ion chromatography and gel chromatography were used for the preparation of homogeneous insulin derivatives. In order to prove the intramolecular cross linkage of insulin its derivate was split by mercaptoethanol and trypsin. TLC, GC and EP showed that crosslinked insulin gives only one product after splitting of the disulfide bonds by thiol and the peptide bond B22-B23 by trypsin. In both cases native insulin yields two products: A- and B-chains (thiol) or des-octapeptide (B22-B30) insulin and the octapeptide.

ORD investigation showed that the cross linkage results in a slight destabilization of the secondary structure of the hormone (Table 1), pointing to the optimal length of the crosslinking agent that is equal to the A1, B29 distance of insulin (4).

Nevertheless the cross linkage of the hormone results in some stabilization of its secondary structure with respect to the destabilizing influence of urea and of the basic pH. The crosslinked insulin exhibits full biological activity. However, its immunoreactivity is only 44 % of that of the native hormone (Table 2).

Biologically active A1, B29 crosslinked insulin was used as a macromolecular initiator of the polymerization of N-vinylimidazole, N-vinylpyrrolidone, acrylamide and acrylic acid. The polymerization was carried out in water at 55 to 65°C. Polymerization reactions result in the formation of A1, B29 polymeric derivatives of insulin. Molecular masses of polymeric conjugates of insulin were determined by the GC (5) and the Lowry (6) method.

Table 1. ORD data of insulin derivatives.

N	Insulin derivate	pH 1.5		pH 7.5		pH 11.0	
		c	% α-helicity	c	% α-helicity	c	% α-helicity
1	Insulin	272	42	270	41	250	27
2	Al,B29 cross-linked insulin	265	38	259	34	246	24
3	PVI-insulin (1500)*	257	32	-	-	250	27
4	PVI-insulin (5000)	257	32	-	-	258	33
5	PVI-insulin (14000)	257	32	-	-	257	32
6	PVP-insulin (3000)	257	32	-	-	250	27
7	PVP-insulin (6000)	257	32	-	-	258	33
8	PVP-insulin (14000)	257	32	-	-	257	32
9	PAA-insulin (4000)	254	30	-	-	242	22

* - molecular mass of modifying polymer, PVI : polyvinylimidazole, PVP:polyvinylpyrrolidone, PAA: polyacrylic acid

ORD studies showed additional destabilization of the structure of insulin up to 32 % of α-helicity (Table 1). However, the investigation of the dependence of specific rotation on pH showed that OVI and PVI stabilize the hormone in the range of basic pH and stabilization increases at higher molecular masses of modifying polymers. In contrast, polyacrylic acid destabilizes the structure of insulin, both, in the acid and in the basic pH range.

The investigation of biological activity showed that PVI conjugates are generally more active. However, the chemical structure of the modifying polymer influences the biological acti-

Table 2. Biological properties of insulin derivates.

N	Insulin derivate	Mol.mass of polymeric component	Biol.activity, % glucose in rabbits	Immunoreactivity %
1	Native insulin	-	100	100
2	Al,B29-cross-linked insulin	-	100	44
3	PVI-insulin	1500	150	1.5
4	PVI-insulin	5000	56	-
5	PVI-insulin	14000	45	0.5
6	PVI-insulin	34000	25	-
7	PVP-insulin	3000	70	2.0
8	PVP-insulin	6000	50	-
9	PVP-insulin	14000	0	0
10	PA-insulin	2500	56	3.5
11	PA-insulin	7000	49	-
12	PA*-insulin	12000	20	0
13	PAA-insulin	4000	80	2.5
14	PAA-insulin	6000	59	-
15	PAA-insulin	10000	48	0.2

* polyacrylamide

vity of the conjugate up to 4000 - 5000 molecular masses of the modifying polymer. At higher molecular masses only the molecular mass and the position of the modified amino acid of the hormone influence its activity.

The immunoreactivity of polymeric derivatives falls more sharply than the biological activity (Table 2).

It is difficult to determine exactly the mechanism of the effect of the polymeric component on the activity of modified hormone. The polymer can influence the stage of the formation of the productive hormone-receptor complex by shielding the active site of the hormone. It can also affect the stage of signal generation by "freezing" the hormone in an unfavourable, conformation, thus preventing the generation of the signal.

References

1. Brandenburg, D., Gattner , H., Wollmer,A.: Z.Physiol.Chem. 353, 599-617 (1972).
2. Vlasov, G., Glushencova, V., Bolotina, I., Glikina, M., Maltzev, V.: Z.Allgemein.Chem.42, 1128-1132 (1972) (Russ.).
3. Vlasov, G., Izvarina, N., Illarionova, N.: Biochemistry 46, 942-951 (1981 (Russ.).
4. Blundell, T.L., Dodson, G.G., Hodkin, D.C., Mercola, D.A.: Adv.Prot.Chem.26, 279-402 (1972).
5. Granath, K.A., Kvist, B.E.: J.Chromatog.28, 69-81 (1967).
6. Lowry, O., Rosenbrough, N., Tarr, A., Randall, R.: J.Biol. Chem.350, 717-721 (1951).

BACTERIAL CELL WALL GLYCOPEPTIDES: SYNTHESIS, CONFORMATION AND ANTITUMOR ACTIVITY

Tatyana Andronova, Larisa Rostovtseva, Irina Sorokina, Veronika Mal'kova, Vadim Ivanov

Shemyakin Institute of Bioorganic Chemistry, USSR Academy of Sciences, Moscow, USSR

Introduction

In recent years natural synthetic glycopeptides modelling the microbial cell wall segments have been shown to have interesting immunomodulating and/or antitumor activity. The immune response can be enhanced by a small molecule, N-acetylmuramyl-alanyl-D-isoglutamine, which is a part of the monomeric unit of the bacterial peptidoglycan (1, 2). Several hundred analogues and derivatives of MDP have been prepared to date (3). In 1975 glycopeptides isolated from partial hydrolysate of the *Lactobacillus bulgaricus* cell wall (m.m. 1000-10000) (a preparation now called "Blastolysin") were found to induce a specific antitumor effect (necrosis)of some well developed tumors, for example sarcoma 180. A chemical formula for one of these glycopeptides has been proposed (4):

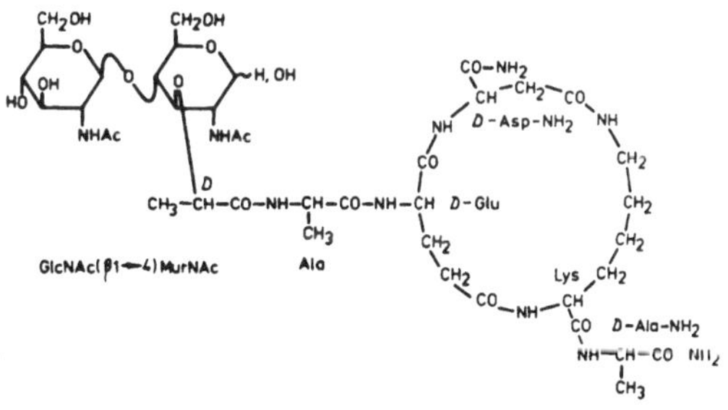

Chemistry of Peptides and Proteins, Vol. 1

Although this structure did not find a synthetic confirmation (5) there is little doubt that among the active components are species with the GlcNAc(β1→4)MurNAc disaccharide or its oligomers as a glycan part.

Results

As a part of structure-functional studies of disaccharide containing glycopeptides we obtained glycopeptide preparations from lysozyme hydrolyzates of *E. coli* and *Micrococcus lysodeikticus* cell walls. In spite of notable structural differences in the cell walls (see Fig. 1), and therefore in the components of their hydrolyzates, all three bacterial preparations showed commensurate antitumor activity (see below, Table 1), indicating that antitumor activity should not be considered as a unique property of preparations from *L. bulgaricus*; it rather belongs to a broad spectrum of glycopeptides. With this in mind we undertook synthesis of a series of model glycopeptides containing the GlcNAc(β1→4)MurNAc unit.

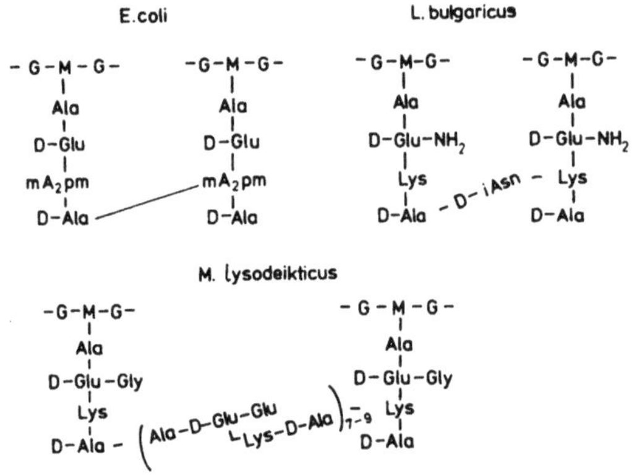

Fig. 1. Glycopeptides of bacterial cell walls.

Table 1. Antitumor activity of glycopeptides and their derivatives.

	Compound	Dose mg/kg	Necrotic reaction vs. total number of mice	Tumor growth inhibition %
1.	GlcNAc-MurNAc-Ala-D-Glu-NH_2(GMDP)	200	4/10	0
		100	8/10	70
		100	10/10	35
		50	9/10	35
		5	8/10	23
		0.5	6/10	16
2.	GlcNAc-MurNAc-Ala	100	3/5	19
		50	2/5	17
3.	MurNAc-Ala-D-Glu-NH_2(MDP)	100	4/5	0
		50	1/5	16
4.	GlcNAc-MurNAc-Ala-Glu-NH_2	100	0/5	16
		50	0/5	0
5.	GlcNAc-MurNAc-D-Ala-D-Glu-NH_2	100	3/5	19
		50	2/5	17
		5	0/8	40
6.	GlcNAc-MurNAc-Ala-D-Glu	100	3/5	45
		50	4/10	28
7.	GlcNAc-MurNAc-D-Ala-D-Glu	100	5/5	44
		50	3/5	38
8.	GlcNAc-MurNAc-Ala-D-Asp-NH_2	100	2/10	0
		50	0/10	0
9.	MurNac-Ala-D-Asp-NH_2	100	0/10	0
		50	0/10	0
10.	GlcNAc-MurNAc-GlcNAc-MurNAc Ala-D-Glu-NH_2 Ala-D-Glu-NH_2	50	3/5	66
11.	GlcNAc-MurNAc-Ala-D-Glu-D-Asp-NH_2 Lys-D-Ala-NH_2	100	3/5	41
		50	3/5	23
12.	GlcNAc(β1→4)MurNAc	250	0/8	0
		100	0/8	0
13.	[GlcNAc(β1→4)MurNAc]$_2$	250	0/10	0
		100	0/10	0
14.	Ala-D-Glu-NH_2	100	0/10	+28
15.	Ala-D-Glu-D-Asp-NH_2 Lys-D-Ala-NH_2	250	0/5	0

16. Clycopeptides from *L. bulgaricus* (Blastolysin)	250	10/10	33
	100	10/10	64
	50	10/10	57
17. Glycopeptides from *M. lysodeikticus* m.m. 1000-10000	250	10/10	73
	100	10/10	40
18. Glycopeptides from *E. coli* m.m. 1000-10000	250	2/5	-

Usual approaches to such derivatives involve multistep synthetic schemes with a number of protection-deprotection steps of the sugar moiety. Such strategy was employed in the syntheses of compound 1 which appeared after completion of this work (6, 7). In this work we were able to couple the unprotected di- and tetrasaccharides with peptides, with the aid of Woodward reagent K. Glycopeptides 1-10 were synthesized according to schemes 1-6. They were purified by ion exchange chromatography on Dowex 1x8 and characterized by TLC, elementary, amino acid and carbohydrate analysis as well as by PMR spectroscopic data. More details of the synthesis are given in (8). Below are presented the first results of biological studies of the glycopeptides obtained. Their antitumor properties in sarcoma 180 are summarized in Table 1. The effect was visually detected in tumor tissue following 24 hrs after single intravenous injection to the mice at the 7-8th day after tumor transplantation (well developed tumor). Inhibition of tumor growth was measured from the mean tumor weight at the 7th day after application of the glycopeptide.
The best studied preparation 1 showed rather high activity, maximal at 50-100 mg/kg and clearly seen even at 0.5 mg/kg. Interestingly, the effect is less pronounced at 200 mg/kg. The antitumor spectrum of 1 is shown in Fig. 2 (8). Pronounced antitumor activity of 1 correlates with its remarkable activity as an adjuvant (8, 9).
The following regularities can be deduced from the data of Table 1. Aminosugars and unprotected peptides alone do not

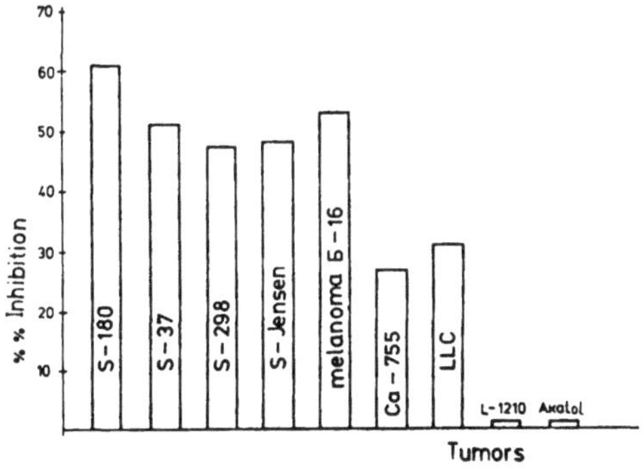

Fig . 2. Antitumor activity of GMDP.

cause the necrotic effect (slight stimulation in the case of 14 obviously requires further confirmation). Shortening of the peptide 2 or sugar moiety 3 lowers the activity. Elongating of the peptide 11, 5 or the carbohydrate chain 10 leads to active preparations. It is worthwhile noting that maximal effect (100% necrosis accompanied with considerable growth inhibition) was found with *M. lysodeikticus* cell wall lyzate apparantly also containing longer than disaccharide and/or dipeptide carbohydrate and peptide moieties. Ala→D-Ala or Glu(NH_2)→Glu replacements or both these modifications taken together give rise to active preparations 5-7 , while D-Glu→Glu or D-Glu→D-Asp replacements, both in the GlcNAc-MurNAc (compounds 4, 8) and MurNAc series (compound 9) result in sharp fall in the activity.

NMR spectra of GMDP and its analogue 8 were studied in an attempt to find out whether conformational properties are affected at the D-Glu→D-Asp replacement. As seen in Table 2, the δ and J values of the NH signals are rather similar for the two compounds while notable differences in the $\Delta\delta/\Delta T$ values were found for the acetyl glucosamine and acetylmuramyl NH moieties. The low values in GMDP ($1.5 \cdot 10^{-3}$ and $1.4 \cdot 10^{-3}$ ppm/^{o}C) point to

Schemes 1-6.

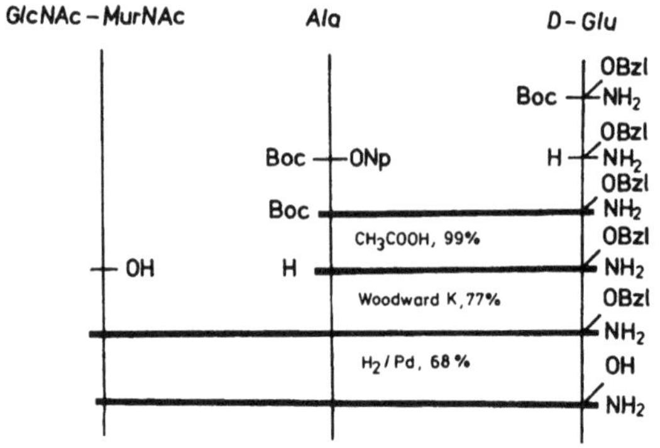
GlcNAc-MurNAc
Ala
D-Glu
OBzl
Boc
NH2
OBzl
Boc
ONp
H
NH2
OBzl
Boc
NH2
CH3COOH, 99%
OBzl
OH
H
NH2
Woodward K, 77%
OBzl
NH2
H2/Pd, 68 %
OH
NH2

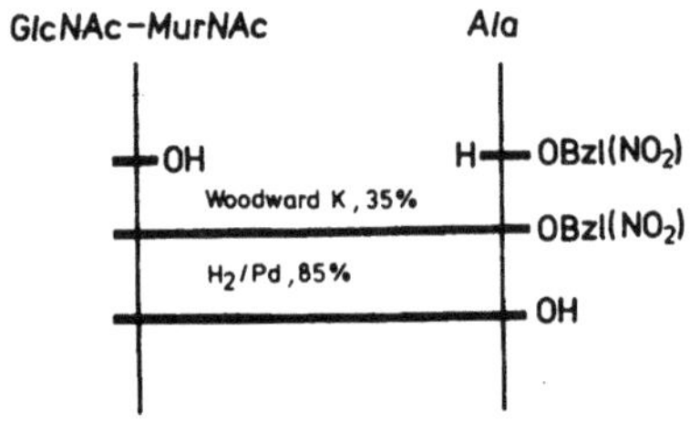
GlcNAc-MurNAc
Ala
OH
H
OBzl(NO2)
Woodward K, 35%
OBzl(NO2)
H2/Pd, 85%
OH

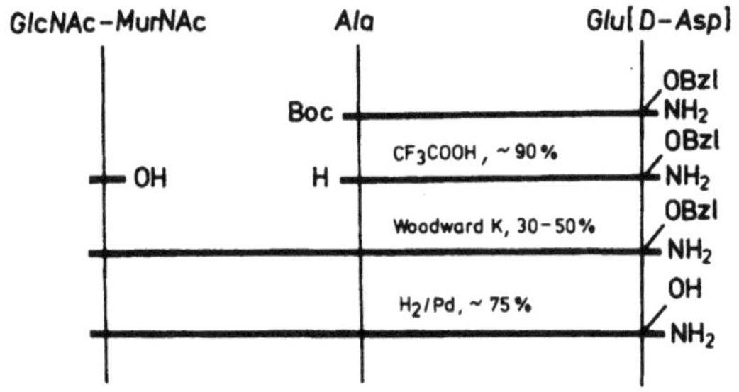
GlcNAc-MurNAc
Ala
Glu[D-Asp]
OBzl
Boc
NH2
CF3COOH, ~90%
OBzl
OH
H
NH2
Woodward K, 30-50%
OBzl
NH2
OH
H2/Pd, ~75%
NH2

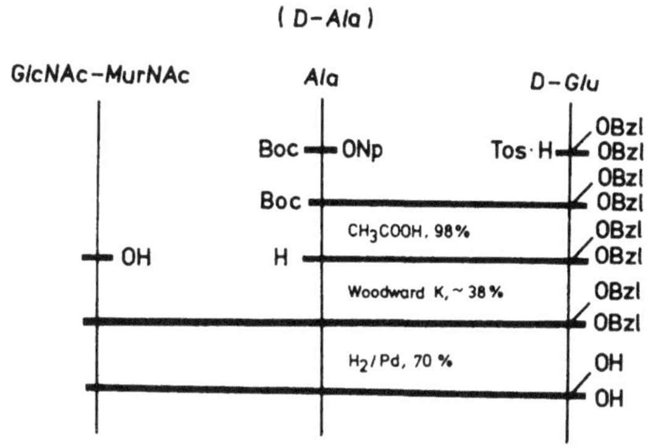
(D-Ala)
GlcNAc-MurNAc
Ala
D-Glu
Boc
ONp
Tos·H
OBzl
OBzl
Boc
CH3COOH, 98%
OH
H
Woodward K, ~ 38 %
H2/Pd, 70 %
OH

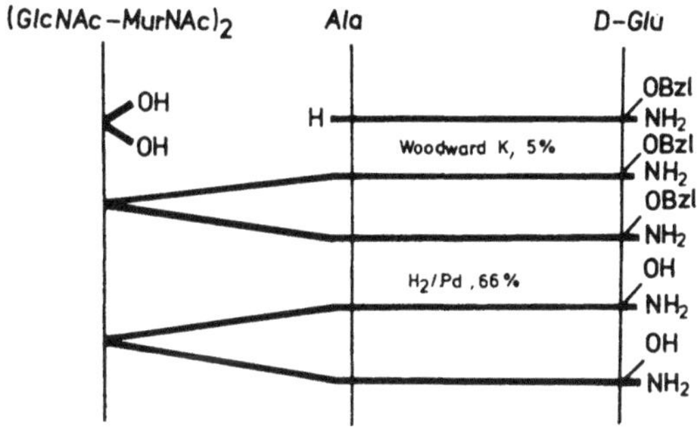
(GlcNAc-MurNAc)2
Ala
D-Glu
OH
H
OBzl
NH2
Woodward K, 5%
H2/Pd, 66 %
OH

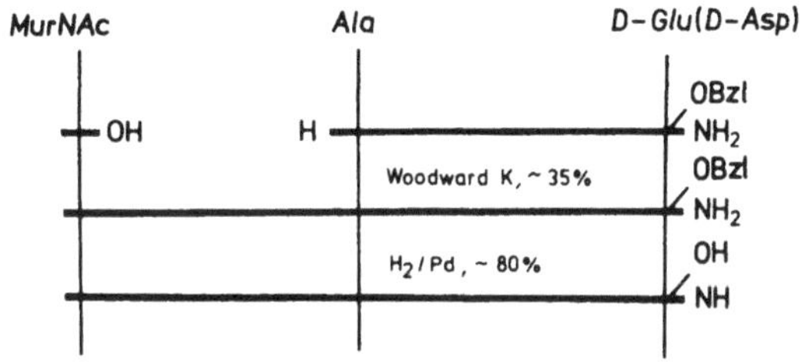
MurNAc
Ala
D-Glu(D-Asp)
OH
H
OBzl
NH2
Woodward K, ~ 35%
H2/Pd, ~ 80%
OH
NH

Table 2. Chemical shifts (δ, ppm), vicinal spin-spin coupling constants ($^3J_{HN-C^{\alpha}H}$, Hz) and temperature dependencies ($k = \Delta\delta/\Delta T \cdot 10^3$) of the NH proton signals in GlcNAc and glycopeptides.

Compound		GlcNAC	MurNAc	Ala	Glu(NH_2) or Asp(NH_2)
GlcNAc-MurNAc	δ	8.169	7.785	-	-
	J	4.4	8.8	-	-
	k	5.3	4.6	-	-
GlcNAc-MurNAc-	δ	8.137	7.850	8.476	8.674
-Ala-D-Glu-NH_2	J	6.6	8.5	5.2	7.9
(1)	k	1.5	1.4	7.8	7.6
GlcNAc-MurNAc-	δ	8.396	7.795	8.330	8.603
-Ala-D-Asp-NH_2	J	5.8	8.1	8.4	8.4
(8)	k	7.4	3.9	7.2	5.8

participation of the NH groups in intramolecular hydrogen bonding, probably with the glutamic γ-carboxyl and the lactyl carbonyl, as shown in Fig. 3. In the Asn derivative with its shorter side chain the former H bond is more difficult to form which in turn destabilizes the second H bond. This explains the

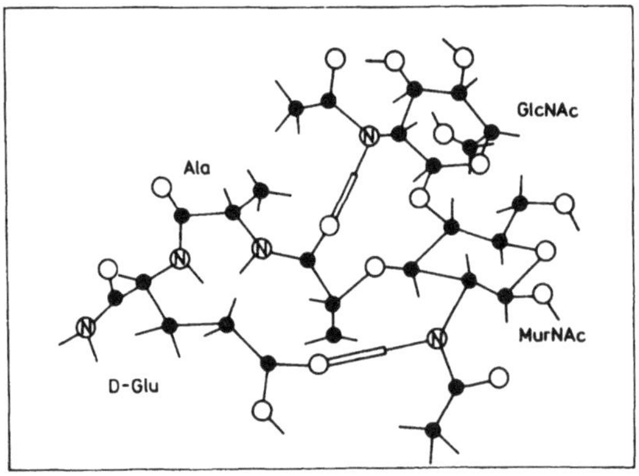

Fig. 3. Tentative conformation of GMDP.

high $\Delta\delta/\Delta T=7,4\cdot 10^{-3}$ and $3,9\cdot 10^{-3}$ ppm/^{o}C values found in the analogue (8). It is only natural to assume that structures of the type presented in Fig. 3 participate in biological interactions.

References

1. Ellouz, F., Adam, A., Ciorbaru, R., Lederer, E.: Biochem. Biophys. Res. Commun. 59, 1317 (1974).
2. Kotani, S., Watanabe, Y., Kinoshita, F., Shimono, T., Morisaki, I., Shiba, T., Kusumoto, S., Tarumi, Y., Ikenaka, K.: Biken. J. 18, 105 (1975).
3. Lederer, E.: J. Med. Chem. 23, 819 (1980).
4. Bogdanov, I.G., Dalev, P.G., Gurevich, A.I., Kolosov, M.N., Mal'kova, V.P., Plemyannikova, L.A., Sorokina, I.B.: FEBS Letters 57, 259 (1975).
5. Andronova, T.M., Rostovtseva, L.I., Dobrushkina, E.P., Gavrilov, Yu.D., Deshko, T.N., Ivanov, V.T.: Bioorg. Khim. (Russian) 6, 1830 (1980).
6. Kusumoto, S., Yamamoto, K., Shiba, T.: Tetrahedron Lett. 45, 4407 (1978).
7. Durette, P.L., Meitzner, E.P., Shen, T.Y.: Carbohydr. Res. 77, C1 (1979).
8. Rostovtseva, L.I., Andronova, T.M., Mal'kova, B.P., Sorokina, I.B., Ivanov, V.T.: Bioorg. Khim. (Russian) in press.
9. Nikitina, L.E., Maslennikova, L.K., Bichurina, M.A., Fridman, E.A., Noskov, F.S., Rostovtseva, L.I., Andronova, T. M., Jurovskaya, A.A., Ivanov, V.T.: Immunol. (Russian) in press.

USE OF DIPEPTIDE SUBSTRATES IN STUDIES OF SPECIFIC FEATURES OF THROMBIN CATALYSIS

V.K.Kibirev and S.B.Serebryany
Institute of Molecular Biology and Genetics, Academy of Sciences of the Ukrainian SSR, Kiev, USSR

Introduction

Thrombin, a blood-clotting enzyme, is a key factor of the coagulation process and a highly specific trypsin-like serine proteinase. It shows the sequence homology and similarity with trypsin in the primary specificity (1,2). But contrary to trypsin, the effect of thrombin on macromolecular substrates is very restricted. The aim of the present work is a comparative study of esterolytic activity of these serine hydrolases on dipeptide esters I which differ in the nature of the amino acid residue at P_2 according to Schechter and Berger (3).

Results

To investigate the relationship between the chemical structure of thrombin substrates, their conformation and reactivity, a series of dipeptides I has been synthesized, CD-spectra in different solvents studied and kinetic parameters of their hydrolysis by the afore-mentioned enzymes determined.

$$R\text{-}C_6H_4\text{-}SO_2\text{-}X\text{-}Arg\text{-}OCH_3 \qquad (I)$$
$$R{=}CH_3\text{-},\ CH_3O\text{-},\ X{=}Asn,\ Ala,\ Val,\ Phe$$

The substrates were synthesized by two methods: (i) condensation of tosyl amino acids with nitroarginine methyl ester followed by hydrogenation of protected dipeptides; (ii) in-

Chemistry of Peptides and Proteins, Vol. 1

teraction of N-hydroxysuccinimide esters of different tosyl amino acids with unprotected arginine in the dioxane-water mixture followed by treatment of the product with $SOCl_2$ in methanol. Peptide synthesis without protection of guanidinium group, which produces zwitterion with carboxylate, is the most suitable method of obtaining these compounds. The synthesized peptides were purified by the ion-exchange chromatography on CM-cellulose at pH 6.0.

In order to study the conformational lability of dipeptides I, their CD-spectra were studied in solvents of different polarities: in water, alcohol as well as in water-alcohol and water-dioxane mixtures. The data thus obtained indicate that the alteration of polarity in peptides, having LL-configuration, results in significant change of chiroptic characteristics. Such a situation, therefore, may suggest an idea that the compounds mentioned, possess, for the most part, the extended conformation, which contributes to complementary binding of side chains of amino acids at P_1 and P_2 with the subsites S_1 and S_2 of the thrombin active site. It facilitates to estimate which of the contacts between the enzyme and substrate are favourable for catalysis.

The rate of dipeptide ester hydrolysis was measured on pH-stat under steady-state conditions at pH 6.0-9.5 and temperatures 8-40°C. The values of the kinetic parameters k_{cat} and K_m were determined by a progressive curve method. The results of the pH-dependence of enzyme reaction demonstrated that during hydrolysis of Tos-Val-Arg-OCH_3 by thrombin, the two ionizing enzyme groups with pK 7.5 and 8.8 are catalytically essential for esterolysis. The former value is consistant with pK of histidine residue of the thrombin active site. The dependence of the catalysis on the group with pK 8.8 may be explained by the breakdown of the "salt bridge" between the α-amino group of N-terminal isoleucine and Asp 194 of the enzyme active site (4). Table 1 gives the kinetic parameters of dipeptides I hydrolysis by thrombin at the optimum value of pH.

Table 1. Kinetic parameters of synthetic substrates hydrolysis by thrombin and trypsin at pH 8.5 and 25°C [a].

Substrate	$10^5 K_m$ (M)	k_{cat} (c^{-1})	$10^{-5} k_{cat}/K_m$ ($M^{-1}c^{-1}$)
P_3 P_2 P_1		Thrombin	
Tos-Asn-Arg-OCH_3	-	-	-
Tos-Ala-Arg-OCH_3	1.43	11.0	7.7
Tos-Phe-Arg-OCH_3	0.24	0.86	3.6
Tos-Val-Arg-OCH_3	0.45	2.56	5.8
CH_3O-C_6H_4-SO_2 - Val-Arg-OCH_3	0.56	3.02	5.5
Tos-Arg-OCH_3	3.3	2.5	0.76
		Trypsin	
Tos-Val-Arg-OCH_3	0.52	39.6	85
CH_3O-C_6H_4-SO_2- Val-Arg-OCH_3	0.39	36.0	97
Tos-Phe-Arg-OCH_3	0.37	45.0	122
Tos-Arg-OCH_3	0.67	70.0	105

[a]Reaction with thrombin was carried out in 0.15 M KCl (2-3) $\times 10^{-8}$ M enzyme; hydrolysis by trypsin was carried out in 0.1 M NaCl containing 0.002 M $CaCl_2$ at 3×10^{-9}M trypsin concentration; substrate concentrations were $(3-7)\times10^{-5}$M; given are the average values from 4 runs, standard deviation is $\pm$10 %.

The above data substantiate the idea that the nature of the amino acid at P_2 affects considerably the substrate hydrolysis by thrombin. The compound, having at P_2 the alanine residue, undergoes the most effective cleavage: compared to tosyl-L-arginine methyl ester which, at P_2, has a tosyl group, the k_{cat}/K_m value of alanine-containing dipeptide grows 10-fold. For thrombin catalysis the presence of a hydrophobic amino acid at P_2 is of great importance which has favourable contacts with tryptophan residue in S_2 of the enzyme. Hydrophobic increase of the amino acid side chain at P_2 results in the improvement of thrombin binding of dipeptides, while the decrease of the amino acid residue size stimulates the growth of catalytic constants. So, the alanine introduction at P_2, instead of valine residue, leads to a 4-fold growth of k_{cat}. It is worth mentioning, that the replacement of hydrophobic amino acid residue by asparagine with hydrophilic properties results in loss of substrate sensitivity to thrombin. In contrast to this the hydrolysis of such compound by trypsin continues according to some tentative data in a conventional manner. The study of the temperature dependence of the enzyme reactions (Table 2) suggests that deacylation of acyl-thrombins is accompanied by much more unfavourable entropy changes than the hydrolysis of the same compounds by trypsin.

Table 2. Activation parameters of synthetic substrate hydrolysis by thrombin and trypsin at pH 8.5 and 25°C.

Substrate	ΔG kcal/mol	ΔH kcal/mol	ΔS e.u.
		Thrombin	
$CH_3-C_6H_4-SO_2$-Arg-OCH_3	17.6	10.3	-24.4
$CH_3O-C_6H_4-SO_2$-Val-Arg-OCH_3	16.9	11.0	-19.6
		Trypsin	
$CH_3-C_6H_4-SO_2$-Arg-OCH_3	15.0	9.9	-17.2
$CH_3O-C_6H_4-SO_2$-Val-Arg-OCH_3	15.3	12.4	- 9.8

The activation parameters were determined from Arrhenius plots, which appeared broken for thrombin; the results are average from 3 experiments, standard deviation is $\pm$ 10-20 %. Elongation of a peptide decreases the unfavourable entropy change, obviously due to additional interactions in thrombin subsites S_2 and S_3. The peculiar feature of the thrombin binding site lies in the fact that it catalyzes most effectively the cleavage of those substrates which have at their P_2 the hydrophobic amino acid with a relatively small side chain.

The proline residue may apparently correspond to these steric and lipophilic characteristics. In this connection, it should be emphasized that in such physiological thrombin substrates as prothrombin and some other proteins cleaved by the enzyme, there is a sequence of amino acids -Pro-Arg-Ser(Gly)- containing the proline residue at P_2.

The above investigation has shown that in spite of the similarity between thrombin and trypsin, the thrombin active site is considerably different from the respective trypsin region and that for effective thrombin catalysis, a more precise corresponding and fitting of the substrate side chains to the thrombin binding site is required.

References

1. Magnusson, S.: The Enzymes 3, 277-321, Academic Press, New York, 1971
2. Curragh, E.F., Elmore, D.T.: Biochem.J. 93, 163-171 (1964).
3. Schechter, I., Berger, A.: Biochem.Biophys.Res.Commun. 27, 157-162 (1967).
4. Magnusson, S., Petersen, T.E., Sottrup-Jensen, L., Claeys, H.: Proteases and Biological Control, Cold Spring Conf.Cell Proliferation 2, 123-149 (1975).

STUDIES ON THE EFFICACY OF PREPARATIVE AND ANALYTICAL HPLC IN PEPTIDE CHEMISTRY

Walter Göhring, Manfred Gemeiner, Luis Moroder, Rolf Nyfeler and Erich Wünsch

Max-Planck-Institute of Biochemistry, Department of Peptide Chemistry, 8033 Martinsried / Munich, FRG

Introduction

In the last years the high performance liquid chromatography (hplc) has found widespread application as a new efficient analytical and preparative chromatographic method for various classes of natural products. With great efforts this technique has also been adapted to chromatographic purposes in the field of peptides and proteins. First positive results compelled us to ulteriorly investigate the resolution power and consequently, the usefullness of hplc for protected and "free" synthetic peptide preparations.

Hplc of protected peptide derivatives

In the case of protected peptide derivatives the main difficulties derive from the solubility factor which strongly may limit the choice of eluent systems and of stationary phases. In fact, only for low molecular weight derivatives the chromatography can be performed in reversed phase systems, whereas for high molecular weight fully protected peptide derivatives silica gel columns have to be used. Additionally, latter peptide derivatives are usually not soluble in common eluent systems. This difficulty, however, is bypassed by dissolving the samples in trifluoroethanol. As shown in fig. 1, following this procedure it has been possible to analyze a fully protected dihydrosomatostatin preparation, i.e. Boc-Ala-Gly-Cys(StBu)-Lys(Adoc)-Asn-Phe-Phe-Trp-Lys(Boc)-Thr(tBu)-Phe-Thr(tBu)-Ser(tBu)-Cys(StBu)-OtBu (1) on a μ-Porasil column using the eluent system 1,2-

Chemistry of Peptides and Proteins, Vol. 1

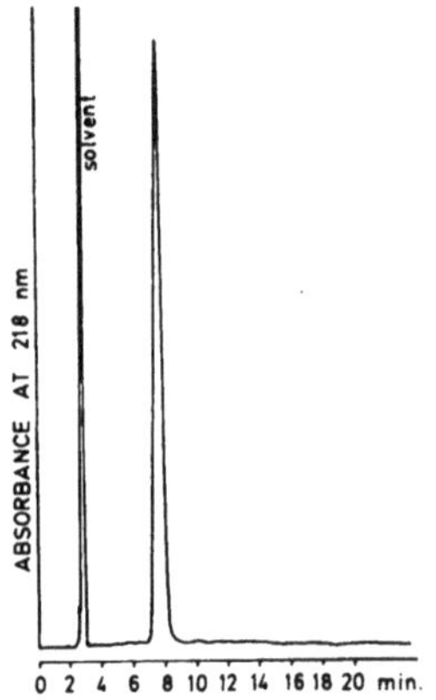

fig. 1 : Hplc of fully protected linear somatostatin; sample dissolved in $CHCl_3/CH_3OH$, 8:1; column: µ-Porasil, 4mm x 30cm; eluent: 68.3% ethylenechloride (50% saturated with water)/24.4% n-heptane/7,3% methanol, v/v; flow rate: 1,2 ml/min.

dichloroethane/n-heptane/methanol (68.3:24.4:7.3). Similarly, a crude condensation product of our Nle^{15}-human little gastrin I synthesis (2), isolated upon acylation of H-[Glu(OtBu)]$_5$-Ala-Tyr(tBu)-Gly-Trp-Nle-Asp (OtBu)-Phe-NH_2, H-[6-17]-NH_2, with H-Pyr-Gly-Pro-Trp-Leu-OH, H-[1-5] -OH, via DCC/HONSu was analyzed, again on a µ-Porasil column upon dissolving it in trifluoroethanol (fig. 2). These analytical conditions were thus transferred to a preparative scale as shown in fig. 2, yielding a fast and highly efficient purification procedure: the fully protected heptadecapeptide amide, i.e. H-Pyr-Gly-Pro-Trp-Leu-[Glu(OtBu)]$_5$-Ala-Tyr(tBu)-Gly-Trp-Nle-Asp(OtBu)-Phe-NH_2 was isolated in 50% yield at a high degree of purity as judged by silica gel tlc ($CHCl_3$/trifluoroethanol/80% acetic acid; 5:1:1), amino acid analysis (Asp 1.00, Glu 5.82, Pro 0.97, Gly 1.97, Ala 1.02, Leu 0.94, Nle 1.01, Tyr 1.02, Phe 1.00, Trp 1.91) and by elemental analysis. Unreacted amino component (30 to 40% on weight bases) was additionally recovered by this procedure as homogeneous and utilizable material.

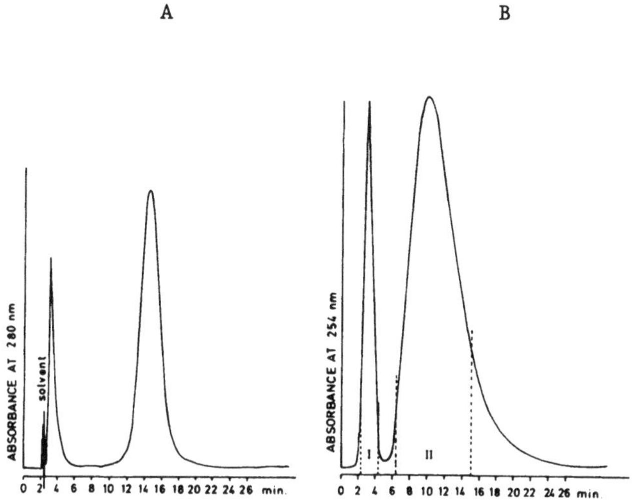

fig. 2 : A) Analytical hplc of fully protected crude Nle^{15}-human-little gastrin I; column: μ-Porasil, 4mm x 30cm; eluent: ethylenechloride (50% saturated with water)/isopropanol/acetic acid 65:35:0,45, v/v; flow rate: 1.5ml/min.

B) Preparative hplc of fully protected crude Nle^{15}-human-little-gastrin I; column: Prep Pak silicagel; eluent: ethylenechloride.(50% saturated with water) / isopropanol/acetic acid 65:35:0,45, v/v; flow rate: 300ml/min; sample: 700mg dissolved in trifluoroethanol; yield: 360mg pure product.

Hplc of peptides

Even for unprotected peptides their solubility properties may limit the choice of optimal chromatographic conditions. On the other side, however, the resolution power of hplc strongly depends from the composition of the solvent mixture used as eluent, since minimal changes in the eluent systems may produce noticeable effects as exemplified in the case of our synthetic somatostatin preparation (1): upon lowering the acetonitrile content of 2% a "one-peak-material" is resolved into three components (fig. 3). Thus, for every synthetic preparation the most favorable conditions have to be elaborated by optimizing retention time and resolution.

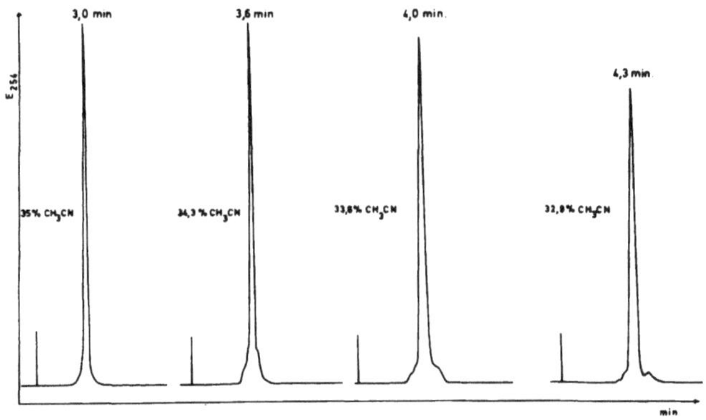

fig. 3 : Hplc of somatostatin; column: μ-Bondapak C 18, 4mm x 30cm; eluent: x % CH_3CN/(100-x)% 0,01M NH_4OAc, pH 4.0, v/v; flow rate: 1.7ml/min.

Additional difficulties in the hplc of peptide preparations often derive from irreversible absorption of products and byproducts to different extents, as observed in the case of crude somatostatin preparations. Isocratic analytical hplc of this materials yields elution profiles characterized by one main peak with minor contaminants (fig. 4). On tlc, however, a group of substances with higher molecular weight than somatostatin - the different polymeric forms - are detected. In this case the hplc clearly produced a falsified picture about the degree of homogeneity of the analyzed material. It follows that hplc chromatograms may reflect only <u>partially</u> the actual state of homogeneity of peptide materials. Semipreparative hplc with recovery determination, however, should guarantee unequivocal interpretation of elution profiles. Such quantitative hplc-analysis on purified somatostatin preparations indicated no irreversible absorption of this cyclic tetradecapeptide under the conditions reported in fig. 4.

A possible source of byproducts in synthetic peptide preparations is the racemization, which may occur to grater extents during the condensation steps. In most of the cases it is difficult to detect and to separate

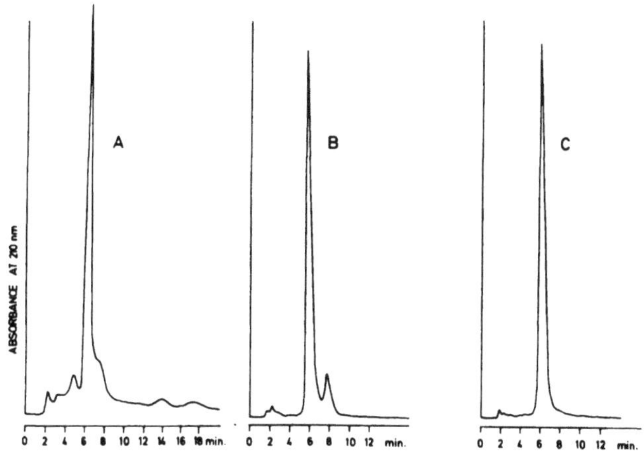

fig. 4: Hplc of somatostatin; A: crude product; B: after LH 20 gelfiltration; C: after Biogel P 6 gelfiltration of B; column: µ-Bondapak C18, 4mm x 30cm; eluent: 32,5% CH_3CN/67,5%; 0.01M NH_4OAc, pH 4.0, v/v; flow rate: 1.4ml/min.

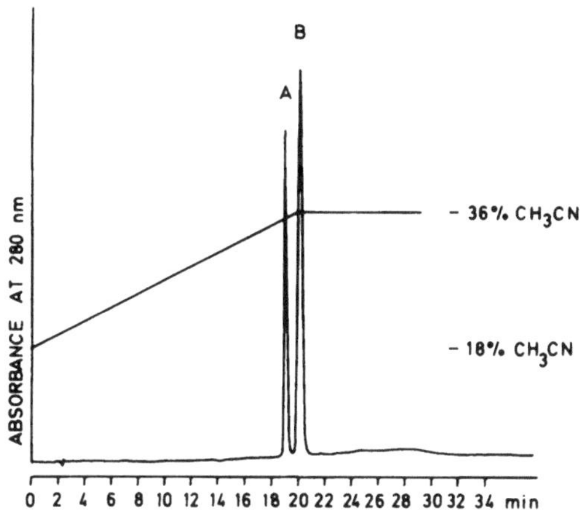

fig. 5: Hplc of gastrins; A: D-Leu^5, Nle^{15}-human-little-gastrin I; B: Nle^{15}-human-little-gastrin I; column: µ-Bondapak C 18, 4mm x 30cm; eluent: 0.01M NH_4OAc, pH 4.0 linear gradient (20min), 18%-36% CH_3CN, v/v; flow rate: 1.7ml/min.

such byproducts, the hplc, however, may represent a valuable analytical tool for this purpose. During the last fragment condensation step of our Nle^{15}-human little gastrin I synthesis (2), reported above, considerable racemization took place at the level of leucine in position 5. Upon deprotection we succeeded to separate by partition chromatography the two diastereomers. An artifical mixture of D-Leu^{5}- and L-Leu^{5}-gastrin analogue is well resolved by gradient elution on a µ-Bondapak column, thus allowing a fast and quantitative analysis of such gastrin preparations (fig. 5).

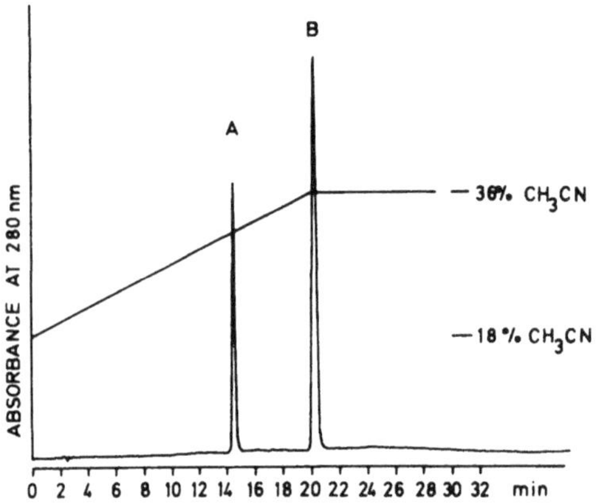

fig. 6 : Hplc of gastrins; A: Nle^{11}-human-minigastrin I , B: Trp $(1'\text{-tBu})^{10}$, Nle^{11}-human-minigastrin I; column: µ-Bondapak C 18, 4mm x 30cm; eluent: 0.01M NH_4OAc, pH 4.0, linear gradient (20 min), 18%-36% CH_3CN, v/v; flow rate: 1.7ml/min.

In the course of the synthesis, especially in the deprotection step chemical modifications at reactive side chains may also occur. Recently we succeeded to characterize one of such side reactions involving the tryptophan side chain, i.e. a tert-butylation at the indole group in different positions (3,4), particularly in position 1. An artificial mixture of Nle^{11}-human minigastrin I and its Trp$(1'\text{-tBu})^{10}$-derivative (2) is well resolved by gradient elution from a C^{18}-column (fig. 6), whereby the retention time of the N^{in}-tert-butylated material is about 6 min greater than that of the unmodified peptide. Hplc of such synthetic gastrin preparations not only allows to detect minimal amounts of this side product, but on preparative scale should also represent a highly efficient purification procedure.

Conclusion

The examples illustrated above clearly indicate that hplc represents a powerful analytical technique for purity control of peptides and peptide derivatives, when optimal chromatographic conditions were elaborated. But the method has also its limitations and thus, as all other analytical procedures, it does not represent an absolute criterion of purity. It gives, however, important information, complementary to that derived from other assays. For its more general application as a purification procedure of synthetic peptides in larger scale additional experiences have to be collected.

References

1. Moroder, L., Gemeiner, M., Göhring, W., Jaeger, E., Thamm, P., Wünsch, E.: Biopolymers 20, 17-37 (1981).
2. Wünsch, E.: unpublished results.
3. Wünsch, E., Jaeger, E., Kisfaludy, L., Löw, M.: Angew. Chem. 89, 330-331 (1977).
4. Jaeger, E., Thamm, P., Schmidt, I., Knof, S., Moroder, L., Wünsch,E.: Hoppe Seyler's Z.Physiol.Chem. 359, 155-164 (1978).

A METHOD FOR AMINO ACID SEPARATION WITH A MICROCOLUMN CHROMATOGRAPH

E.Ya.Kreindlin, V.S.Onoprienko, O.V.Evseeva, L.B.Kaminir
Institute of Molecular Biology, USSR Academy of Sciences, Moscow 117984, U.S.S.R.

Developments of biochemical investigations which are closely connected with the progress in fields such as molecular biology, cytology, pharmacology, medicine etc. urgently necessitate the advancement of microscale determinations. The introduction of ultramicroanalysis into biochemical studies is of high expedience and advantage for the reasons given below. First of all, very often biological samples for such investigations are very expensive since to obtain the necessary amounts of the substance, a complex multistage and labour-consuming processing of extremely large quantities of the initial material is needed. Hence, the consumption for the analysis of a smaller volume of test substance is undoubtedly a great advantage of a microscale method. The second reason is that in many cases the initial amounts of the substance are very scarce and cannot be increased.

A chromatographic method of amino acid analysis has been developed using a model of a microcolumn chromatograph manufactured in the Special Design Bureau of Analytical Instrumentation of the Scientific Technical Union, Acad.Sci.USSR. The instrument is designed for ultramicroanalysis of amino acids. Its design and analysis technique envisage maximum reduction of all volumes in the fluid lines after the column (fluid space of the reaction capillary, connections and the cuvette). Utilization of microcolumn chromatography for the needs of amino acid analysis is somehow complicated by the

Chemistry of Peptides and Proteins, Vol. 1

necessity to carry out the colour reaction prior to the detection stage. The reaction tube additionally increases dispersion of chromatographic zones of the separated components. The set-up of the instrument is basically the same as that of the conventional type, except that before the chromatographic separation all buffer solutions are combined in the predetermined sequence in one teflon tube of a very small diameter.

Two 310 mm long glass columns with 0.5 mm and 1 mm internal diameters were employed for the chromatographic separation of the standard amino acid mixture. The columns were packed with ion-exchange resin DC-4A (Durrum, U.S.A.). Better separation was obtained with the column of 1 mm diameter. Pressure drop in the column was substantially reduced thus allowing a twofold increase in the volumetric rate of the eluent. As a result, the time for analysis was reduced by ca. 50% and the life time of the column became significantly longer. Moreover, better resolution was gained for the pair of amino acids: threonine-serine. The selectivity coefficient calculated for separation of the above amino acid pair on the narrower column (0.5 mm diameter) was within the range from 1.028 to 1.055 (in some cases its value was equal to 1.077). In the case of the column with 1 mm internal diameter, this value amounted to 1.073 - 1.11. The resolution in these two cases was within the limits of 0.78 - 0.88.

It is essential that sensitivity of the chromatograph remains the same for the 0.5 and the 1 mm diameter column. The chromatographic peak is known to widen in proportion to the increase in the cross-sectional area of the column (1). However, the final width of the peak is the sum of peak dispersion resulting from chromatographic separation and dispersive effects due to diffusion and mixing in the reactor, the cuvette of the photometer, capillaries and places of connections. The total volume of the latter elements is commensurate with the column volume and their contribution to peak dispersion is very significant. Therefore, in this case the increase in the column diameter did not affect sensitivity. The above considerations account for the selection of the 310 mm x 1 mm column. This does not, however, mean that columns with diameters less than 1 mm are unapplicable for ultramicroanalysis. In order to use those for this purpose, one should introduce design innovations and take some other measures for reducing the volumes causing additional dispersion and for improving sample input, specify in full detail the analysis conditions, select the appropriate resin and elaborate the technique for packing the microcolumns with ion-exchange resin.

Column packing. Since the amino acid analysis is performed in the system of three buffer solutions, the column is packed with resin in the high molar buffer solution. This permits to prevent decreasing of the resin bed volume and compress the latter during running the column. Before packing, the resin is suspended in the buffer solution, and the suspension formed is poured into a reservoir. The reservoir is connected to a column and a pump, and the suspension is pumped at the maximum possible packing flow rate into the column(3).

Buffer solution. For elution of amino acids, sodium citrate buffer solutions are used. Acidic amino acids are eluted with 0.2 N buffer at pH 3.25, neutral amino acids with 0.2 N buffer at pH 4.25, and the basic ones with 1.1 N buffer at pH 7.0. To remove ammonium from the two former buffers, these are pumped through the column packed with DC-3 resin (Durrum, U.S.A.).

Preparations for instrument operation include regeneration and equilibration of the column, and preforming a set of the sample and buffer solutions .

Column regeneration and equilibration are performed by pumping 120 μl of 0.2 N NaOH and 250-300 μl of the first buffer solution respectively. The buffer solutions and the sample are delivered by means of a micropump. The teflon tube is disconnected from the column and its end is successively placed into the reservoirs with the buffer solutions and the sample. The buffers and the sample are introduced into the tube in the reverse order of elution. In order to run one chromatographic separation, the following amounts are sufficient: 760 μl of the first buffer, 240 μl of the second, and 960 μl of the third. The sample volume is 4 μl. After the tube is filled, it is reconnected to the column.

During chromatographic separation the buffers are delivered at the flow rate of 8 μl/min, and ninhydrin at the rate of 4 μl/min. The colour reaction of amino acids with the ninhydrin reagent proceeds at 110°C. Separation starts at the column temperature of 55°C. After alanine is eluted, the temperature is increased up to 68°C and thereafter maintained constant.

Fig. 1a shows the chromatogram of the standard mixture consisting of 18 amino acids of 2×10^{-10} mole each (the standard mixture obtained from Pierce, U.S.A.).

The drift of the base line, the unidentified peak 16 and the hump in the region of the ammonia peak are due to contaminations in the buffer solutions what can be clearly seen in Fig. 1b. It is evident that with increased sensitivity of the analysis, the requirements for the purity of all components of the buffer solutions should become higher. The reproduction error for the analysis of the standard amino acid mixture does not exceed 7%.

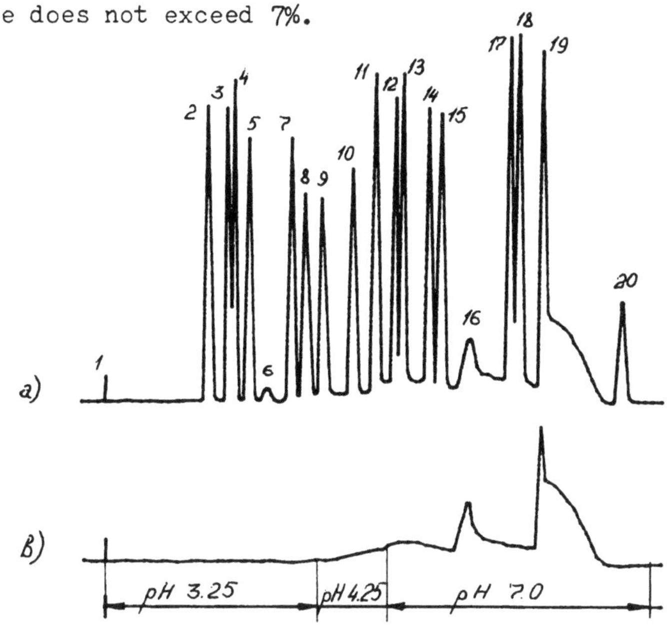

Fig.1 Chromatogram of: a) the standard amino acid mixture: 1-sample input; 2-aspartic acid; 3-threonine; 4-serine; 5-glutamic acid; 6-proline; 7-glycine; 8-alanine; 9-cystine; 10-valine; 11-methionine; 12-isoleucine; 13-leucine; 14-tyrosine; 15-phenylalanine; 16-unidentified peak; 17-histidine; 18-lysine; 19-ammonia (as $(NH_4)_2SO_4$); 20-arginine;

b) the buffer solutions (without input of the standard mixture).

References

1. Hamilton, P.B.: Advances in Chromatography 2, 3-62 (1966) Marsel Dekker, New York.

2. Devenyi, T., Gergely, J.: Amino Acids, Peptides end Proteins (1974) Elsevier Scientific Publishing Company, Amsterdam-London-New York.

3. Webber, T.I.N., McKerrell, E.N.: J.Chromatography 122, 243-258 (1976).

AN EXPRESS METHOD FOR DETERMINATION OF ABSOLUTE CONFIGURATION AND QUANTITY OF AMINO ACID ENANTIOMERS IN MIXTURES

N.A. Voskova, V.V.Romanov, G.A. Korshunova, Yu.P.Shvachkin
Moscow State University, Moscow, USSR

Introduction

The elaboration of an effective method for determination of absolute configuration of amino acid stereoisomers in mixtures is a quite actual problem. No less actual is also the problem of development of a simple and reliable method of quantitative determination of individual enantiomers of amino acids in reaction mixtures without preliminary fractionation of the latter.

Results

We have developed a new method for determination of absolute configuration and quantitative content of amino acid stereoisomers in situ which consists in treating the mixture to be analyzed with o-phthalaldehyde in the presence of mercaptoethanol and subsequent measuring circular dichroism (CD) spectra and UV absorption spectra of the chromophore derivatives formed thereby. The absolute configuration of amino acid stereoisomers is determined from the sign of a CD peak in the region of 340 nm and their quantitative content from UV absorption intensity in the same region.

The method is simple in performance and permits a rapid and unambiguous determination of absolute configuration and quantitative content of stereoisomers of amino acids in various

Chemistry of Peptides and Proteins, Vol. 1

mixtures. The sensitivity of the method is 2.10^{-5}. A necessary premise for the development of this method was a preliminary study of regularities of interaction between individual enantiomers of different amino acids and o-phthalaldehyde and mercaptoethanol performed earlier by us. o-Phthalaldehyde in the presence of mercaptoethanol is known to react with amino acids to form chromophore derivatives with an intensive absorption maximum at 340 nm (1,2). The mechanism of this reaction was recently studied (3) and on its basis a method for quantitative determination of amino acids was proposed (4).

However, up till now no data on optical characteristics of the products of this reaction were available in the literature. We have studied CD spectra of the products of the reaction of the known L- and D-isomers of different amino acids with o-phthalaldehyde in the presence of mercaptoethanol and found that the resulting derivatives of the L-isomers exhibit a positive CD peak in the region of 340 nm. Within the experimental error the CD curves of the D-isomer derivatives are mirror images of the respective CD curves of the L-isomer derivatives.

In the group of protein amino acids exceptions from this rule are the derivatives of L-alanine, L-tryptophan and L-aspartic acid which show a negative Cotton effect in this region as well as the L-histidine derivative for which no distinct dichroic absorption was observed in the 340 nm region. It should be noted that an anomalous behaviour of L-alanine was also observed for its chromophore derivatives with fluorescamine(5) and 2-methoxy-2,4-diphenyl-3-oxohydrofuran (6). Molar absorption coefficients ϵ_{334} of the protein amino acid derivatives have values of the order 6 000 - 7 000. The ϵ_{334} values for cystine and lysine are equal to 9 200 and 10 000, respectively. Molar differential dichroic absorption coefficients $\Delta\epsilon_{334}$ for the chromophore derivatives which are formed on reaction of the protein amino acids with o-phthalaldehyde and mercaptoethanol lie within a range from + 0.60 to + 6.00.

The absorption coefficients and differential dichroic absorption values of the protein amino acid chromophore derivatives linearly change with the concentration within an interval from 2.10^{-5} to 2.10^{-4}M.

The concentration of a predominant isomer in the case of a mixture of L- and D-isomers of a particular amino acid was estimated according to the formulae:

$$C_L = \frac{1}{2}\left(\frac{\Delta D}{\Delta \epsilon_L} + \frac{D}{\epsilon}\right), \qquad (1)$$

$$C_D = \frac{1}{2}\left(\frac{D}{\epsilon} - \frac{\Delta D}{\Delta \epsilon_L}\right), \qquad (2)$$

where C_L and D_C are molar concentrations of L- and D-isomers, ϵ -molar absorption coefficient of the amino acid derivative, $\Delta\epsilon_L$-molar differential dichroic absorption of the amino acid derivative, D - measured value of the UV absorption, ΔD-measured value of the circular dichroism differential absorption.

The proposed method is applicable for the determination of absolute configuration and quantitative content not only of the protein amino acids but also of various nonproteinogenic amino acids in particular, nucleoamino acids (7).

The absolute configuration of an amino acid stereoisomer and its quantitative content can be determined by this method practically simultaneously without a preliminary separation of the amino acid chromophore derivative from the reaction mixture. Such being the case this method affords an advantageous means for a rapid analysis of the mentioned and similar systems.

The developed method has been used:

1. For determination of the absolute configuration and optical purity of amino acids obtained by the microbiological synthesis,

2. for determination of the reaction rate and stereospecificity of the fermentative hydrolysis of N^{α}-acyl derivatives of D,L-nucleoamino acids under the effect of the pig kidney acylase I, E.coli penicillamidase and Aspergillus oryzae acylases. The experimental value $\Delta\epsilon = \Delta D \cdot \epsilon / D$ equal to the $\Delta\epsilon$ value of a pure isomer of amino acid can be regarded as a criterion of stereospecificity of the enzymatic reaction.

References

1. Roth, M.: Anal.Chem. 43, 880-882 (1971).
2. Schwabe, C., Catlin, J.C.: Anal.Biochem. 61, 302-304 (1974).
3. Simons, S.S., Johnson, D.F.: J.Org.Chem.43, 2886-2891(1978).
4. Shviadas, V.-Yu.K., Galaev, I.Yu., Berezin, I.V.: Bioorgan.Chem.4, 19-25 (1978).
5. Toome, V., Wegrzynski, B., Reymond, G.: Biochem.Biophys. Res.Commun. 69, 208-211 (1976).
6. Toome, V., Reymond, G.: Biochem.Biophys.Res.Commun. 66, 75-80 (1975).
7. Shvachkin, Yu.P.: J.Gen.Chem. 49, 1157-1161 (1979).

III Structural Features of Proteins

STRUCTURE-ACTIVITY STUDIES ON NEUROTOXIN 1 FROM SEA ANEMONE RADIANTHUS MACRODACTYLUS *

Modification of Tryptophan, Lysine, Arginine

Emma Kozlovskaya, Helen Vozhova, Georgy Elyakov
Pacific Institute of Bioorganic Chemistry, Far East Science Centre, USSR Academy of Sciences, Vladivostok, 690022,USSR

Introduction

Neurotoxin 1 from sea anemone Radianthus macrodactylus (order - Actiniaria, family - Stoichactiidae) is a basic polypeptide, consisting of 47 amino acids, molecular weight - 5000 /1/. As the other representatives of anemonotoxin class /2/,neurotoxin 1 does not contain free sulfhydryl groups, the six cysteine residues form three intermolecular disulfide bridges /1/. This paper describes the determination of the role that tryptophan, lysine and arginine play in physiological activity of neurotoxin 1 from R.macrodactylus, using the method of specific chemical modification of amino acid side chains. According to amino acid analysis, the polypeptide molecule contains one tryptophan residue, two - arginines and three - lysines /2/.

Results

Triptophan residue modification. N-bromosuccinimide (NBS) and Koshland reagent, 2-hydroxy-5-nitro-benzylbromide (HNB-Br)

* Earlier, the sea anemone had been identified as Homostichantus duerdemi by mistake.

Chemistry of Peptides and Proteins, Vol. 1

are mostly used for modification of tryptophan residues in proteins and peptides. NBS is considered to be the most reactive and selective reagent acting upon tryptophan at pH 4.0. Spande and Witkop showed that the 10-20-fold excess of NBS led to the modification of tyrosine and histidine residues. Titration of neurotoxin 1 (RTX-1) was performed in acetate buffer (pH 4.0) in the presence and absence of denaturants - urea and guanidine chloride. The calculation of modified tryptophan residues was carried out by the formula, suggested by Spande and Witkop /3/. The reaction was monitored spectrophotometrically by the absorption decrease at 280 nm (Fig.1). Simultaneously, the toxicity was estimated by intraperitoneal injections in strainless white mice. The results are given in Table 1.

Table 1. Modification of RTX-1 with NBS.

Molar excess	$\frac{\text{RTX-1 oxid.}}{\text{RTX-1}}$	$\frac{LD_{100}\ \text{mod.}}{LD_{100}\,\text{native}}$
0.8	0.175	1.6
1.2	0.39	2.0
1.6	0.7	2.4
2.0	0.92	4.5
2.4 (6 M guanidine chloride)	1.05	10.0
2.8 (8 M urea)	1.17	12.0

A two-fold molar excess of NBS was used for reaction, which allowed to suggest the modification of the only tryptophan residue. This suggestion was confirmed by the amino acid analysis data and UV-spectroscopy. The differential UV-spectra, taken during the titration of neurotoxin 1 with NBS, are represented in Fig.2. These spectra have negative maxima at 284 and 292 nm and a wide peak at 310 nm, which is characte-

ristic of tryptophan residue oxidation.
The analysis of CD-spectra, molecular weight determination and the analysis of N-terminal amino acids of modified (by the tryptophan residue) neurotoxin 1 showed that the modification was not accompanied by the cleavage of peptide bond. It should be noted that the modification of tryptophan residue in the presence of denaturants induced considerable alteration in the toxin activity. The decrease in activity can be supposed to be connected with the change in the toxin molecule conformation or with the partial modification of another functionally important residue.

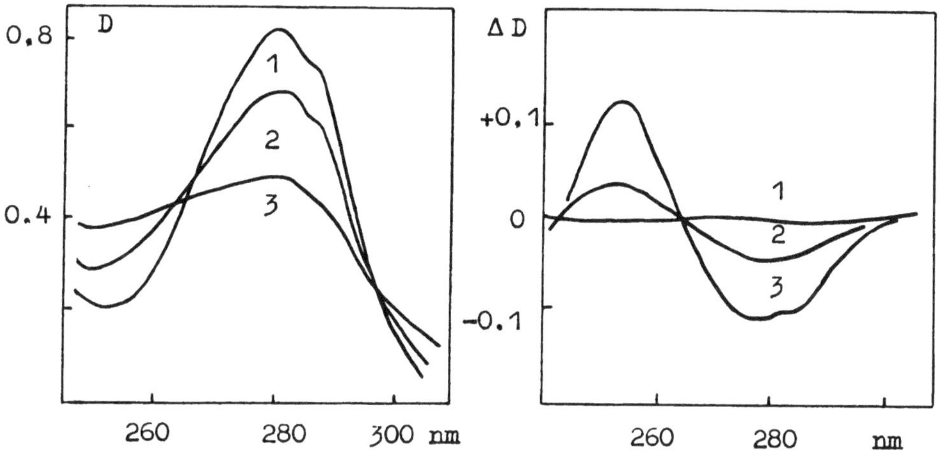

Fig.1. UV-spectra of neurotoxin 1 and neurotoxin 1 modified with NBS (1 - RTX-1; 2 - RTX-1 + 0.5 molar excess of NBS; 3 - RTX-1 + 2 molar excess of NBS)

Fig.2. Differential spectra of RTX-1 and the modified RTX-1 with NBS (designations are the same).

The Koshland /4/ reagent (HNB-Br) is considered to be another highly selective modifying tryptophan reagent. Titration of neurotoxin 1 with HNB-Br was performed in acetate buffer at pH 4.0. The calculation of modified residues was carried out using the formula suggested by Norton and Koshland /5/. The reaction was monitored by the formation of oxynitrobenzyl chromophore, which gave the pH-dependent increase in absorption at 410 nm (Fig.3). Simultaneously, the toxicity determination was carried out. The results are given in Table 2. The absence of by-modifications was shown by the amino acid analysis data. As is seen from Table 2, even at a large excess of the reagent about 50% of activity was still preserved.

Table 2. Modification of RTX-1 with HNB-Br.

Molar excess of HNB-Br	RTX-1 mod. / RTX-1	LD_{100} mod. / LD_{100} native
0.6	0.10	1.0
1.5	0.32	1.66
5.0	0.42	1.8
10.0	0.75	2.0
15.0	1.1	2.2

The modification of tryptophan residue with the Koshland reagent and NBS led to insignificant loss of the toxin activity. Thus, the data obtained permit to conclude that tryptophan residue was not essential for the toxicity of RTX-1 but its presence in RTX-1 is probably essential for RTX-1 - receptor interaction, as the alteration in neurotoxin activity takes place.

Lysine residue modification. Acylation with anhydrides and arylation with substituted nitrobenzenes are the mostly used methods of modifying lysine residues in proteins. No modification type is entirely specific, as these reagents easily react with tyrosine, histidine, cysteine, threonine, serine and methionine /6/. Gilbert et al. /7/ showed that in contrast to acylating and alkylating reagents 2,4-pentanedione was the specific reagent for modifying lysine residues, arginine was the only exception. However, the reaction of 2,4-pentanedione with the primary amines was 20 times faster than that with guanidine at pH 9.0.

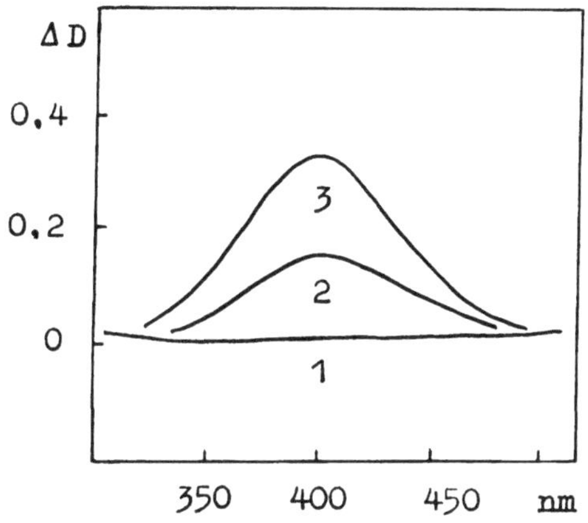

Fig.3. UV-spectra of RTX-1 modified with HNB-Br. (1 - RTX-1; 2 - RTX-1 + 5 molar excess of HNB-Br; 3 - RTX-1 + 15 molar excess of HNB-Br).

That is why almost at any time it is possible to choose the reaction conditions at which the protein molecule can be modified by the lysine residues only. The reaction of neurotoxin 1 with 2,4-pentanedione was performed in sodium phosphate buffer (pH 7.0) or in potassium bicarbonate buffer (pH 9.0). The number of modified lysine residues was calculated by the formula suggested by Gilbert /7/. The extent of lysine modification was monitored spectrophotometrically by the absorption increase at 310 nm, as the forming enamine had the absorption maximum at this wave length ($\varepsilon_{310} = 2.1 \times 10^{-4}\ M^{-1}\ cm^{-1}$), and the native protein did not practically absorb at this region. Simultaneously the toxicity was determined in the usual way.
The results are given in Table 3 and Fig. 4.

Table 3. Modification of RTX-1 with 2,4-pentanedione

RTX-1 mod. / RTX-1	LD_{100} mod. / LD_{100} native
0.6	1.0
1.6	3.3
2.7	10.0

Fig.4. UV-spectra of RTX-1, modified with 2,4-pentanedione (1 - RTX-1, pH 9; 2-reaction 22h, pH 9; 3-100 h, pH 9, 20°).

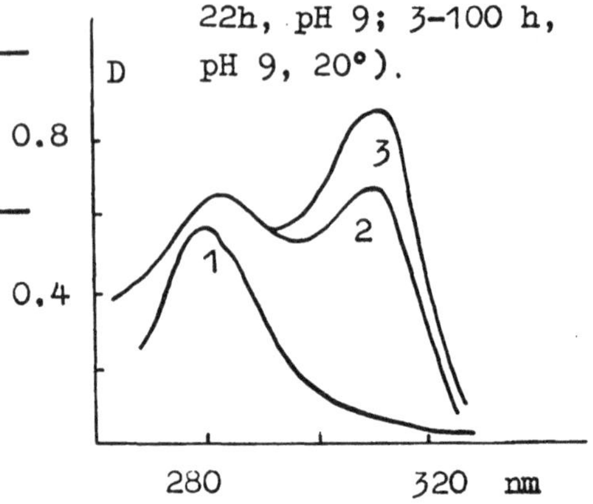

Lysine amino group was regenerated by treatment with 0.1 M hydroxylamine at pH 6.0. The toxin activity at such a treatment was completely restored and the absorption at 310 nm, corresponding to the modification product, completely disappeared. It should be noted that the modification of arginine residue of neurotoxin 1 with 2,4-pentanedione was not observed even at pH 9.2 and a 100 hours incubation at 20°C. The analysis of CD spectra of free amino group modified neurotoxin 1 showed that conformation did not dramatically change. The observed decrease in neurotoxin activity as a result of modification of the three lysine residues could be explained by the alteration of the molecule charge.

<u>Arginine residue modification</u>. Several reactions, based on the condensation of the guanidino group with different diketones and aldehydes were suggested to modify arginine residues in proteins /8-11/. The method of modification arginine residues with malonic aldehyde attracted our attention by its completeness of transformation and specificity of reaction. Besides, this modification permits to obtain a derivative which is electrically neutral in the greater part of pH range

in contrast to the native toxin. The reaction of neurotoxin 1 with 50-fold molar excess of malonic dialdehyde was carried out in 10 N HCl for an hour at 4°C. Malonic dialdehyde was formed from tetraethoxypropane (TEP) directly introduced into reaction mixture. The process of reaction was monitored by the formation of N^{δ}(2-pyrimidinyl)-ornithine, which absorbed at 315 nm; native toxin did not absorb in this region (Fig.5). The determination of toxicity was carried out in white mice in the usual way. The completeness of transformation of arginine residues into residues of a pyrimidine derivative was shown by the absence of a specific coloration at the treatment of the product with Sakaguchi reagent. The modified toxin completely loses activity while the native one keeps 100% activity under the reaction conditions ($\frac{LDmod.}{LDnat.} > 20$). The absence of cleavages in the polypeptide chain of modified neurotoxin 1 was shown by the analysis of the N-terminal amino acids and molecular weight determination.

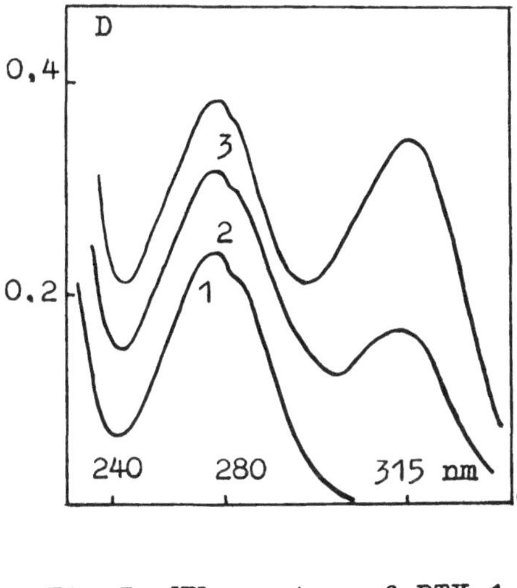

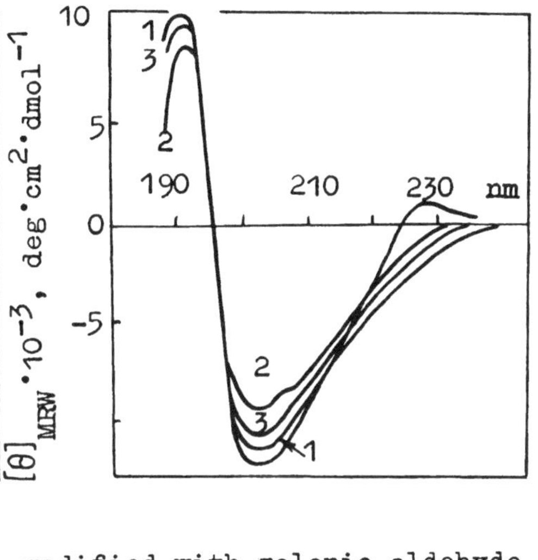

Fig.5. UV-spectra of RTX-1, modified with malonic aldehyde (1-RTX-1; 2-RTX-1 + 5 molar excess of TEP; 3-RTX-1 + 50 molar excess of TEP).

Fig.6. CD-spectra of RTX-1 and RTX-1, modified with NBS (1), TEP (2), 2,4-pentanedione (3).

The analysis of CD spectra (Fig.6) showed that modification of arginine residue does not lead to the total change in molecule conformation, which permits us to assume an important role of arginine residues for the physiological activity of neurotoxin 1.

Thus, this work allowed us to come to two conclusions: first, the native structure of neurotoxin 1 from R.macrodactylus is kept intact regardless of the used modification method; second, the arginine residues are functionally important for the physiological activity of neurotoxin. It should be noted that toxicological experiments on animals are not sufficient to come to the conclusion of the appurtenance of this or that amino acid to toxophore centre. The toxic effect, even in the simpliest case, is achieved as a result of some stages of the toxin transport to the affected tissue, binding it with the corresponding receptor and its own toxic action. The direct study on the interaction of toxins with their molecular receptors is required.

References

1. Zykova, T.A., Kozlovskaya, E.P., Elyakov, G.B.: V Vsesoyuzny sympozium po khmii i phyzike belkov i peptidov. Baku, st. 70 (1980).
2. Rathmayer, W.: Advances in Cytopharmacology, 3, edited by B.Ceccarelli and F.Clementi, Raven Press, New York, 335-344 (1979).
3. Spande, T.F., Witkop, B.: Methods in Enzymology, 11, Academic Press, New York, 498-532 (1967).
4. Norton, H.R., Koshland, D.E. Jr. : Methods in Enzymology, 11, Academic Press, New York, 556-565 (1967).
5. Norton, H.R., Koshland, D.E. Jr. : J. Amer. Chem. Soc., 86, 1126-1130 (1965).
6. Means, G.E. and Feeney, R.E.:Chemical Modification of Proteins.San Francisco,Calif., Holden Day 1971

7. Gilbert, H.F. and O'Leary, M.H.: Biochemistry 14, 5194-5199 (1975).
8. King, T.R.: Biochemistry 5, 3454-3459 (1966).
9. Riordan, J.F.,: Biochemistry 12, 3915-3923 (1974).
10. Patthy, L., Smith, E.L.: J. Biol. Chem., 250, 557-564 (1975).
11. Toi, K., Byhum, E., Norris, E. and Itano, H.A.: J. Biol. Chem., 242, 1036-1043 (1967).
12. Muranov, A.V., Modyanov, N.N.: Bioorganicheskaya Khimia, 5, 210-216 (1979).

COBRA VENOM NEUROTOXINS: CONFORMATION AND INTERACTION WITH THE ACETYLCHOLINE RECEPTOR

Victor Tsetlin, Evert Karlsson*, Yuri Utkin, Alexandr Arseniev, Kirill Pluzhnikov, Alexandr Surin, Vladimir Pashkov, Vladimir Bystrov, Vadim Ivanov

Shemyakin Institute of Bioorganic Chemistry, USSR Academy of Sciences, Moscow, USSR

*Institute of Biochemistry, Uppsala, Sweden

Introduction

Neurotoxins from the cobra venom served as invaluable tools for isolating the acetylcholine receptor (AchR) of the postsynaptic membrane in a pure form (1,2). A lot of information is currently available for the neurotoxins of postsynaptic action as to their chemical and spatial structure in solution and in a crystal form. In conjunction with data available on the biological activity of the modified neurotoxin derivatives, it allowed to formulate some ideas on the nature of the neurotoxin "active site". Knowledge of the neurotoxin spatial structure has made it possible to gain an insight into certain features of the AchR topography.

Most of the detailed information on the solution conformation of the neurotoxins was provided by NMR spectroscopy. With proteins, it is rarely possible to get a fairly subtle picture when studying the object under consideration alone - the necessary assistance is rendered by the studies on homologous proteins and chemically modified derivatives.

Results

At the Shemyakin Institute of Bioorganic Chemistry, USSR Academy

Chemistry of Peptides and Proteins, Vol. 1

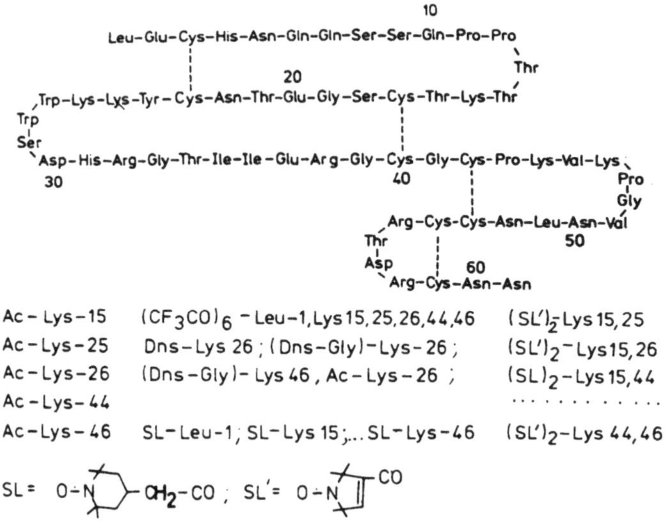

Fig. 1. Amino acid sequence of neurotoxin II *Naja naja oxiana* (3) and its chemically modified derivatives.

of Sciences, we have been studying the spatial structure of neurotoxin II (NT-II) from the venom of the Central Asian cobra *Naja naja oxiana*. Along with the native toxin, its chemically modified derivatives (Fig. 1) have been analyzed. This allowed a wider use of various NMR techniques and has rendered assistance to obtain quantitative results by EPR and fluorescence spectroscopy.

In respect of the mono-acetylated derivatives (Fig.1) it was found that blocking the lysine positive charges does not affect the neurotoxin conformation as follows from CD and fluorescence data (4,5). Biological activity is essentially preserved as indicated by the toxicity assays in mice and electrophysiological tests on the frog muscle. The acetylated derivatives have made possible the signal assignments in the ^{19}F NMR spectra of trifluoroacetylated NT-II, which in turn provided the data on the microenvironment of the amino groups (4). The radiationless energy transfer efficiency from the tryptophan residues onto the dansyl group measured by fluorescence spectroscopy for dansylated derivatives, allowed us to elucidate mutual disposition of the respective groupings (5, 6). At present all the other α- or ε-amino singly dansylated derivatives of NT-II are being investigated.

Extremely valuable proved to be the spin labeled derivatives. By varying the reaction conditions it was possible to achieve a predominant formation of either mono- or di-labeled species and then to separate them (5, 7), as illustrated for the former in Fig. 2.

Localization of the label in each spin labeled derivative was accomplished by reduction, carboxymethylation and tryptic hydrolysis followed by the EPR identification of the spin labeled peptide in the fractionated hydrolysate.

The mono-spin-labeled compounds were studied by ^{1}H NMR spectroscopy. The signals of protons in spatial proximity to the unpaired electron of spin label experience broadening, from the extent of which the respective distances can be calculated (Fig. 3). For example, a difference in the distances to the ortho and meta protons of Tyr 24 gives an idea of the phenol

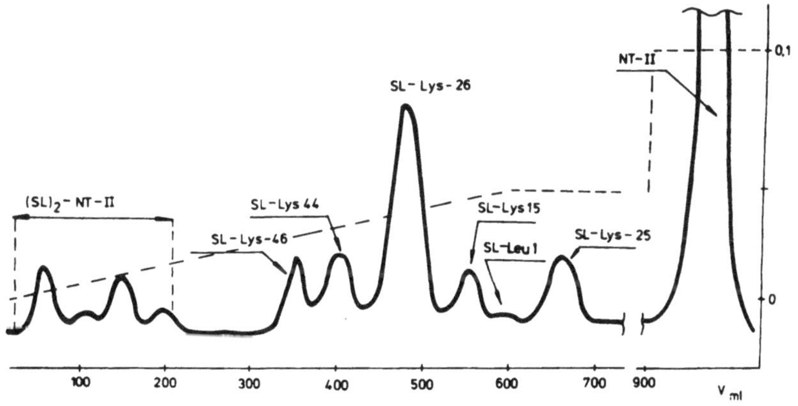

Fig. 2. Separation of NT-II spin labeled derivatives by Bio-Rex 70 column chromatography.

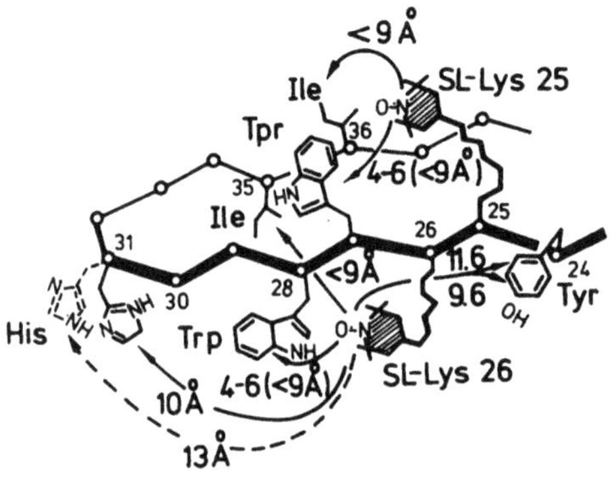

Fig. 3. Intramolecular distances in the central loop fragment derived from the ^{1}H NMR and fluorescence studies on mono-spin-labeled derivatives.

ring orientation in respect to the spin label. An increase in distance to His 31 from 10 Å at neutral pH to 13 Å at acidic pH is a manifestation of molecular dynamics: the imidazole ring moves upon protonation. The spin label at Lys 25 is close to the Trp 27 and Ile 36 side chains, whereas the label at Lys 26 lies in proximity to the Trp 28 and Ile 35 side chains. Such an alternating disposition of the side chains is indicative of an antiparallel β-structure at this segment of the neurotoxin molecule.

A comparative fluorescence study on NT-II and its singly acetylated or spin labeled derivatives unambiguously proved the proximity of the Trp 28 indole ring and Lys 26 amino group (5). The spin label bound to Lys 26 completely quenches the fluorescence of Trp 28, whereas the spin label at Lys 25 exerts the same effect on the Trp 27 fluorescence. These facts mean that the distances from the indole rings to the unpaired electrons of the respective spin labels are less than 4-6 Å. Noteworthy, in this case fluorescence spectroscopy permitted more accurate determination of the space between the groupings than did ^{1}H NMR.

In total, spectroscopic studies on native NT-II and its congeners illuminated the microenvironment of many amino acid side-chains. The disclosed interactions between the residues lying apart in the amino acid sequence allow us to trace the folding of the polypeptide backbone. For example, the proximity of the Lys 26 spin label to Val 43 and Val 49 (Fig. 4) and that of Lys 46 to Trp 28, His 31 and Ile 35 (as inferred from the ^{1}H NMR studies of the mono-spin-labeled derivatives) clearly shows that the two disulfide-bridged loops 17-40 and 42-53, which in a conventional "two-dimensional" picture of neurotoxins are depicted as being the most remote (see Fig. 1), are factually in proximity. Such an arrangement is confirmed by the closeness of trifluoroacetyl groups at Lys 26 and Lys 46 found earlier by ^{19}F NMR (4), and a short distance (~12 Å) measured by EPR in the Lys 26, Lys 46 di-spin-labeled derivative. The data on the NT-II solution structure are in general agreement with the X-ray structure of erabutoxin *b* (8-10); in particular it concerns the closeness of the two loops depicted in Fig. 4. The homologous neurotoxins results are usually

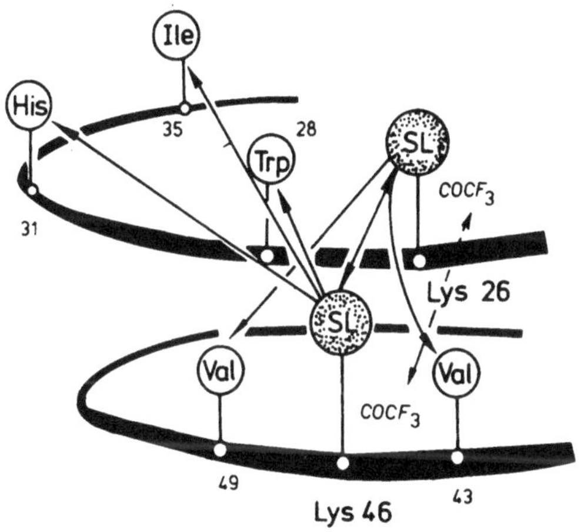

Fig. 4. Mutual dipposition of NT-II fragments 26-35 and 43-49.

interpreted as the evidence for a close similarity of the solution and crystal structures. As a rule, the comparisons are only qualitative and reveal a proximal or distant relation of residues, but not in terms of intramolecular distances. Noteworthy, from the distances measured between the spin label pairs in ten di-labeled derivatives of NT-II (7), a consistent picture of the arrangement of spin labels (and, consequently, lysine side chains) could be derived (Fig. 5). These distances were also realized in the model of the NT-II three-dimensional structure built on the basis of the erabutoxin *b* coordinates (10).

For this purpose it was necessary, however, to make some sort of alteration in the Lys 15 side chain orientation; this may be the consequence of a small conformational difference between the two toxins, or between their solution and crystal structures. In spite of the marked differences either in conformational stability or microenvironment of some residues for various neurotoxins (4), the X-ray structure of erabutoxin *b*

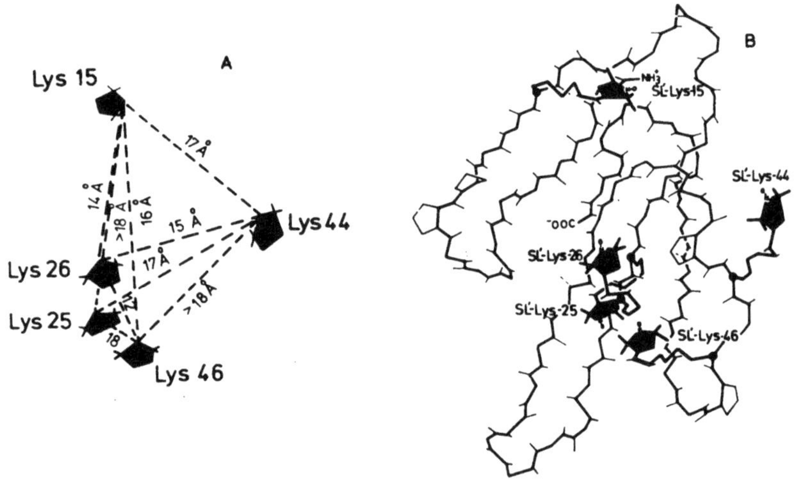

Fig. 5. Distances between labels in NT-II di-spin-labeled derivatives (A) and disposition of the labels in the model of NT-II spatial structure (B).

offered an adequate description of the spatial structure in solution for the whole family of short neurotoxins.

Incorporation of labels into various sites of the NT-II molecule persued a dual purpose: we have been able, for the first time, using EPR and fluorescence spectroscopy, to obtain direct information on the neurotoxin regions involved in binding its target, the acetylcholine receptor of the postsynaptic membrane (11, 12). The receptor was isolated from the electric organs of *Torpedo marmorata* using affinity chromatography on neurotoxin-Sepharose in nonionic detergents.

Binding of dansylated neurotoxins to AchR is accompanied by an increase in the fluorescence intensity and the blue shift of

Table 1. EPR data for NT-II spin labeled derivatives and their AchR complexes[a].

Labeled residue	Correlation time τ_{obs}, nsec[b]		Broadening by paramagnetic probe, $k_w/2k_p$[c]			
			$K_3Fe(CN)_6$		$Ni(CH_3COO)_2$	
	free	bound	free	bound	free	bound
Leu 1	1.1	>10	1.5	2.3	1.2	7.5
Glu 2	3.8	>20	0.5	>4	1.3	1.7
Lys 15	0.8	2.0	1.3	4.4	0.9	1.1
Lys 25	2.0	>20	2.3	4.2	1.5	2.8
Lys 26	2.0	>20	2.9	>10	1.5	6.2
His 31	0.8	>20	0.5	>10	1.0	2.4
Lys 44	0.5	0.9	1.0	1.2	1.2	0.8
Lys 46	0.9	2.0	1.1	10	0.9	7.0

[a]Spectra taken in 0.05 M Tris-HCl, pH 7.5, 0.025% Triton X-100, 0.02% NaN_3, at 21°C;

[b]Calculated in terms of isotropic model

[c]Describes relative microviscosity around a spin label, higher values corresponding to lower accesibility for paramagnetic probes

the emission maximum. When the spin labeled neurotoxins combine with AchR, their signals in the EPR spectra broaden to a various degree. The dissociation constants and 2:1 neurotoxin-AchR stoichiometry were determined in titration experiments (12).
By adding paramagnetic probes to the solutions of neurotoxin-AchR complexes, it is possible to distinguish whether a spin label is accessible and, consequently, exposed into solution. The EPR spectra of the free neurotoxin labeled at Lys 44 and its AchR complex are equally broadened by the probes (Table 1) indicating the label noninvolvement in binding. For all the rest of the spin labeled derivatives a decrease in the accessibility or complete inaccessibility indicates that the labels are contacted to the receptor or are embedded into it to a various degree.
The results obtained allowed us to suggest a model for the neurotoxin-AchR binding (Fig. 6). The interaction is of a multipoint character: the labels at Leu 1, His 31, Lys 15, 25 26 and 46 are all involved in the binding, the most pronounced for the latter two. Obviously, the binding of a label does not prove the role of the group to which it is attached; incorpora-

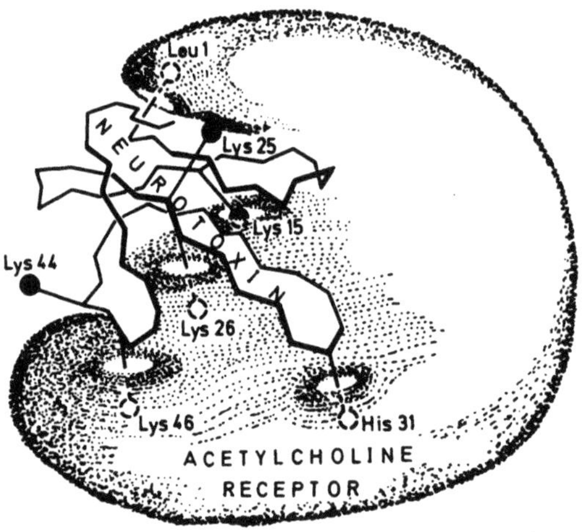

Fig. 6. A model for neurotoxin-AchR binding .

tion of the spin labels into lysine side chains abolishes a positive charge that might be of importance for binding. The role of such a charge is most noticeably seen in a decrease both in the toxicity and efficacy of AchR binding upon modifying Lys 26. The incorporated labels seem to be screened from the medium due to the AchR binding of some nearby non-modified groupings of the neurotoxin.

Our data show that the neurotoxin covers the AchR with at least three of its four disulfide-confined loops, and the interaction is not reduced to the penetration of the central loop into a hypothetical cleft as proposed in (8). Our data are in better agreement with the idea of the neurotoxin active site, including two loops, formulated by Low and co-authors (10).

Snake venom neurotoxins, interesting as they are by themselves, play a subordinating role as tools in elucidating the mechanisms of nerve impulse transmission and other problems of neurobiology. Fortunately, labeled neurotoxin derivatives provided new information about binding surface not only of the neurotoxin, but of AchR as well. Since the neurotoxin conformation is essentially preserved in solutions of different pH and polarity or upon label incorporation, one may assume that the inter-label distances, measured for di-spin-labeled compounds or deduced from the X-ray structure, are essentially preserved in complexes with AchR. It allows us to assess the dimensions of the AchR site where neurotoxins are bound . Experimentally determined distances between the lysine residues range from 12 to 20 Å, while the distance between the most far separated Leu 1 and His 31 labeles is about 30 Å as deduced from the erabutoxin *b* X-ray structure (10). Since the diameter of one AchR molecule composed of 5 subunits is about 75-80 Å, it is quite evident that a neurotoxin cannot accomodated within one α-subunit (whose function is to bind the natural neurotransmitter, acetylcholine), but should also contact other subunits. The spin labels at Lys 15, Lys 26, Lys 46 and His 31 are situated on the same side of the disk-like neurotoxin molecule and probably interact with one subunit, whereas oppositely directed

Leu 1 and Lys 25 labels may interact with the neighboring subunits. Hopefully, further studies utilizing neurotoxins with spectral and/or photosensitive (1) labels will clarify the spatial organization of AchR.

References

1. Karlsson, E.: in Handbook of Experimental Pharmacology (Lee, C.Y. ed.) vol. 52, pp 159-212, Springer Verlag, Berlin (1978).

2. Heidmann, T., Changeux, J.-P.: Annu. Rev. Biochem. 47 317-358 (1978).

3. Grishin, E.V., Sukhikh, A.P., Lukjanchuk, N.N., Slobodyan, L.N., Lipkin, V.M., Ovchinnikov, Yu.A., Sorokin, V.M.: FEBS Lett. 36, 77-78 (1973).

4. Tsetlin, V.I., Arseniev, A.S., Utkin, Yu.N., Gurevich, A.Z., Senyavina, L.B., Ivanov, V.T., Bystrov, V.F., Ovchinnikov, Yu.A.: Eur. J. Biochem. 94, 337-346 (1979).

5. Surin, A.M., Utkin, Yu.N., Pluzhnikov, K.A., Efremov, E.S., Tsetlin, V.I.: Bioorgan. Khim. 7, 342-356 (1981).

6. Utkin, Yu.N., Pashkov, V.S., Surin, A.M., Tsetlin, V.I., Ivanov, V.T.: in Peptides 1978, pp 397-400. Wrocław University Press, 1979.

7. Ivanov, V.T., Tsetlin, V.I., Karlsson, E., Arseniev, A.S. Utkin, Yu.N., Pashkov, V.S., Surin, A.M., Pluzhnikov, K.A., Bystrov, V.F.: in Natural Toxins (Eaker, D., Wadström, T. eds) pp 523-530, Pergamon Press, Oxford and New York (1980).

8. Tsernoglou, D., Petsko, G.A.: FEBS Lett. 58, 1-4 (1976).

9. Low, B.W., Preston, H.S., Sato, A., Rosen, L.S., Searl, F.E., Rudko, A.D., Richardson, J.S.: Proc. Natl. Acad. Sci. USA 73, 2991-2994 (1976).

10. Kimball, M.R., Sato, A., Richardson, J.S., Rosen, L.S. Low, B.W.: Biochem. Biophys. Res. Commun. 88, 950-959 (1979).

11. Tsetlin, V.I., Karlsson, E.: in Proceedings of the Third Symposium on Plant, Animal and Microbial Toxins, London 1978 (Habermehl, G., Mebs, D., eds) p. 106, THD Schriftenreihe Wissenschaft und Technik 14, 1979

12. Tsetlin, V.I., Karlsson, E., Arseniev, A.S., Utkin, Yu.N., Surin, A.M., Pashkov, V.S., Pluzhnikov, K.A., Ivanov, V.T., Bystrov, V.F., Ovchinnikov, Yu.A.: FEBS Leet. 106, 47-52 (1979).

13. Hucho, F.: FEBS Lett. 103, 27-32 (1979).

THE STUDY OF THE SCORPION NEUROTOXIN MEMBRANE RECEPTORS

Nickolai Soldatov, Victor Kovalenko, Eugenii Grishin
Shemyakin Institute of Bioorganic Chemistry, USSR Academy of Sciences, Moscow, USSR

Introduction

The ascertainment of the molecular mechanisms of the nerve impulse propagation is going to be one of the most intriguing problem of the modern bioorganic chemistry. Now the leading role of the fast sodium channels in the realization of excitation phenomena is well known. Up today some fundamental parameters of the functioning of sodium channels are received but its molecular characteristics are still under question (1). On the other hand now it is evident that sodium channel is a combination of receptors for at least three groups of natural neurotoxins (2). The use of these substances led to the distinction of some functional fragments of sodium channel. For example, tetrodotoxin (TTX) interacts with carboxyl group near the outside of the channel thereby blocking the ionic current (3). Aconitine and other steroidal neurotoxins change the properties of the selective filter (4). Scorpion as well as sea anemone toxins slow down the sodium channel inactivation upon interaction with potential-dependent gating mechanism (5). So the isolation and following study of these toxin receptors could bring some essential information about the molecular organization of sodium channel.

Results

In this work the mice neurotoxin isolated from the Central Asian scorpion *Buthus eupeus* has been used as an instrument for the study of fast sodium channels. The amino acid sequence of this toxin was established by means of combined automatic and manual Edman degradation (6):

```
                                      10
Val-Arg-Asp-Gly-Tyr-Ile-Ala-Asp-Asp-Lys-Asp-Cys-Ala-Tyr-Phe-
             20                                            30
Cys-Gly-Arg-Asn-Ala-Tyr-Cys-Asp-Glu-Glu-Cys-Lys-Lys-Gly-Ala-
                                      40
Glu-Ser-Gly-Lys-Cys-Trp-Tyr-Ala-Gly-Gln-Tyr-Gly-Asn-Ala-Cys-
                   50                                      60
Trp-Cys-Tyr-Lys-Leu-Pro-Asp-Trp-Val-Pro-Ile-Lys-Gln-Lys-Val-

Ser-Gly-Lys-Cys-Asn .
```

The polypeptide chain of 65 amino acid residues comprises 4 disulfide bridges and a high quantity of lysine and aspartic acid residues, but lacks threonine, methionine and histidine. The affinity constant of this toxin in respect to nerve membranes is in the range from 10 to 100 nM, and it is important that the depolarization of the membrane reduces the affinity (5). That is why the radioactively labeled photosensitive derivative of this toxin has been used for the identification of the molecular organization of its receptor in sodium channel. The dinitrofluorophenylazide has been used for the labeling of lysine-57 of M_{10} (7). This reagent is much more reactive with respect to free protein amino groups, than traditionally used NAP (8). The monosubstituted toxin derivative (DNPA-M_{10}) possessed the characteristic absorptions at 355 and 415 nm and so it could be photoactivated at wavelengths apart from the absorption of common aromatic protein groups.
The last step of the modification was radioiodination, and the obtained labeled photosensitive derivative of the scorpion neurotoxin possessed the high level of biological activity, that has been shown by a number of voltage clamp experiments with neuroblastoma cell membranes (7). Thus far it was found that toxin derivative irreversibly modified sodium channels upon UV-light irradiation. In other words nitrenes of DNPA-M_{10}

interacted directly with gating structure.

The estimation of the in vitro specific binding of ^{125}I-DNPA-M_{10} has been done in competition with native scorpion toxin for the receptors in rat brain synaptosomes (9). The irradiation was performed at wavelengths higher than 320 nm that didn't induce any change in properties of sodium channels. It was found that the increase of the native toxin concentration reduced the bindung of ^{125}I-DNPA-M_{10} up to 50 % (Fig.1). Sea anemone toxin ATX II, which has been kindly supplied by Dr.L. Beress, also inhibited the binding of ^{125}I-DNPA-M_{10} by rat brain synaptosomes in competitive manner with an apparent constant of about 200 nM (Fig.2). Ca^{++} ions, too, effectively suppressed the reception of scorpion toxin. But at the same time, these ions enhance the non-specific binding of ^{125}I-DNPA-M_{10}. This effect could be due to the increased non-specific interactions of the toxin with lipid phase, the state of which is known to be modified by Ca^{++} ions.

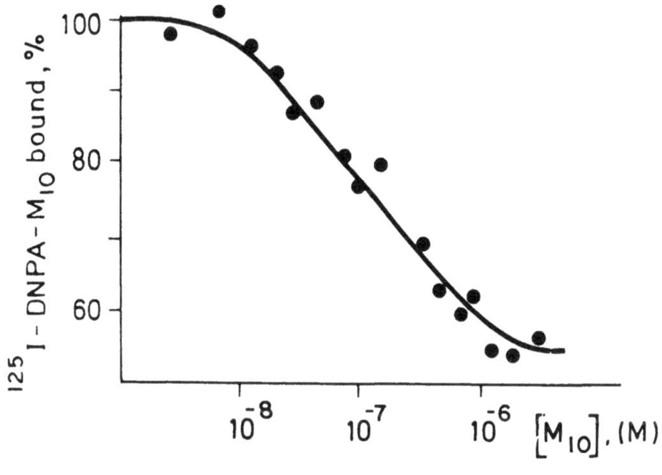

Fig. 1. Competitive interaction of ^{125}I-DNPA-M_{10} and M_{10} upon photo-induced modification of rat brain synaptosomes. For M_{10} $K_{0.5}$ ca. 100 nM, number of binding sites ca. 100± 20 fmol/mg protein (9).

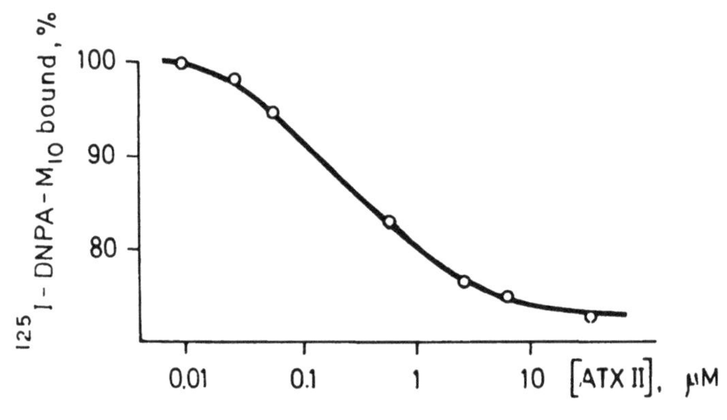

Fig. 2. Inhibition of ^{125}I-DNPA-M_{10} photolabeling of synaptosomes by ATX II (9).

Thus far the direct evidences have been received that labeled toxin derivatives do interact with the receptor that belongs to the sodium channel. So we studied this receptor and carried out the comparative analysis of the respective binding components of the excitable membranes from different sources.
The identification of the receptor membrane components has been carried out by the difference analysis of the radioactivity in sections of the SDS-PAAG electropherogram which was received in the absence or in the presence of the excess of native toxin to suppress the specific covalent binding of ^{125}I-DNPA-M_{10} in the control run. Fig. 3a shows the results obtained in case of rat brain synaptosomes. The above curve is the densitogram and lower curve represents the profile of the specifically bound ^{125}I-DNPA-M_{10}. The main zone of radioactivity is present in the molecular weight region up to 20 KD, and its level is a high as tenths of pmoles per mg of synaptosomal protein. The extraction of modified membranes by different

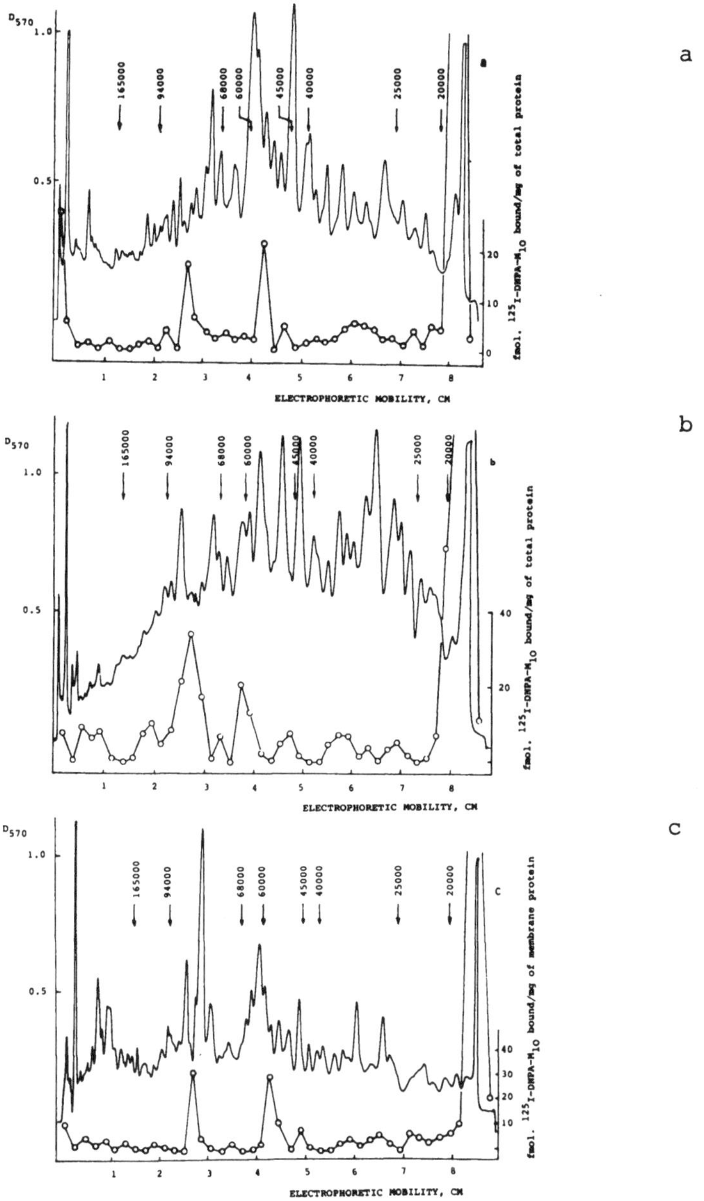

Fig. 3. SDS-electrophoretic analysis of ^{125}I-DNPA-M_{10} binding components in a) rat brain synaptosomes, b) neuroblastoma cells and c) crab nerve plasma membranes from walking legs of *Paralithodes kamtschatica*. The arrows show the mol. weights of reference proteins (9).

organic solvents didn't reduce significantly the radioactivity in this zone. So it could be suggested that the toxin in this part of the gel is bound to the lipid components of the membrane. This type of interaction could be responsible for the high level of non-specific binding of scorpion toxin to membrane.

Nevertheless the presence of two radioactive peaks was detected that could be suppressed by the excess of native toxin. Mol. weights of the respective components were 84 and 58 KD. Both peaks comprised the specifically bound toxin in the level that was very close to the binding capacity of the membrane. The true mol. weights of the receptor components were equal to 77 and 51 KD in the assumption that the binding of ^{125}I-DNPA-M_{10} to receptor is equimolar and the toxin mol. weight is near 7 KD.

Fig. 3b,c show the results that were obtained for neuroblastoma cells in culture and crab nerve plasma membranes. In these cases the irradiation also led to the covalent binding of ^{125}I-DNPA-M_{10} to a couple of membrane components of the similar mol. weights. In all cases the assumed error for mol. weight estimations makes up near 3 KD. So it became evident that the presence of two specifically binding components for ^{125}I-DNPA-M_{10} is typical for nerve membranes of both mammalian and crustacean. These components possess the apparent mol. weights of 76 and 51 KD and are suggested to belong to the fast sodium channel that most probably has a similar organization for excitable membranes of different types of animals. The unique modification of two membrane components could be accounted for their spatial proximity with respect to the point of fixation of the toxin molecule in membrane.

Crab nerve plasma membranes were subjected to covalent modification by ^{125}I-DNPA-M_{10} upon irradiation and than membranes were solubilized by Lubrol PX. The following sucrose density gradient sedimentation has shown that covalent toxin-receptor complex was present mainly in two fractions that possess the coefficients of 9,5 and 16 S. In separate experiments it was found that only these fractions contain also the solubilized tetrodotoxin receptor.

The covalently labeled 77 KD component was identified in both fractions upon SDS electrophoresis. This individual protein has been isolated by means of large-scale electrophoretical technique and its amino acid composition includes (in mol. %): 14-18 Asp, 4,55 Thr, 6,49 Ser, 12,05 Glu, 8,64 Pro, 8,73 Gly, 7,60 Ala, 5,53 Val, 1,33 Met, 3,34 Ile, 7,07 Leu, 2,13 Tyr, 5,27 Phe, 5,17 His, 3,85 Lys and 4,03 Arg. N-Terminal amino acid residue was Ala. The amino acid composition is typical for an amphiphilic membrane protein and so far it might be attributed to the transmembrane protein structure of the channel type that possesses spacious regions exposed, both, into aqueous and lipid phases.

We hope that the forthcoming investigation of described membrane components as well as its reconstruction will further clear up the questions of the molecular organization and function of the sodium channels of excitable membranes.

References

1. Hille, B.: Biophys. J. 22, 283-294 (1978).
2. Catterall, W.A.: Ann. Rev. Pharmacol. Toxicol. 20, 15-43 (1980).
3. Sigworth, F.J., Spalding, B.C.: Nature (L) 283, 293-295 (1980).
4. Mozhayeva, G.N., Naumov, A.P., Negulyaev, Yu.A., Nosyreva, E.D.: Biochem. Biphys. Acta 466, 461-473 (1977).
5. Mozhayeva, G.N., Naumov, A.P., Nosyreva, E.D., Grishin, E.V.: Biochem. Biophys. Acta 597, 587-602 (1980).
6. Grishin, E.V., Soldatova, L.N., Shakhparonov, M.I., Kazakov, V.K.: Bioorg. Khim. 6, 714-723 (1980).
7. Grishin, E.V., Soldatov, N.M., Ovchinnikov, Yu.A., Mozhayeva, G.N., Naumov, A.P., Zubov, A.N., Nisman, B.Ch.: Bioorg. Khim. 6, 724-730 (1980).
8. Wilson, D.F., Miyata, Y., Erecinska, M., Vanderkooi, J.M.: Arch. Biochem. Biophys, 171, 104-107 (1975).
9. Grishin, E.V., Soldatov, N.M., Ovchinnikov, Yu.A.: Bioorg. Khim. 6, 914-922 (1980).

DEHYDRATION OF IONOGENIC GROUPS IN PEPTIDE LIGANDS ON RECEPTOR SURFACE

G.V.Nikiforovich, S.A.Rozenblit, G.Chipens
Institute of Organic Synthesis, Latvian SSR Academy of Sciences, Riga, USSR

The space structure studies on oligopeptides are rather common today and are generally aimed at the establishment of conformation-function relationship in peptides. Implicit in the realization of this relationship is the assumption that a conformation complementary to the specific receptor surface, i.e. a "biologically active" conformation, is present among the set of stable molecular structures of peptide ligands (cf.e.g. 1,2). Unfortunately, the results of physico-chemical studies on peptide conformations in solution probably fail to provide a detailed description of the set of possible stable oligopeptide structures since extraction from the averaged values of experimentally measured parameters of the contribution due to any particular conformer is hardly feasible today. Nevertheless, the methods of semi-empirical conformational analysis (3,4) providing each possible molecular structure with their conformational energy estimation allow not only to reduce significantly the number of accessible oligopeptide conformations (in particular, the types of peptide backbone structures), but also to attain, in some cases, descriptions of "biologically active" conformations of peptide ligands (5-9).
The results of semi-empirical conformational analysis are generally well correlated with the experimental evidence derived from physico-chemical studies, although there are some peculiarities characteristic of three-dimensional structures the presence of which is unequally assessed by computation or in

Chemistry of Peptides and Proteins, Vol. 1

the experiment. One is to mention here the close spacing of alternatively charged ionogenic groups in low-energy conformations of oligopeptides that could be deduced from the calculations performed for enkephalin (5), angiotensin (7), bradykinin (10), tuftsin (11), bradykinin-potentiating peptides (12) and other molecules. In contrast, the experimental evidence (mainly NMR spectroscopy data) rejects such a possibility for bradykinin (e.g., 13) and tuftsin (14,15) in aqueous solution.

The observed discrepancy is probably due to the fact that the semi-empirical conformational analysis conducted with in vacuo approximation fails to take into consideration the effect of solvent on the molecule structure. The calculations reported in (10,11) were carried out bearing in mind the ionization of ionogenic groups in the corresponding peptides (the situation similar to that found in the aqueous solution under physiological conditions), however the values of dielectric constant ε were chosen equal to 3.5, i.e. the value characteristic of compact packing of amides (16). In aqueous solution, however, the hydration shells would prevent effective electrostatic interaction of ionogenic groups (that is responsible for their close spacing) almost leading to its complete inhibition. It is noteworthy that according to the data (17), interaction of ionic pairs in water must be assessed using $\varepsilon \sim 81$ (i.e., the value of water) starting with interionic distances 5-7 Å.

The above assumption was put to trial by comparing the results of two variants of energy calculations performed on tuftsin molecule. An oligopeptide immunostimulator with the amino acid sequence Thr-Lys-Pro-Arg, tuftsin contains four ionogenic groups: basic α-amino group, ε-amino group of the Lys residue and guanidyl group of Arg, and an acidic C-terminal carboxyl. In the first variant of calculations, all these groups were taken completely ionized, in the other variant-deionized. The computation procedure involves the analysis of 16 possible types of peptide backbone structures (combinations of local energy minima B ($\varphi \sim -140^{\circ}$, $\psi \sim 140^{\circ}$) and R ($\varphi \sim -60^{\circ}$, $\psi \sim -60$) for Thr and Pro residues (where angle φ is fixed at -60°) and B,L are minima ($\varphi \sim 60^{\circ}$, $\psi \sim 60^{\circ}$) for Arg and Lys residues) with concurrent refinement of side chain spacing at angles Θi according to a specially designed algorithm (19). The potential functions of interatomic interactions were identical with those described in (18); transconformation was assumed for the Lys-Pro peptide bond (14).

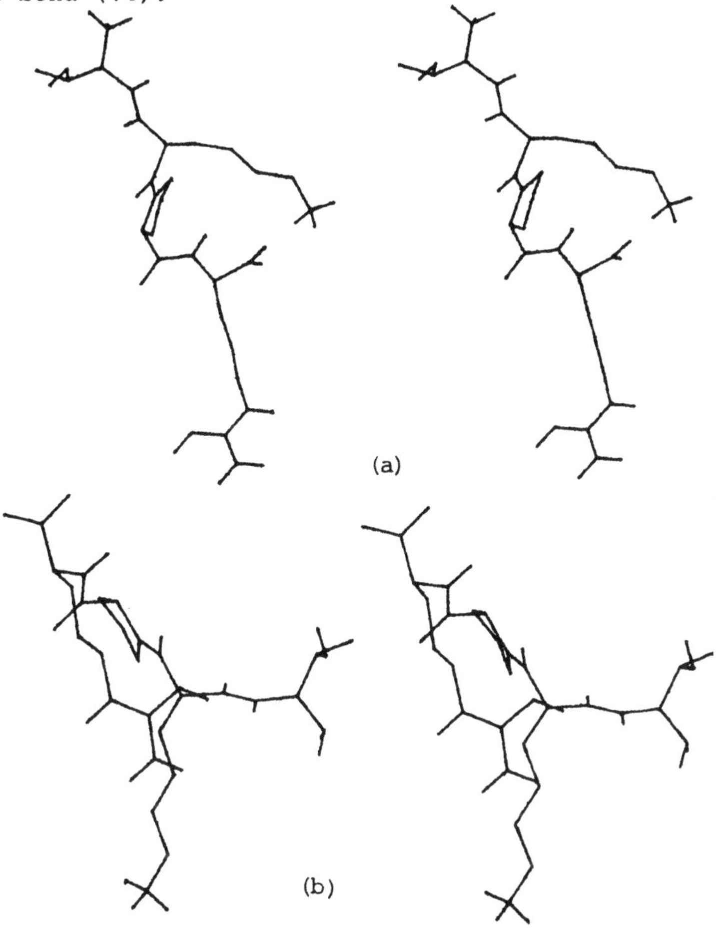

Fig.1. Lowest-energy structures of tuftsin molecule. Calculation with ionized(a) and deionized(b) ionogenic groups.

Table 1. Low-energy structures of tuftsin molecule

Residue	Angle*	Ionogenic groups ionized					Ionogenic groups deionized			
		BBRB	BBBB	BLRB	BBBL	BBRL	BBBB	BBRL	BLBB	BBRB
		1	2	3	4	5	1	2	3	4
Thr	φ	-117	-117	-117	-117	-117	-141	-146	-143	-146
	ψ	150	150	149	150	152	138	139	136	139
	χ_1	47	47	46	47	47	53	54	53	52
	χ_2	161	164	161	161	160	136	154	136	139
Lys	φ	-126	-126	55	-125	-125	-127	-126	52	-127
	ψ	121	121	110	115	129	134	127	105	129
	χ_1	-146	-160	-142	189	-158	-72	-159	-159	-76
	χ_2	-143	-98	-139	-103	-134	178	-178	-88	180
	χ_3	180	162	-179	170	163	-179	178	177	180
	χ_4	-157	-163	-153	-177	-164	180	-76	-179	180
Pro	ψ	-51	152	-48	111	-22	112	-41	120	-37
Arg	φ	-160	-112	-152	36	67	-132	48	-132	-128
	ψ	131	135	132	86	137	139	98	139	138
	χ_1	-159	-160	-159	-162	-165	-69	-68	-70	-70
	χ_2	180	-110	180	-178	-117	174	173	176	176
	χ_3	-179	73	-179	178	72	-174	-178	-175	-177
	χ_4	97	-106	97	96	-101	97	95	96	94
$U-U_{min}$ kcal/mole		0.0	1.7	4.7	4.9	7.2	0.0	2.7	3.5	4.1
Distance N^{ε}_{Lys}-COO, Å		3.1	5.4	3.2	3.0	8.3	6.8	7.3	6.5	12.2

* Internal rotation angle values according to (21).

The results of calculations are listed in Table 1, where 5 and 4 types of peptide backbone structures are described, respectively, which have been found to satisfy the requirement $U-U_{min} \leqslant 7.5$ kcal/mole, with the ionogenic groups in the molecule being either ionized or deionized. The interatomic distances N^{ε}_{Lys}-COO are also presented demonstrating that 3 out of 5 structures shown in the first part of Table 1 permit close spacing of the corresponding functional groups (distances ~ 3Å) which is not observed in any of the structures presented in the second part of Table 1. It can be noted that the set of structures in the second part of Table 1 agrees well with the set found for the tuftsin molecule by (19), where the authors have also suggested deionization of ionogenic groups. Similar results have been also obtained using an additional variant of calculations when the ionogenic groups have also been assumed ionized but the ε value was taken as 81.

Thus, it can be claimed that the assumed deionization of ionogenic groups provides a good model for the situation existing in aqueous solution. On the other hand, the assumed ionization of these groups as seen from Table 1 leads to a more distinct energy differentiation of the peptide backbone structures bringing about a certain restriction of the possibilities of space organization. E.g., the structures 1,3 and 4 in the first part of Table 1 are virtually identical as to the peptide backbone structure, the variation being reduced only to the orientation of the planes of peptide bonds between the Thr-Lys and Pro-Arg residues. Both these facts (transition from an disorderly to a more orderly three-dimensional structure and restriction of conformational mobility) are easily observable by means of NMR spectroscopy upon transferring the tuftsin molecule from water to a less polar solvent like DMSO (14,15). Upon transition to DMSO, a sharp decline in the temperature dependence of the chemical shift of amide proton in the Arg residue is observed indicating the predominance of structures in which the proton is shielded from solvent (14,15). These particular structures appear to possess the low energies as evidenced by the calculation results based on the assumption that ionogenic groups are ionized in the molecule (cf. Fig. 1a and b).

Consequently, it can be rightly suggested that calculations based on the assumed ionization of ionogenic groups provide a more adequate description of space molecular structures existing in media less polar than water. As a logical continuation of this assumption, it follows from the data given in Table 1 that in less polar media close spacing of ionogenic groups in the tuftsin molecule occurs leading to the formation of quasi-cyclic structures of the type depicted in Fig. 1a, a situation which is more probable than the variation in the sets of peptide backbone structures in media characterized by different polarity. However, the biophase of receptor is characterized by less polar conditions as compared to the aqueous environment. Hence, the binding of tuftsin molecule to receptor can lead to the formation of quasi-cyclic structures in the ligand as a result of close spacing of the lysine ε-amino group and the C-terminal carboxyl. The resulting quasi-cyclic structures can be regarded as "biologically active" molecule conformations.

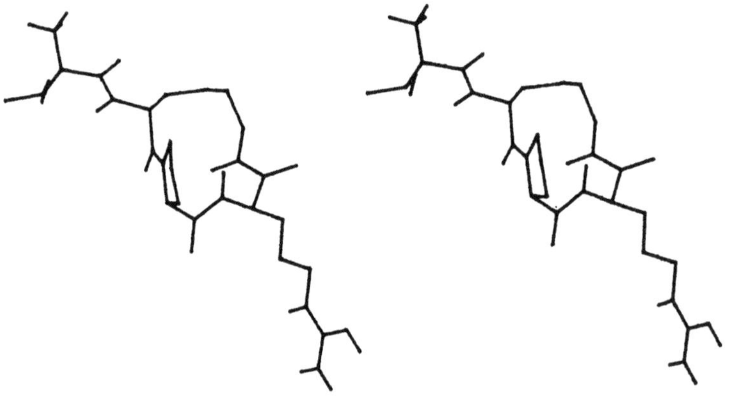

Fig.2. Lowest-energy structure of cyclic tuftsin analogue.

If this assumption is true, the cyclic analogue of tuftsin obtained by covalent bonding between the ε-amino group of lysine and the C-terminal carboxyl (Thr-Lys-Pro-Arg) should bind effectively to tuftsin receptors.

Such analogue has been synthesized and has been shown to possess a high phagocytosis-stimulating activity comparable with the activity of its natural counterpart (20); on the other hand, the results of semi-empirical conformational analysis have confirmed the occurrence of only 3 peptide backbone structures of the types 1,3 and 4 described in the first part of Table 1: the lowest-energy conformation of the cycloanalogue is depicted in Fig.2.

Thus, it can be apparently stated that quasi-cyclic structures the existence of which has been predicted by semi-empirical conformational analysis based on the assumed ionization of ionogenic groups are indeed characteristic of peptide ligands during their binding to receptors. Apart from tuftsin, active cycloanalogues have been found for bradykinin (20) in which the predicted close spacing of ionogenic groups (10) was fixed. This phenomenon may be due, for instance, to the disturbance of hydration shells around ionogenic groups in the ligand upon entry into the lipophylic "pocket" on the receptor surface, the state of their ionization being retained.

References

1. Walter R., Schwartz I.L., Darnell J.H., Urry D.W.: Proc. Natl. Ac. Sci. USA 68, 1355-1359 (1971).
2. Nikiforovich G.V., Vegners R.E., Balodis Yu.Yu., Shenderovich M.D., Chipens G.I.: In: Epigenetical information transfer in biorecognition (J.G.Vassileva-Popova, J.R.Tata eds.) Plenum Press, in press
3. Scheraga H.A.: Adv. Phys. Org. Chem. 6, 103-183 (1968).
4. Ramachandran G.N., Sasisekharan V.: Adv. Prot. Chem. 23, 283-437 (1968).

5. Balodis Yu.Yu., Nikiforovich G.V., Grinsteine I.V., Vegner R.E., Chipens G.I.: FEBS Lett. 86, 239-242 (1978).

6. Manavalan P., Momany F.A.: In: Peptides. Structure and Biological Function,(E.Gross, J.Meienhofer eds.) Pierce Chemical Company, Rockford, 893-896, 1980

7. Chipens G.I., Ancan Yu.E., Nikiforovich G.V., Balodis Yu.Yu., Makarova N.A.: In: Peptides-1978, Wrocław University Press, (G.Kupryszewski, I.Z.Siemion eds.) Wrocław, 415-419 (1979).

8. Nikiforovich G.V., Balodis Yu.Yu., Chipens G.I.: In: 16 Eur. Pept. Symp., Abstracts. Scriptor Copenhagen, 152 (1980).

9. Nikiforovich G.V., Shenderovich M.D., Balodis Yu.Yu.: Bioorgan. Khim., 7, 179-188 (1981).

10. Galaktionov S.G., Sherman S.A., Shenderovich M.D., Nikiforovich G.V., Leonova V.I.: Bioorgan. Khim., 3, 1190-1197 (1977).

11. Nikiforovich G.V.: Bioorgan. Khim., 4, 1427-1430 (1978).

12. Sevastianova N.N., Lipkind C.M., Archipova S.F., Popov E.M.: Bioorgan. Khim., 3, 473-484 (1977).

13. London R.E., Stewart J.M., Cann J.R., Matwiyoff N.A.: Biochemistry 17, 2270-2277 (1978).

14. Blumenstein M., Layne P.P., Najjar V.A.: Biochemistry 18, 5247-5253 (1979).

15. Sekacis I.P., Liepinsh E.E., Veretennikova N.I., Chipens G.I.: Bioorgan. Khim., 1617-1622 (1979).

16. Bass S.J., Nathan W.I., Meighan R.M., Cole R.N.: J. Phys. Chem. 68, 509-515 (1964).

17. Finkelstein A.V.: Mol. Biol., 11, 811-819 (1977).

18. Nikiforovich G.V., Leonova V.I., Galaktionov S.G., Chipens G.I.: Int. J. Pept. Prot. Res., 13, 363-373 (1979).

19. Fitzwater S., Hodes Z.I., Scheraga H.A.: Makromolecules, 11, 805-811 (1978).

20. Chipens G., Nikiforovich G., Mutulis F., Veretennikova N., Vosekalna I., Sosnov A., Polevaya L., Ancans Y., Mishlyakova N., Liepinsh E., Sekacis I., Breslav M.: In: Peptides. Structure and Biological Function, (E.Gross, J.Meienhofer eds.) Pierce Chemical Company. Rockford, 567-570, 1980

21. IUPAC-IUB Commission on Biochemical Nomenclature: Pure Appl. Chem. 40, 291-308 (1974).

THE STUDY OF THE STRUCTURAL AND FUNCTIONAL ORGANIZATION OF THE SOMATOTROPIN MOLECULE

Yu.A.Pankov, A.A.Bulatov, Yu.M.Keda, T.A.Osipova, V.I.Pozdnyakov
Institute of Experimental Endocrinology and Hormone Chemistry, USSR Academy of Medical Sciences, Moscow

Introduction

Pituitary protein hormones: somatotropin and prolactin play an essential role in the biological regulation. Apart from controlling growth, somatotropin exerts influence upon many branches of metabolism. Prolactin controls lactation, reproduction and possesses multiple biological activity. Somatotropin and prolactin isolated from pituitaries of many species are represented by families of proteins which being combined constitute a superfamily. To contribute to the investigation of the superfamily of somatotropin and prolactin we inquired into the primary structure of somatotropin isolated from the sey whale (Balaenoptera borealis) pituitary which had not been studied yet. We have also carried out a comparative analysis of the somatotropin and prolactin amino acid sequences of various species and studied the biological activity of some fragments of the somatotropin molecule.

Results

Somatotropin was extracted from the pituitaries, precipitated by salt and ethanol and purified by the ion exchange and exclusion chromatography. The hormone was cleaved by cyanogen bromide, trypsin, chymotrypsin and pepsin, the resulting mix-

Chemistry of Peptides and Proteins, Vol. 1

tures of peptides being separated and purified by the ion exchange chromatography, gel filtration on Sephadexes, partition paper chromatography and high voltage paper electrophoresis. The routine Edman-dansyl procedure was used to study the amino acid sequences of the peptides. The glutamic and aspartic acid residues and their amides were identified as phenylthiohydantoins. The study of overlapping peptides has brought about the primary structure of sey whale somatotropin which comprises 190 amino acid residues (Fig.1). The hormone structure shows many common features with that of somatotropins of other species but differs from them in many amino acid residues.

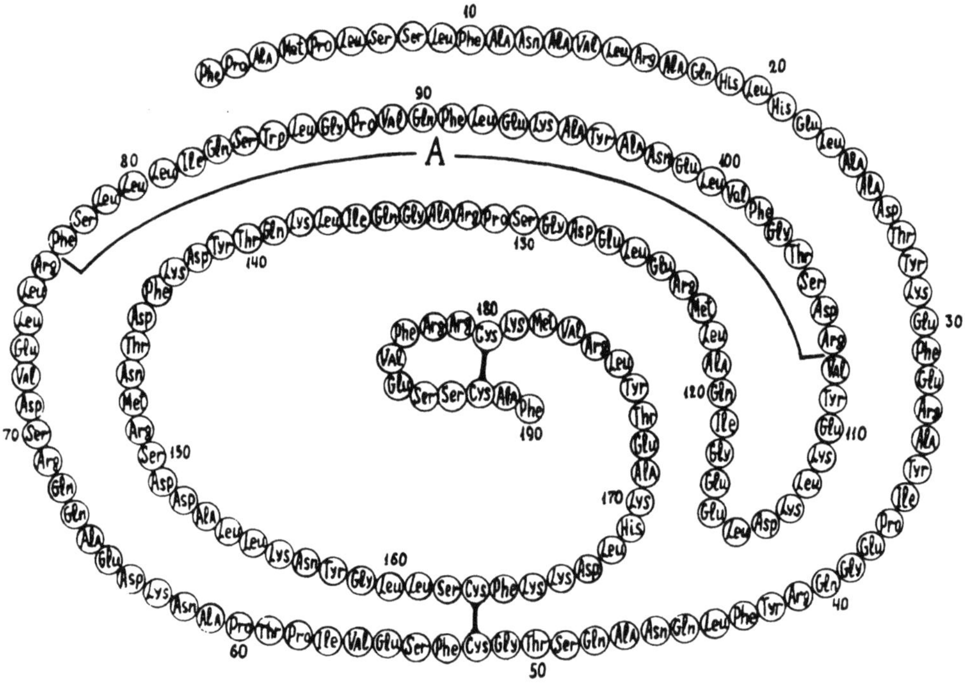

Fig.1. The primary stucture of sey whale somatotropin. Section A - biologically active somatotrophic fragment.

The comparative analysis of the structure and biological activity of similar hormones from various species or homologous hormones which differ in their functions but belong to the same superfamily of proteins provides an insight into their structure and function relationship. To simplify the approach to the problem we have developed mathematical methods of comparative analysis of the protein primary structures and the appropriate computer programs. The methods have been presented in a series of papers (1-4). They allow to present the results of the primary structure comparative analysis in the form of the curves which characterize the local similarity along the sequences under investigation. Some curves of the somatotropin and prolactin similarity are presented on Fig.2. They show that there are evolutionary stable, moderately changeable and variable segments in the peptide chains of the hormones which play different roles in the mechanism of hormonal action. This concept has been outlined more thoroughly in the review (4).

Recently Chou and Fasman (5) have suggested numerical potentials reflecting probable participation of each amino acid in α- and β-structures of proteins. This scoring system has simplified the computer realization of their method. The curves of the probable distribution of α-helix in the protein hormones are shown at the bottom of Fig.2.The profiles of α-helix distribution for somatotropin and prolactin nearly coincide. The probable α-helix location seems to be very similar in both hormones and coincide qualitatively with the curve of their local similarity. This coincidence suggests that the basic curve of local similarity between somatotropin and prolactin amino acid sequences should reflect the similarity of the spatial arrangement of these hormone molecules. In spite of the fact that there are no X-ray cristal data on somatotropin and prolactin those curves suggest that there are at least four separate portions of supersecondary structure in their molecules. Divergency of the similarity curves calculated for somatotropins and prolactins of different species

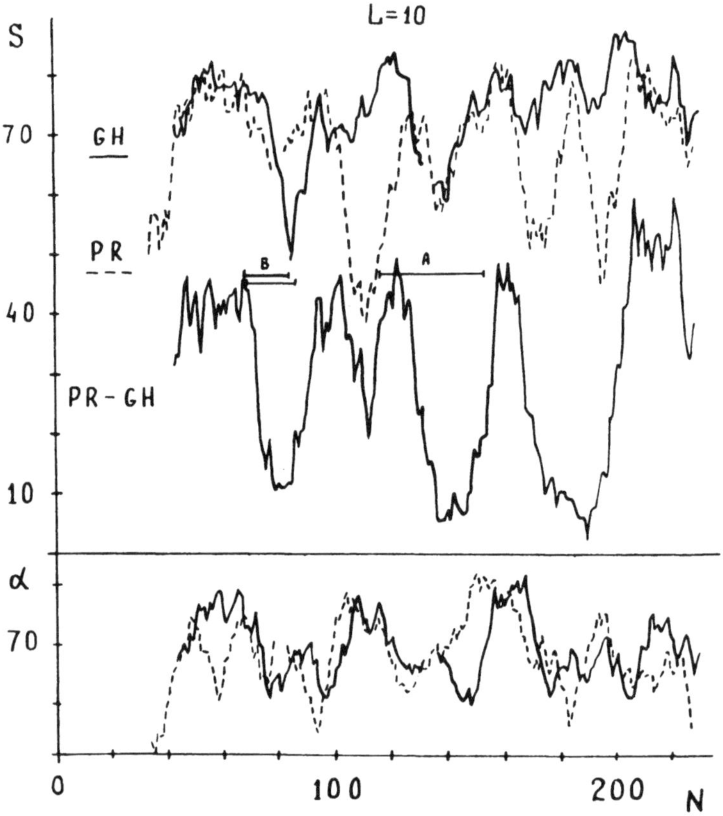

Fig.2. Profiles of local similarity S and probable helicity α along somatotropin (human, ovine, bovine, horse and sey whale) and prolactin (human, ovine, bovine, porcine and rat) amino acid sequence families. Relative values S and α are placed in the middle of each region of L=10 amino acids long throughout the superfamily alignment and calculated as follows:

within families- $$S = \sum_{N\leq k\leq N+L-1} \sum_{i<j\leq n} \frac{2\cdot s(a_k^i, a_k^j)\cdot 100\%}{s_m\cdot L\cdot n\cdot(n-1)},$$

$$\alpha = \sum_{N\leq k\leq N+L-1} \sum_{i\leq n} \frac{\alpha(a_k^i)\cdot 100\%}{\alpha_m\cdot L\cdot n};$$

between families- $$S = \sum_{N\leq k\leq N+L-1} \sum_{i_1\leq n_1, i_2\leq n_2} \frac{s(a_k^{i_1}, a_k^{i_2})\cdot 100\%}{s_m\cdot L\cdot n_1\cdot n_2},$$

where s_m and α_m are, respectively, maximum values in the matrix of amino acid similarity $s(a^i,a^j)$ (6) and in the vector of amino acid α-potentials $\alpha(a^i)$ (5); i_1,n_1 and i_2,n_2 are the index and the number of sequences in the families 1 and 2; N - index of the first position of the region. Designations. Solid and dashed lines correspond to similarity and α-helicity for somatotropin (GH) and prolactin (PR) families, respectively; solid line PR-GH is similarity between the families. Section A - somatotrophic fragment of whale somatotropin, B - lipotropic fragment of human somatotropin and deletion in its modified form.

from the basic profile seems to be the result of the functional specificity of the hormones. Therefore this divergency may indicate the location of the most probable segments which constitute zones of "growth promoting" or "lactogenic" activities in the spatial structure of the hormones.

In connection with this problem various fragments of somatotropin possessing hormone-like activity have been obtained by many authors by means of the chemical synthesis or enzymatic degradation of the native hormone. After the tryptic hydrolysis of the sperm whale somatotropin we have also isolated a hydrophobic peptide corresponding to 77-107 of its sequence. The peptide contains 16% of the hormone amino acid sequence and possesses about 5% of its growth promoting activity as determined by the conventional tibia test on hypophysectomized rats. In comparison with similar peptides described in the literature this fragment exceeds them in biological activity and has lower molecular weight. The position of the peptide in the amino acid sequence of the hormone is indicated by segment A in Fig.1 and 2. One can see there is only a partial coincidence of the peptide with one of the similarity peaks of somatotropins. If it were possible to obtain a peptide coinciding better with the peak of similarity it would likely possess a higher biological activity.

There is another interesting fragment of somatotropin which we have obtained after pepsin hydrolysis of the hormone. This fragment extends from position 31 to 44 in the molecule and possess a very high lipotropic but no growth-promoting activity (7-10). The effect of the peptide greatly exceeds that of the intact hormone and is manifested very fast without a 3 hour lag period nessesary for displaying the effect of native somatotropin. The 31-44 segment is thus considered to be responsible for the lipotropic (or adipokinetic) activity of the hormone, and manifestation of this effect in tissue should be preceeded by the enzymatic cleavage of somatotropin and the release of a fragment with high lipotropic activity. The position of the peptide in the amino acid sequence and its correspondence to the similarity profiles of different hormones are indicated on Fig.2 by segment B. It implies that the peptide coincides with a minimum of local similarity between somatotropin and prolactin families, but overlaps the right side of the similarity peak when somatotropins of different species are compared. The results suggest that this peptide is important for some biological effects of somatotropin and located between the two portions of α-helix of the hormone molecule.

A new light has been recently thrown upon this peptide in view of the data obtained by others. Lewis et al.(11), for example, have found out in the human pituitary along with normal somatotropin a hormone lacking 32-46 sequence, the residue 31 in its molecule binding directly to residue 47. The modified hormone amounts to 15% of the content of normal somatotropin in human pituitaries and possesses ,both, growth-promoting and lactogenic activities but lacking lipotropic and some other side activities (12). Since our fat mobilizing peptide coincides well with the 32-46 sequence this zone of the somatotropin molecule is thought to be responsible for the adipokinetic effect of the hormone. At present it is difficult to account for the existence of the two types of somatotropins in human pituitaries. Findings by Fiddes et al. (13) may to a certain extent provide an insight into the problem. These authors have shown that the natural somatotropin gene contains segments that code the amino acid sequence of the protein - the so-called exons, which are separated by several non-coding intervening sequences, or introns. It is interesting that the second intron is located between the codons of amino acid residues 31 and 32 - precisely the position of the start of the deletion in the modified somatotropin molecule. Therefore the existence of the two types of somatotropin is supposed to be accounted for by a shift of the border between the second exon and the second intron in the two natural gene copies or during two ways splicing mRNA of human somatotropin. The lengthening or shortening of the exons may induce the synthesis and release of several types of somatotropin with somewhat different biological activities. The changeable part of the second exon coincides with the nucleic acid sequence that codes the structure of the peptide responsible for the lipotropic activity of human somatotropin.

References

1. Pankov Yu.A., Pozdnyakov V.I., Tumanyan V.G.: Moleculyarnaya Biologiya 10, 423-436 (1976).
2. Pozdnyakov V.I., Pankov Yu.A.: Doklady Akad. Nauk SSSR 246, 1, 232-236 (1979).
3. Pozdnyakov V.I., Pankov Yu.A.: Moleculyarnaya Biologiya 13, 899-911 (1979).
4. Pankov Yu.A., Pozdnyakov V.I.: Biochimiya 45, 195-209 (1980).
5. Chou P.Y., Fasman G.D.: Adv. Enzymol. New York 1, 45-148 (1978).
6. McLachlan A.D.: J. Mol. Biol. 64, 417-437 (1972).
7. Keda Yu.M., Sinicyna A.L., Osipova T.A., Pankov Yu.A.: Biochimiya 38, 659-663 (1973).
8. Yudaev N.A., Pankov Yu.A., Keda Yu.M., Shvachkin Yu.P., Ryabcev M.N., Chukashev S.G.: Biochimiya 41, 843-846 (1976).
9. Yudaev N.A., Pankov Yu.A., Keda Yu.M., Shvachkin Yu.P., Ryabcev M.N.: Biochimiya 41, 1484-1487 (1976).

10. Yudaev N.A., Pankov Yu.A., Keda Yu.M., Shvachkin Yu.P., Ryabcev M.N.: Biochimiya 44, 1779-1785 (1979).

11. Lewis U.J., Bonewald L.E., Lewis L.J.: Biochem. Biophys. Res. Commun. 92, 511-516 (1980).

12. Friger L.G., Peterson S.M., Lewis U.J.: Biochem. Biophys. Res. Commun. 91, 778-782 (1979).

13. Fiddes J.C., Seeburg P.H., deNoto F.M., Hollewell R.A., Baxter J.D., Goodman H.M.: Proc. Natl. Acad. Sci. USA 76, 4294-4298 (1979).

STRUCTURAL FEATURES OF Bacillus thuringiensis CRYSTAL PROTEIN

Valentin M.Stepanov, Galina G.Chestukhina, Igor' A.Zalunin, Lubov' I.Kostina, Alla L.Mikhailova
Institute of Genetics and Selection of Industrial Microorganisms, II3545, Moscow, USSR

Introduction

B.thuringiensis preparations are applied in large scale in agriculture and forestry to control pest insects. The activity of these preparations is mainly dependent on the presence of insecticidal δ-endotoxin that appears within the sporulating B.thuringiensis cells as crystalline proteinaceous inclusions. Entomocidal activity of B.thuringiensis crystal is known to exhibit large variations which apparently reflects certain differences in the structure of crystal-forming proteins. Nevertheless, up to now, no data are available on the chemical structure of the crystal protein. Moreover, even the data on the number and the size of subunits that form the crystal were extremly contradictory, e.g. for the molecular weight values ranging from 230 000 to 1 500 were published (1-6). The data presented below allowed us to disclose the cause of these discrepancies and to gain a new concept on molecular organization of B.thuringiensis crystal protein.

Results

It had been found that the crystal preparations contain a mixture of endogenous proteinases tenaciously kept by the surface or internal defects of the crystal structure. According to their action on synthetic substrates as well as on their

Chemistry of Peptides and Proteins, Vol. 1

inhibition pattern these enzymes might be referred to as serine and metalloproteinases as well as aminopeptidases (7). These proteinases are capable of degrading proteins. whereas the intensity of this hydrolysis is sharply enhanced under conditions favourable for denaturation of the protein. The extent of the degradation and, consequently, the number, size and relative amount of resulting fragments depend on a number of factors.

The development of the methods of selective inactivation of the proteinases permitted us to show that each of the studied serotypes of B.thuringiensis (I-IV, VI, VIII-XII) with the only exception of serotype V produced crystals composed of only one protein with the molecular weight of 145 000, 135 000 or 130 000 (the latter value is dependent on the strain (8,9)). The crystals formed by strains of serotype V were shown to contain two proteins with molecular weights of 135 000 and 130 000, some of these strains revealed simultaneously the pre-

Table. Molecular weight of proteins forming the crystal of various strains of B.thuringiensis.

145 000	135 000	130 000	135 000 + 130 000 or 135 000, 130 000 + 65 000
thuringiensis (I)	alesti (III)	finitimus (II)	galleriae (V)
	kenyae (IV)	kurstaki (III)	
insectus (I)	dendrolimus (IV)	galleriae (V)	
		tolworthi (IX)	
	morrisoni (VIII)	thompsoni (XII)	
	darmstadtiensis (X)		

sence of the polypeptide chain with the molecular weight of 65000.

To unveil the molecular organization of the crystal protein the latter was subjected to limited degradation by a number of proteolytic enzymes - extracellular serine proteinases of B.thuringiensis, subtilisin, trypsin, chymotrpysin, proteinases from gut juice of Galleria mellonella larvae. This lead practically in all cases to the disappearance of the high molecular weight protein. Simultaneously the fragments with the molecular weight of about 70 000 emerged, and accumulated being apparently stable to further proteolysis. This indicates the presence of a compact domain within the crystal protein. Variation in trypsin concentration and duration of the treatment allowed us to trap apparently less stable intermediate degradation products of molecular weight 110 000 and 90 000. Remarkably, differences in the molecular weights of undegradated proteins (145 000, 135 000, 130 000) reappears in the stable fragments and in the intermediate products. Hence, it appears that from all crystal proteins a few fragments of equal size were split off in the course of the limited proteolysis.

These observations indicate a certain similarity in the molecular organization of crystal proteins produced by various B.thuringiensis strains. Thus, roughly half of the molecule exists as a compact domain (70 000) fairly resistant to proteolysis. Another half of the molecule might be stepwise degraded by means of cleaving off the fragments that correspond to molecular weights of about 15 000. The eventual structural or functional role of the latter fragments remains to be cleared.

The stable domains (70 000) isolated from the serotype V crystal protein were shown to retain toxicity against the larvae when introduced by feeding or by injecting in haemocoel. Hence, the domains resistant to proteolysis, found by us within the structure of the crystal protein might correspond to a toxin presumably formed from the crystal protein ("protoxin") by enzymes from larvae gut juice. To isolate the stable domains the crystal proteins were incubated with trypsin until the components with molecular weights higher than 70 000 disappeared. The domains were then separated from the low molecular weight degradation products by Sephadex G-75 gel filtration in 2M NaCl to prevent aggregation.

The stable domains were isolated from crystal proteins of the serotype III (mol.weight 70 000), V (65 000) and IX (65 000). The amino acid composition of these domains obtained by the above specified procedure or by SDS gel electrophoresis turned to be similar to initial protein preparations differing only by somewhat lower lysine content. No significant differences were found when the amino acid composition of the domains isolated from proteins of different serotypes were compared. This corresponds to the amino acid composition of initial crystal proteins.

Cyanogen bromide cleavage of serotype III crystal protein was handicapped by its pornounced instability in acid solutions. On the contrary, the domains resistant to proteolysis showed also substantially better stability towards acid cleavage. Initial crystal protein gave after CNBr treatment 3 fragments with molecular weights (as assayed by SDS gel electrophoresis) ranging from 65 000 to 16 000. The amount of 16 000 fragment determined by gel scanning corresponded to about one half of the total protein content that might indicate that this fragment(s) appears repeatedly within the primary structure of the toxin. The amino acid composition of the 16 000 fragment was also rather similar to that of crystal protein (10). CNBr splits the stable (70 000) domain with the formation of the 16 000 fragments. Hence, it might be presumed that this moiety of the toxin gives the 16 000 fragment during CNBr treatment of the crystal protein. CNBr cleavage of serotype V crystal protein gave a substantially different fragmentation pattern that might indicate the difference in the primary structures of crystal proteins produced by the cells of various B.thuringiensis serotypes.

References

1. Bulla, L.A., Jr., Kramer, K.J., Davidson, L.I.: J.Bacteriol. 130, 375-384 (1977).
2. Cooksey, K.E.: Biochem.J. 106, 445-454 (1968).
3. Glatron, M.-F., Lecadet, M.M., Dedonder, R.: Europ.J.Biochem. 30, 330-338 (1972).
4. Herbert, B.H., Gould, H.J., Chain, E.B.: Europ.J.Biochem. 24, 366-375 (1971).
5. Nagamatsu, Y., Tsutsui, R., Ichimaru, T., Nagamatsu, M., Koda, K., Hayashi, K.: J.Invertebr.Pathol. 32, 103-109 (1978).
6. Sayles, V.B., Aronson, J.N., Rosenthal, A.: Biochem.Biophys.Res.Commun. 41, 1126-1133 (1970).
7. Chestukhina, G.G., Zalunin,I.A., Kostina, L.I., Kotova, T.S., Katrukha, S.P., Lyublinskaya, L.A., Stepanov, V.M.: Biokhimiya 43, 857-864 (1978).
8. Zalunin, I.A., Chestukhina, G.G., Stepnov, V.M.: Biokhimya, 44, 693-697 (1979).
9. Chestukhina, G.G., Zalunin, I.A., Kostina, L.I., Kotova, T.S., Katrukha, S.P.,Stepanov, V.M.: Biochem.J. 187, 457-465 (1980).
10. Katrukha, S.P., Chestukhina, G.G., Stepanov, V.M.: Khim. Prir.Soedin. 5, 384-386 (1980).

X-RAY INVESTIGATION OF THREE DIMENSIONAL STRUCTURE OF ACTINOXANTHIN

Vladimir Pletnev, Alexander Kuzin, Sergei Trakhanov, Valery Popovich, Igor Tsigannik

Shemyakin Institute of Bioorganic Chemistry, USSR Academy of Sciences, Moscow, USSR

Introduction

Due to the interesting biological properties of antitumor actinoxanthin protein (1, 2) (isolated from *Actinomyces globisporus*, molecular weight 10,300, 107 amino acid residues) extensive structural and biochemical studies have been undertaken with the purpose of establishing the relationship between its conformation and the action mechanism. The present paper describes the results of X-ray study of the three-dimensional structure of this protein at 2.5 Å resolution.

The protein actinoxanthin crystallizes in space group $P2_12_12_1$ with cell dimensions: $\underline{a}$ = 30.9 Å, $\underline{b}$ = 48.8 Å, $\underline{c}$ = 64.1 Å, and $\underline{Z}$ = 4. Isolation, purification, crystallization and preparation of isomorphous derivatives of actinoxanthin using heavy metal salts $UO_2(NO_3)_2$, $KAuCl_4$, $K_2Pt(NO_2Cl)_2$, $Pb(CH_3COO)_2$, and $HgCl_2$ as well as measurement and processing of experimental data were described in detail elsewhere (3, 4). The total number of reflections from the 2.5 Å range for each derivative used in the calculations was 3,433. The heavy atom binding sites were located by X-ray "direct" methods (5). The selection of correct enantiomorph has been done using anomalous scattering data. The mean figure of merit $< m >$ after last cycle of phase refinement was 0.80.

Chemistry of Peptides and Proteins, Vol. 1

Results

The actinoxanthin backbone folding drawn from C^{α} atom coordinates is shown in Fig. 1.

A distinctive feature of the actinoxanthin structure is the absence of α-helices and the presence of enhanced content of antiparallel β-structure (~55%). The three-dimensional architecture of the molecule has a double part appearance with the two lobes of different sizes (70 and 30%) being separated by a well-defined cavity.

The larger lobe is a regular β-supersecondary structure containing the N- and C-termini.

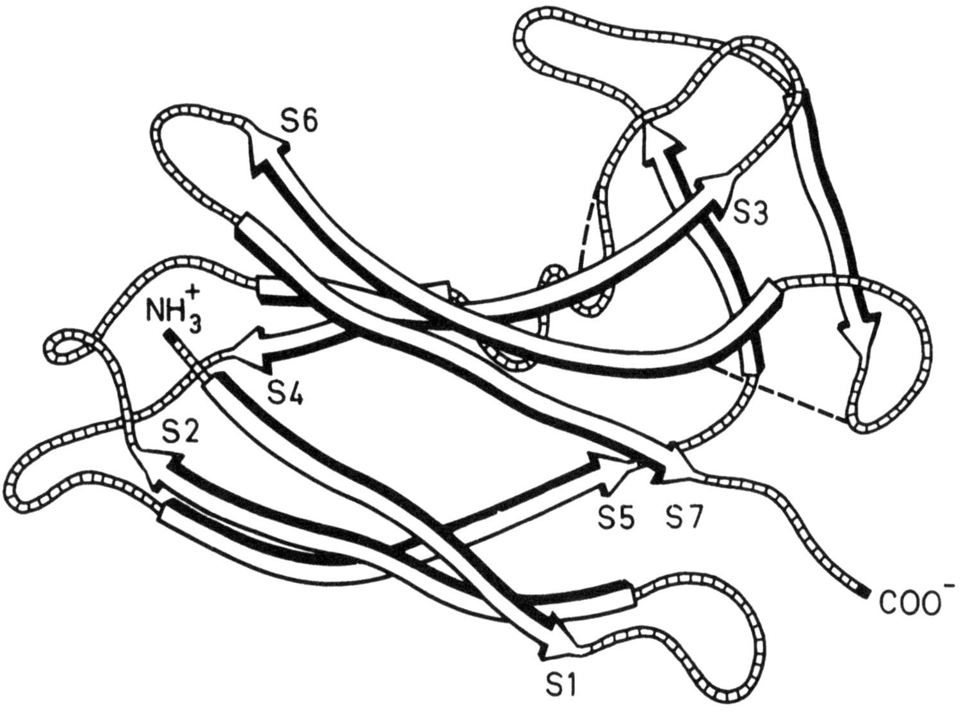

Fig. 1. Topology of actinoxanthin spatial folding. The arrows indicate β-strands, the strings indicate non-regular protein parts.

It is a slightly flattened cylinder formed by two layers of antiparallel β-sheets. The first external layer consists of three strands S_1, S_2 and S_5 and the second one, bounding the molecular cavity, consists of four strands S_4, S_3, S_6 and S_7. β-Strands arrangement in protein architecture has topology of so called "Greek key" (6).
The β-sheets exhibit a right twist when viewed along, and a left twist when viewed across the sheet. The two neighboring β-strands form an angle of 15-20° when looking across the sheet.
The fragments of sharp chain reversal (β-turns) positioned between the β-strands are rich in Ser, Ala and Gly residues. The inner part of the β-cylindrical barrel is tightly packed with hydrophobic side chains oriented towards the axis: Ala(92), Val(6, 18, 20, 64), Phe(4, 51, 61, 63, 106), Leu(90, 99, 104), Ile(31); the Gln(33) side chain is also inside the cylinder. Hydrophilic side chains are arranged mainly on the external surface; the largest group among them is formed of Ser and Thr residues, which are located along the chain alternately with hydrophobic residues. They mainly populate the external three-strands β-sheet S_1, S_2, S_5 and strand S_4.
The cylinder is bounded from one end by turns S_2-S_3, S_4-S_5, S_6-S_7 and the N-terminal and on the other end by turn S_1-S_2, disulfide bond Cys(83)-Cys(88), residue Lys(66) and the C-terminal.
The smaller and less regular lobe of the molecule includes two closely spaced deca- and hexapeptide cycles 34-43 and 83-88 closed by disulfide bonds, curved fragment 43-47 and a β-hairpin shaped part 66-82 on the elongated loop between β-strands S_5 and S_6.
The internal area of the molecular cavity separating the two structural units contains a large number of Gly and Ala residues and disulfide bond Cys(34)-Cys(43). The peripheral part of the cavity is lined by the side chains of Tyr(30), Thr(49, 72), Asn(89, 94), Gln(71), Ser(95) and His(101).
The relative arrangement of the seven β-strands of the super-

secondary actinoxanthin structure shows strong topological similarity with the arrangement of β-strands in C and V domains of immunoglobulins IgG (7), in Bence Jones protein REI (8), in superoxide dismutase subunits (9), and in azurin (10).
Despite the fact that these protein units consist of about the same number of residues and show close resemblance of the spatial folding pattern, the proteins themselves are functionally unrelated and have no evident sequence homology.
Estimation of topological similarity of the actinoxanthin cylindrical subunit with those of immunoglobulin and superoxide dismutase has been performed on a computer by least squares method superimposing of their C^{α}-frames.
Pairwise comparison of the three-dimensional structures shows that the arrangement of practically half of amino acid residues in actinoxanthin and in the two other proteins are topologically equivalent; the root mean-square distances between the residues were less than 2.5 Å. The mean value of base change per codon for comparable protein ranges was about 1.31.
It is unclear whether the observed structural similarity is indicative of divergent evolution of these proteins from a common ancestor or convergent evolution of independent precursors to energetically favorable topological state.
The characteristic cavity found in the actinoxanthin structure undoubtedly has a direct relation to its biological activity. It is the probable binding site for the chromophore in the active complex.

References

1. Khokhlov, A.S., Cherches, B.Z., Reshetov, P.D., Smirnova, G.M., Sorokina, I.B., Koloditskaya, T.A., Smirnov, V.V., Navashin, S.M., Fomina, J.P.: J. Antibiotics 22, 541-544 (1969).

2. Khokhlov, A.S., Cherches, B.Z., Reshetov, P.D., Smirnova, G.M., Koloditskaya, T.A., Sorokina, I.B., Prokoptseva, T.A., Ryabova, I.D., Smirnov, V.V., Navashin, S.M., Fomina, I.P.: Izvestiya AN SSSR (ser. biol.) (Bull. Acad. Aci. USSR,

Russ. ser. biol.) 755-763 (1970).

3. Pletnev, V.Z., Kuzin, A.P., Trakhanov, S.D., Kostetsky, P.V., Popovich, V.A., Tsigannik, I.N.: Biopolymers 20, 679-694 (1981).

4. Pletnev, V.Z., Trakhanov, S.D., Tsigannik, I.N.: Bioorgan. Khim. (Soviet J. Bioorgan. Chem) 5, 1605-1608 (1979).

5. Pletnev, V.Z., Kuzin, A.P., Trakhanov, S.D., Popovich, V.A., Tsigannik, I.N.: Bioorgan. Khim. (Soviet J. Bioorgan. Chem.) 6, 563-569 (1980).

6. Richardson, J.S.: Nature 268, 495-500 (1977).

7. Amzel, L.M., Poljak, R.J.: Annu. Rev. Biochem. 48, 961-997 (1979).

8. Epp, O., Colman, P., Fehlhammer, H., Bode, W., Schiffer, M., Huber, R.: Eur. J. Biochem. 45, 513-524 (1974).

9. Richardson, J.S., Richardson, D.C., Thomas, K.A.: J. Mol. Biol. 102, 221-235 (1976).

10. Adman, E.T., Stenkamp, R.E., Sieker, L.C., Jensen, L.H.: J. Mol. Biol. 123, 35-47 (1978).

ALDIMINE BOND MIGRATION IN THE PHOTOCHEMICAL CYCLE OF BACTERIO-RHODOPSIN

Najmutin Abdulaev, Victor Tsetlin, Alexandr Kiselev, Valdis Zakis, Yuri Ovchinnikov
Shemyakin Institute of Bioorganic Chemistry, USSR Academy of Sciences, Moscow, USSR

Introduction

We found that a migration of the aldimine bond takes place in the course of the bacteriorhodopsin (BR) photochemical cycle. It was also shown that, depending on the $NaBH_4$ reduction of BR or its modified derivatives in the light or in the dark, the retinyl moiety is attached to ε-amino groups of different lysine residues. Such a phenomenon is not an artifact of the reduction conditions, but is a conseguence of the photochemically induced aldimine migration (Fig. 1, 2).

Our experimental approach relies upon analysis of the products formed on limited proteolysis of native (1) or reduced BR catalyzed by chymotrypsin or papain. By polyacrylamide gel electrophoresis two fragments are obtained in both cases: 1-71 and 72-247 (248)-chymotryptic fragments, or 3-65 and 73-230(231)-papainic ones (numbering according to (2), in parenthesis - according to (3)); these fragments are termed as a small and a large one, respectively. Intense fluorescence of the retinyl residue enabled its facile detection in electrophoregrams both in retinyl-bacterioopsin and in the above-mentioned fragments. After $NaBH_4$ reduction of BR in the light (in all experiments a 250 W halogen lamp was used along with 3 cm layer of 10% $CuSO_4$ and yellow-orange glass filter) at pH 9.5 and 20°C (4), retinyl moiety was found in the small and large fragments. Noteworthy, the proton transport across the purple membrane is

Chemistry of Peptides and Proteins, Vol. 1

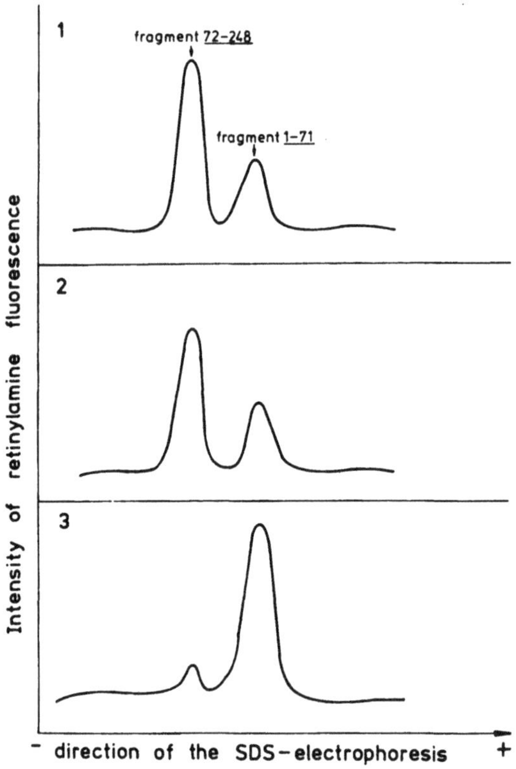

Fig. 1. PAAG-electrophoresis of bacteriorhodopsin fragments obtained by reduction at various conditions followed by chymotryptic hydrolysis. 1) Reduction in the light followed by cleavage; 2) cleavage followed by reduction in the light; 3) cleavage followed by reduction in the dark.

correlated with sequential protonation and deprotonation of the aldimine bond between retinal aldehyde and Lys 41 (see review (5)). The data obtained indicate that the ε-amino group of this residue is not the only site for covalent attachment of retinal to the BR polypeptide chain and that aldimine bond may be formed by a lysine residue in small fragment (apparently Lys 41), or by large fragment residue(s) (designated as Lys x) in position 130, 159, 171(172), or 215(216).
After cleavage with chymotrypsin or papain, BR preparations

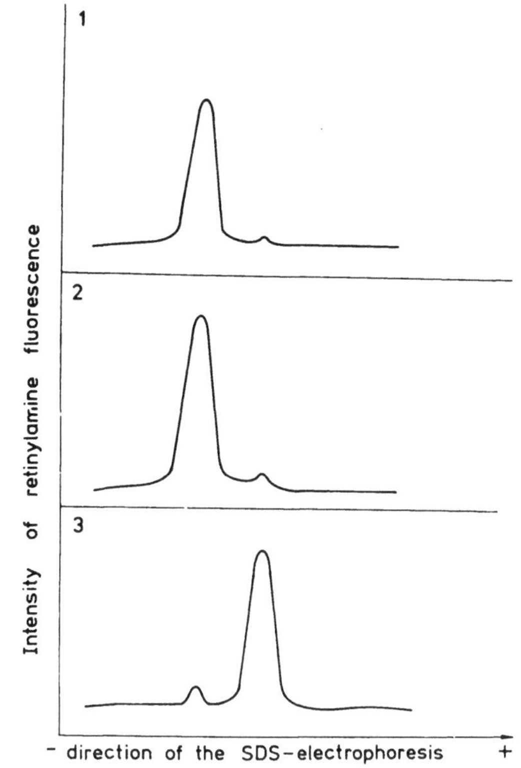

Fig. 2. PAAG-electrophoresis of bacteriorhodopsin fragments obtained by reduction at various conditions followed by chymotryptic hydrolysis. 1) Reduction in the presence of gramicidin S; 2) reduction in the light at 0° and pH 10; 3) reduction in the light at 0° and pH 10.

essentially preserve the spectral and functional properties of the native molecule, however acquire the $NaBH_4$ reducibility. We capitalized on this fact to obtain information about the aldimine localization in dark conditions. In dark preparations the retinyl moiety is associated exclusively with a small fragment. After reduction of native or proteolyzed BR in the light, fluorescence was detected in both fragments. These data testify the N → N retinal migration in the course of photochemical transformations of bacteriorhodopsin.

After dark reduction of proteolyzed BR, preliminary adapted to light or dark, or after dark reduction of preparations regenerated from proteolyzed BR with 13-cis- or all-trans retinal, all the fluorescence was found only in the small fragment. This fact means that in the dark, whatever is the double bond configuration, retinal is attached to Lys 41, and aldimine formation with a new lysine residue (Lys x) in the light is not caused by retinal isomerization during the photocycle.
The BR reduction in the light was also carried out under the conditions wherein an intermediate of the photocycle (so-called M412 form or its congeners) having deprotonated aldimine bond is predominant. Stabilization of such a form is achieved by adding gramicidin S (6) or by lowering the temperature (5, 7). In both cases reduction proceeds at a considerably higher rate and results in exclusive localization of retinyl on Lys x. It might be possible that under these conditions both Lys 41 and Lys x aldimines are present, but only the latter is reducible by $NaBH_4$. However, since Lys 41-aldimine in proteolyzed BR is accessible for $NaBH_4$ in the dark as well as in the light, localization of retinyl moiety only in large fragment when reduction is carried out in the light under the conditions favoring M412 stabilization, is indicative of the absence of Lys 41-aldimine and of complete migration of aldimine onto Lys x. Thus, a correlation exists between the aldimine bond deprotonation and its migration.
The migration may proceed via aldimine hydrolysis and subsequent reaction of released retinal with another lysine amino group (earlier a photoinduced hydrolysis of aldimine bond was discovered and a hypothesis about possible migration was discussed (8)), or via a direct attack of aldimine by a spatially proximal amino group leading to tetrahedral intermediate. We found that trifluoroacetylation or succinylation of the exposed amino groups in BR fails to prevent the aldimine migration, thus allowing a conclusion to be made that Lys x amino group, inaccessible for modification, is a suitable acceptor for migrating retinal, probably being in a hydrophobic environment

and having a low pK.
Collectively, the data on reduction of BR and its modified derivatives clearly show that migration is not a random formation of aldimines with any of accessible amino groups induced by the pH of the medium or by presence of $NaBH_4$. The retinyl residue occupies a chromophoric site in BR when it is bound either to Lys 41 or Lys x, since none of the reduced preparations regenerates with added all-trans retinal. However, marked differences exist in microenvironment of retinyl residues attached to different lysines: a characteristic fine structure and bathochromic shift of the absorption band are seen only for Lys x -retinyl.

A preference for aldimine localization on Lys 41 in the dark and a capability of photoinduced migration onto Lys x are characteristic of native BR and may be lost under certain conditions affecting the relative stability of corresponding aldimines. For example, both, after light and dark reduction retinyl is bound only to Lys 41, when analogs modified at retinal C(4) position are reduced, and only to Lys x when proteolyzed BR is reduced at acidic pH with $NaBH_3CN$.

Aldimine bond migration reported in this communication provides evidence that at least two amino groups, separated in the amino acid sequence and belonging to two different α-helical segments are involved in the BR photocycle. (Localization of these lysines is under way). This finding should be taken into consideration when interpreting the data of physico-chemical studies of BR or mechanisms of its functioning, wherein retinal aldimine is given a key role. A possibility of protonating and deprotonating different aldimines, formed by amino groups of different residues and having different microenvironment, seems very advantageous for realization of a proton transfer across the purple membrane. A photoinduced migration of aldimine bond may be a component part of the mechanism of BR functioning as a proton pump.

References

1. Abdulaev, N.G., Feigina, M.Yu., Kiselev, A.V., Ovchinnikov, Yu.A., Drachev, L.A., Kaulen, A.D., Khitrina, L.V., Skulachev, V.P.: FEBS Letters 90, 190-194 (1978).

2. Ovchinnikov, Yu.A., Abdulaev, N.G., Feigina, M.Yu., Kiselev, A.V., Lobanov, N.A.: FEBS Letters 100, 219-224 (1979).

3. Khorana, H.G., Gerber, G.E., Herlihy, W.G., Gray, C.P., Anderegg, R.J., Nihei, K., Biemann, K.: Proc. Nat. Acad. Sci. USA 76, 5046-5050 (1980).

4. Schreckenbach, T., Walckhoff, B., Oesterhelt, D.: Eur. J. Biochem. 76, 499-511 (1977).

5. Stoeckenius, W., Lozier, R.H., Bogomolni, R.A.:Biochim . et Biophys.Acta 505, 215-278 (1979).

6. Melnik, E.I., Chizhov, I.V., Skopinskaya, S.N., Snezhkova, L.G., Miroshnikov, A.I.: Abstracts of the I Soviet-Swiss Symposium "Biological membranes: structure and function" Tbilisi, p. 75 (1979).

7. Peters, I., Peters, R., Stoeckenius, W.: FEBS Letters 61, 128-134 (1976).

8. Shkrob, A.M., Rodionov, A.V., Ovchinnikov, Yu.A.: Bioorgan. Khim. (Russian) 4, 354-359 (1978).

CHEMICAL MODIFICATION OF PURPLE MEMBRANE

Topography of Bacteriorhodopsin

Horst-Dieter Lemke, Jürgen Bergmeyer and Dieter Oesterhelt
Max-Planck-Institut für Biochemie
D-8033 Martinsried, Germany

Introduction

The primary structure of bacteriorhodopsin (bR) in the purple membrane (PM) is known (1,2).
The disposition of the polypeptide chain first published in (1) was mainly based on surface cleavage sites by proteases (Fig.1).
Favored by the crystalline array of bR in the membrane an electron density map of 7 Å resolution was obtained (3) and the sequence of the polypeptide chain fitted into this map. The arrangements of the seven regions of helical density (A-G in Fig.2) were evaluated based on the criteria of connectivity of the nonhelical links, charge separation and total scattering density per helix (4).
Chemical modification is one appropriate approach to test the proposed arrangement of the polypeptide chain. Additionally it may give more insight into the chromophore structure, formed by retinal and the amino acids near its binding site, as well as into the protein's function as a proton pump. Both features are not explained by the model in Fig. 1 and 2.
The trimeric structure in general was already proven by cross-linking experiments (5). However, the sites of modification were not determined, although the knowledge of neighbouring amino acid residues would represent extremely valuable structural information. In this article we will summarize results of chemical modification of bR and the methods used for the iden-

Chemistry of Peptides and Proteins, Vol. 1

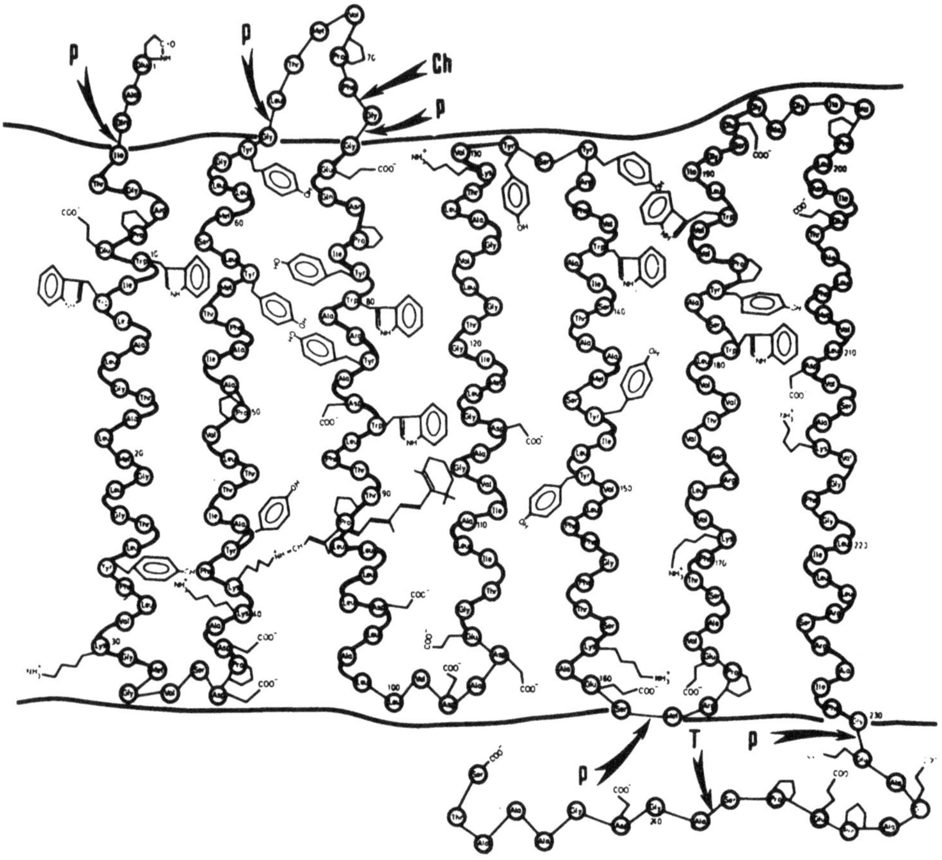

Fig. 1: Primary structure of bacteriorhodopsin and its arrangement in the membrane. Proteolytic cleavage sites of the native purple membrane with different proteases are marked (Ch: Chymotrypsin, P: Papain, T: Trypsin) (1).

fication of modified residues in the sequence.

Results

As a first reaction for chemical modification the nitration of tyrosine residues by tetranitromethane was used. After subti-

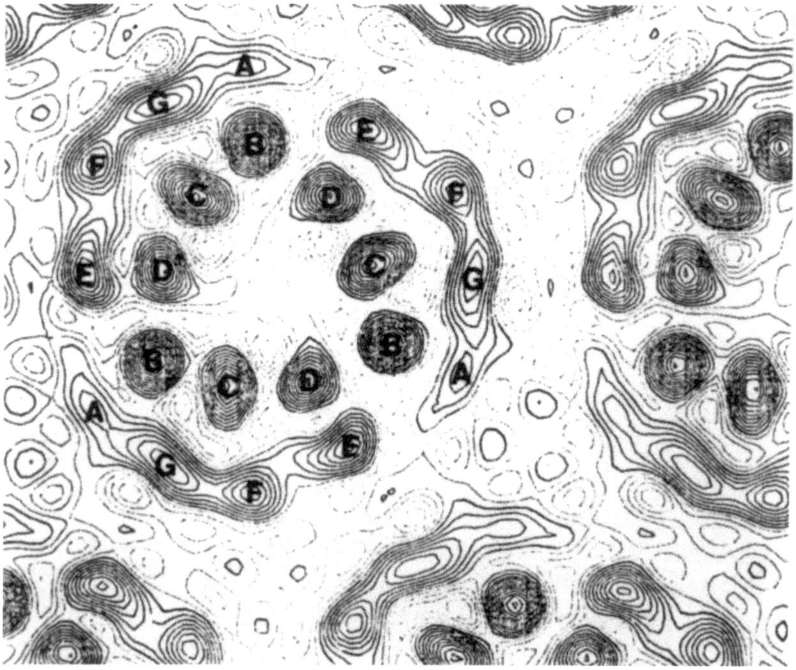

Fig. 2: Suggested arrangement (according to the model in (5)) of the seven helices in the protein's trimeric unit.

lisin digestion of the delipidated protein in suspension a complex mixture of small peptides was obtained. In Fig. 3f a high-performance liquid chromatogram of nitrated peptides (absorbing at 360 nm) of a digest is compared with 3-Nitrotyrosine and the nitrated dipeptides Ser-Tyr(NO_2) (Fig.3b), Gly-Tyr(NO_2) (Fig.3c), Val-Tyr(NO_2) (Fig.3d) and Leu-Tyr(NO_2) (Fig. 3e). Gly-Tyr (positions 63-64), Val-Tyr (positions 130-131) and according to (2) also Ser-Tyr (positions 132-133) are unique in the protein's sequence identifying the tyrosines 64 (1 mol), 131 (0.6 mol) and 133 (0.4 mol) as nitrated. The molar extent of modification of each tyrosine was obtained by integration (Fig. 3f) and comparison with the 3 moles 3-Nitrotyrosine found in the amino acid analysis of the protein's acid hydrolysate (6). For identification of the nitrated positions, from which Leu-Tyr(NO_2) was derived, fragments obtained by CNBr-cleavage of

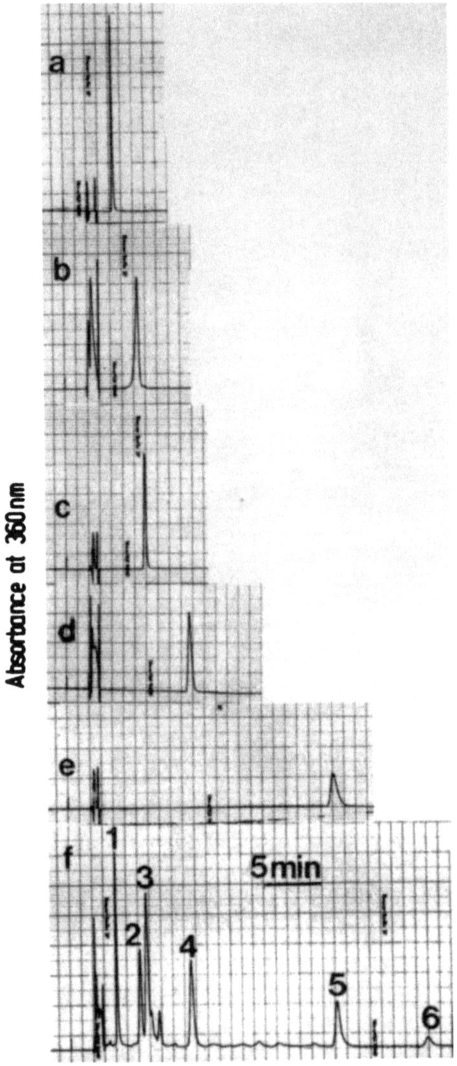

Fig. 3:
Comparison of high performance liquid chromatograms of 3-Nitrotyrosine, nitrated dipeptides and the subtilisin digest of nitrated purple membrane (6).
The following samples were injected separately into an isocratic high performance liquid chromatography system:
a) 3-Nitrotyrosine
b) Ser-Tyr(NO_2)
c) Gly-Tyr(NO_2)
d) Val-Tyr(NO_2)
e) Leu-Tyr(NO_2)
f) subtilisin-digest of nitrated purple membrane.

Peak 6 could be related to a sequence around tyrosine 26 by amino acid composition.

the delipidated protein were analysed. Only fragments II (position 21-32 including Leu-Tyr_{26}), V (positions 61-68 including Gly-Tyr_{64}) and VII (positions 118-145 including Val-Tyr_{131}-Ser-Tyr_{133}) show absorption at 360 nm but not fragment VIII (position 146-163, including Leu-Tyr_{147}-Ile-Leu-Tyr_{150}).

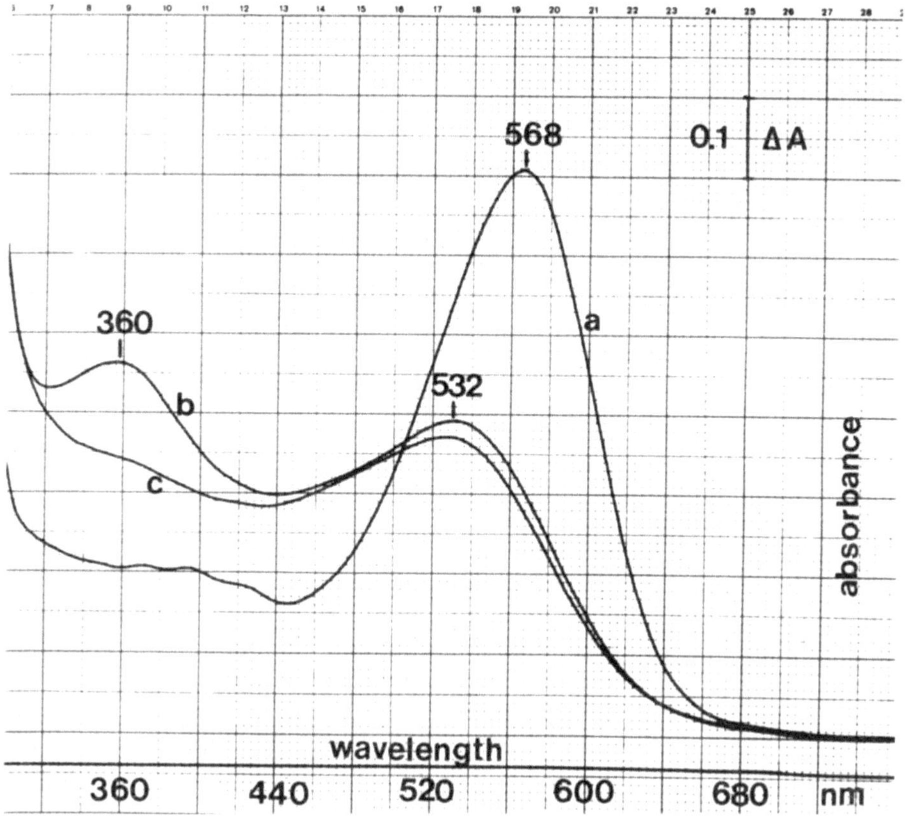

Fig. 4: Spectral changes of purple membrane upon nitration
a) purple membrane b) nitrated membrane
c) nitrated and subsequently reduced membrane .

After reaction of the nitrated membrane with the ionic, water soluble reagent sodium dithionite no peptides absorbing at 360 nm except those derived from tyrosine 26 were detectable any longer indicating a reduction of 3-Nitrotyrosine to the colourless 3-Aminotyrosine. As a consequence Tyr 64, 131 and 133 can be assumed to lie on the surface of the membrane therefore being accessible to the reducing agent. 3-Nitrotyrosine$_{26}$ under these conditions reacts to about 20 %.

In Fig.4 the spectral changes of the purple membrane upon nitration and subsequent reduction are shown. Nitration shifts the

absorption maximum from 568 nm to 532 nm and creates a new maximum at 360 nm. Dithionite reduction only influences absorption at 360 nm, whereas the chromophore absorption at 532 nm remains unchanged. A pH dependent equilibrium of the chromophore with a form absorbing at 510 nm is observed. The pK of this transition is at about pH 9. Therefore and because 3-Nitrotyrosine$_{26}$ does not show absorption at 360 nm, participation of tyrosine 26 in the chromophore structure was proposed (6,8).
Reaction of the purple membrane with the strongly hydrophilic reagent diazotized sulfanilic acid should produce further information about amino acid residues at the surface of the membrane. Modification at pH 8.8, CNBr-cleavage and separation of the resulting fragments by reversed phase HPLC and subsequent ion exchange HPLC revealed Lys 30 (0.5 moles), tyrosine 26 (0.25 moles) both located in CNBr-fragment II and tyrosine 64 (0.75 moles) located in CNBr V as reactive sites.
Reaction of purple membrane-containing vesicles lead to the exclusive modification of tyrosine 64. This clearly demonstrates that all bR molecules are oriented identically and that tyrosine 64 and tyrosine 26/lysine 30 are located on opposite sites of the membrane.
This specific modification can be used to obtain further information about the arrangement of the polypeptide chain in the membrane. Treatment of purple membranes with papain cleaves under certain conditions the carboxy terminal amino acids off the protein. The shortened polypeptide chain migrates differently from the uncleaved chain upon SDS gel electrophoresis.
The relative orientation of tyrosine 64 and the carboxyterminus can therefore be determined by the following experiment: Vesicles coupled with ^{35}S-diazotized sulfanilic acid (DSA) and treated with papain afterwards were prepared. As a prediction radioactivity should be found exclusively in the lower molecular weight band on SDS gels if tyrosine 64 and the carboxyterminus are located on the same side of the membrane but the uncleaved chain should carry the label if they are located on opposite sides.

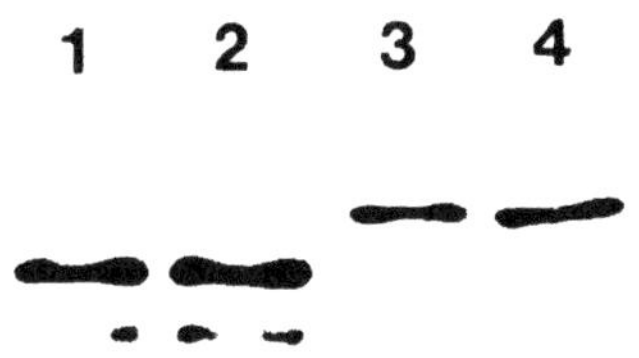

Fig.5: SDS gel electrophoresis after modification of PM-vesicles with ^{35}S-DSA at pH 10 and subsequent treatment with papain (autoradiogram)
1) ^{35}S-methionine PM
2) PM modified with ^{35}S-DSA
3) PM-vesicles modified with ^{35}S-DSA
4) ^{35}S-methionine PM without papain treatment
Samples 1-3 were treated with papain for 6 hours.

Fig.5 shows the result of that experiment. Unmodified ^{35}S-methionine PM serves as a reference for the uncleaved polypeptide chain (trace 4) and the papain cleaved chain (trace 1). Beside the main reaction product (bR minus carboxyterminus) lower molecular weight bands as a result of further papain action are seen. A ^{35}S-DSA modified and papain treated sample of PM shows the same pattern (trace 2). In contrast, only intact ^{35}S-DSA modified chains are found if the bacteriorhodopsin containing vesicles are modified and papain treated (trace 3). This proves the location of tyrosine 64 and the carboxyterminus on opposite sides of the membrane. Since the carboxyterminus was shown to be on the cytoplasmic side of the membrane (7), tyrosine 64 is located on the extracellular side and tyrosine 26/lysine 30 on the cytoplasmic side of the PM.

Discussion

The results of chemical modification of amino acid residues in bR located at the surface of the membrane support fully the arrangement of the polypeptide shown in Fig.1. Only little alterations of the path of the chain are necessary to get lysine 30, tyrosine 64, 131 and 133 in contact with the aqueous phase. Because tyrosine 26 is partially accessible to hydrophilic reagents the retinyliden chromophore itself may be close to the cytoplasmic surface of the membrane. However, based on these results several general questions arise.

None of the modification reactions (with the exception of the nitration at tyrosine 64) is complete as revealed by a reliable quantitation of the extent of modification (6). This non-stoichiometric reaction may be due to a stacking of the membranes in suspension thus excluding the reagent from access to some reactive sites. This complication due to the heterogeneous reaction can only be in part overcome by extensive sonication and has to be considered as a general problem in membrane chemistry. Another reason for non-stoichiometric modification might be the existence of different conformational states of the membrane protein with different reactivity. In the case of bR light-dark adaptation (i.e. cis-trans isomerisation of retinal) could provoke such differences.

Proteases like papain or chymotrypsin with buried active sites can cleave very close to the membrane surface within the small loops shown in Fig.1. No reasonable explanation for these reactions can be found on the basis of the proposed models. Tyrosine 26 and lysine 30 (partially) react with water soluble reagents, the retinyliden chromophore, which is assumed to be close to tyrosine 26, however, only reacts with borohydride or hydroxylamine in the light which induces a structural change of the chromophore. It could well be that tyrosine 26 has a protective function for the C=N double bond in the dark but not at the intermediate stages of the photochemical cycle. In fact, the nitrated membrane reacts with hydroxylamine already in the dark to yield retinaloximes and apomembrane.

References

1. Ovchinnikov, Yu.A., Abdulaev, N.G., Feigina, M.Yu., Kiselev, A.V., Lobanov, N.A.: FEBS Lett. 100, 219-224 (1979).

2. Khorana, H.G., Gerber, G.E., Herlihy, W.C., Gray, C.P., Anderegg, R.J., Niehei, K., Biemann, K.: Proc. Natl. Acad. Sci. USA 76, 5046-5050 (1979).

3. Henderson, R., Unwin, N.T.: Nature 257, 28-32 (1975).

4. Engelman, D.M., Henderson, R., McLachlan, A.D., Wallace, B.A.: Proc. Natl. Acad. Sci. USA 77, 2023-2027 (1980).

5. Dellweg, H.G., Sumper, M.: FEBS Lett. 90, 123-125 (1978).

6. Lemke, H.-D., Oesterhelt, D.: Eur. J. Biochem. 115, 595-604 (1981).

7. Gerber, G.E., Gray, Ch.P., Wildenauer, D., Khorana, H.G.: Proc. Natl. Acad. Sci. USA 74, 5426-5430 (1977).

8. Fischer, U., Oesterhelt, D.: Biophys. J. 28, 211-230 (1979).

FORMATION AND UNUSUAL PROPERTIES OF BACTERIOOPSIN - Ag^+ - 4Di-METHYLAMINOCHALCONE TRIPLE COMPLEX

Almaz Aldashev, Alexander Rodionov, Eugenii Efremov, Alexander Shkrob

Shemyakin Institute of Bioorganic Chemistry, USSR Academy of Sciences, Moscow, USSR

Introduction

Bacteriorhodopsin (BRh) is the main component of purple membranes from *Halobacterium Halobium* (1). Removing of the prosthetic group (retinal aldimine) from BRh transforms the purple membrane into colourless apomembranes containing bacterioopsin (BO) (2). Recombination of BO with, both, retinal and some of its analogs results in formation of aldimine - containing chromoproteins (3). Retinal analogs not only of aldehyde nature, competitively inhibit BO-aldehyde recombination (4). The Ag^+-ions are also the competitive inhibitors of the recombination and destroy the noncovalent complexes formed by BO with retinal analogs (5).

Results

Apomembrane-bound 4-dimethylaminochalcone (DMCh) $(CH_3)_2N-C_6H_4-CH=CH-CO-C_6H_5$ has the bands in the excitation and emission spectra (Fig. 1) at 435 nm and 500 nm, respectively (lifetime in the excited state ca. 2.1 nsec). The binding of the probe results in a large enhance of the quantum yield of fluorescence: this effect enabled us to calculate apomembranc -DMCh dissociation constant ($K_d=3\cdot10^{-7}$ M, see Fig. 2). The bound probe reveals relatively low ellipticity in CD-spectra ($|\Theta|=4,58\cdot10^3$ grad·sm^2/

Chemistry of Peptides and Proteins, Vol. 1

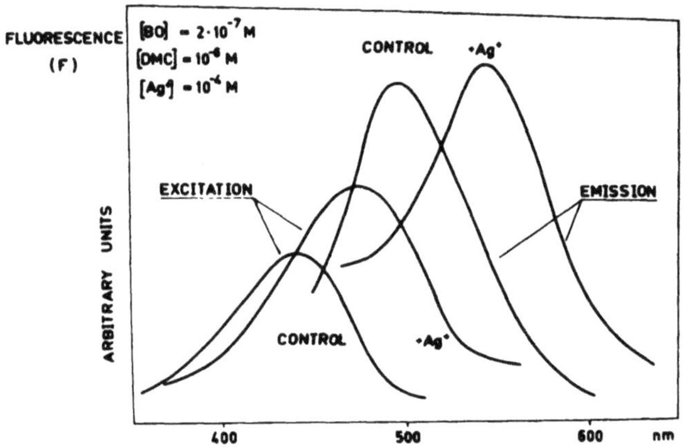

Fig. 1. Excitation and emission fluorescence spectra of DMCh-BO complex before and after Ag^{+}-ions were added.

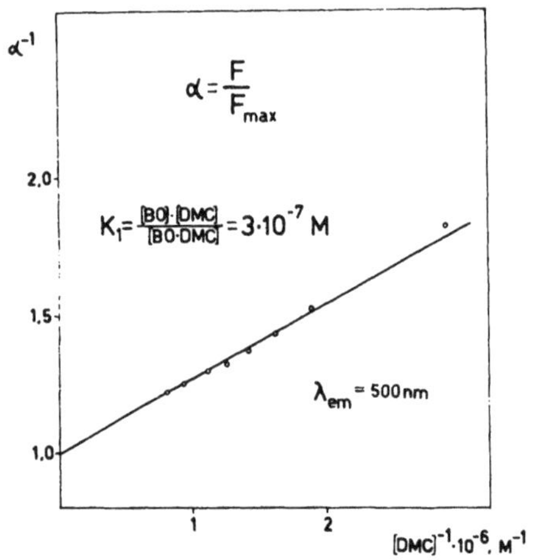

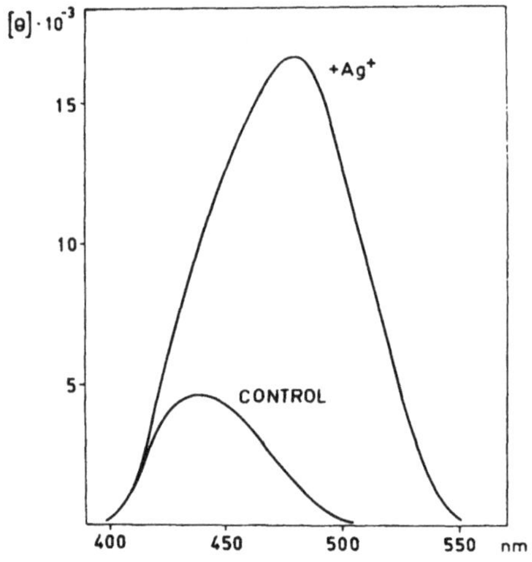

Fig. 2, 3. Determination of equilibrium constant of DMCh-apomembrane binding (C_{BO}=10^{-7} M) and CD-spectra of DMCh in apomembrane without Ag^{+} and in presence of Ag^{+} (C_{BO}=10^{-5} M).

decimole, see Fig. 3). K_d calculated from CD data appeared to be lower than the value obtained fluorometrically.

Taking into account that spectral properties of DMCh in apomembrane are very similar to those of the probe, included in lipid vesicles, we can explain the differences in K_d values obtained by both methods so, that the main contribution in fluorescence enhancement gives the lipid-bound probe, and in CD the protein-bound DMCh. DMCh does not inhibit the formation of BRh from BO and all-trans retinal but suppresses the recombination of BO with 13-cis-retinal (K_i=2,1·10^{-5} M) as well as with tetraenal p-CH_3O-C_6H_4-(CH=CH)$_4$CHO (K_i=1,4·10^{-5} M, Fig. 4). These data demonstrate the differences in the binding sites of isomeric retinals on the stage of the formation of noncovalent complexes.

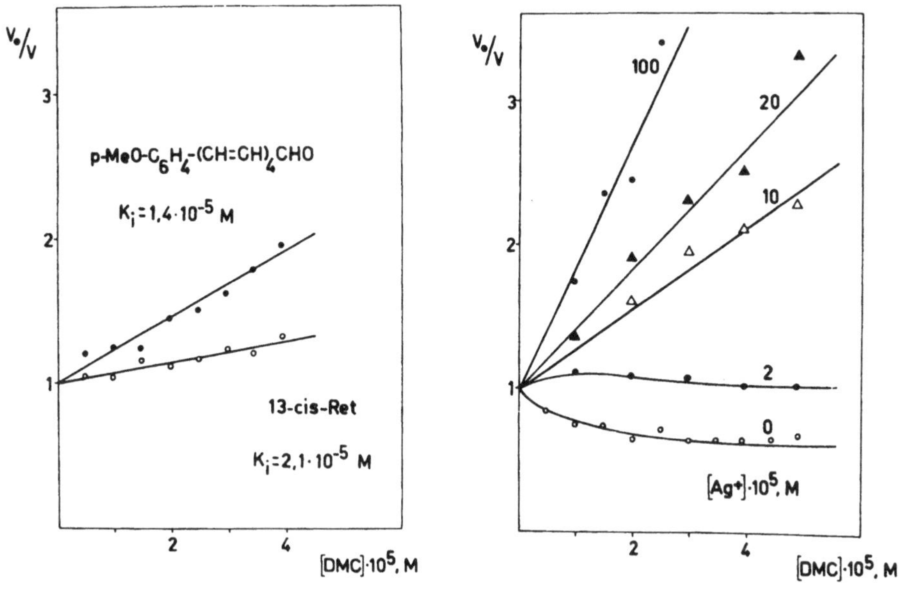

Fig. 4, 5. Competitive inhibition by DMCh of binding to BO (C_{BO}=10^{-6} M) of 13-cis-retinal and tetraenal (left) and all-trans retinal in presence of various concentrations of Ag^+-ions (right).

When Ag^+-ions are present even in negligible concentration, DMCh competitively inhibits the recombination of BO with all-trans retinal (Fig. 5). The Ag^+-ions induce also the large bathochromic shift (40 nm) of the bands in absorption, CD and fluorescence spectra (see Fig. 1, 3).
The halfwidth of the emission band of the BO-DMCh-Ag^+ complex ($\Delta\lambda$=78±2 nm) is similar to that of DMCh in organic solvents. It means that all molecules of the probe are in the homogeneous environment. In addition to the spectral shift, Ag^+-ions induce also the marked rise of DMCh ellipticity (up to $|\Theta|=16,5\cdot 10^3$ grad.sm^2/decimole, Fig. 3), which could be due to increase of asymmetry of the probe environment. Equilibrium constant for reaction BO-DMCh-Ag^+ ⇄ BO-Ag^+ + DMCh ($K_d=3,5\cdot 10^{-7}$ M) is independent of whether it was obtained from CD or from fluorescence data (Fig. 6). Thus, Ag^+-ions in small concentrations

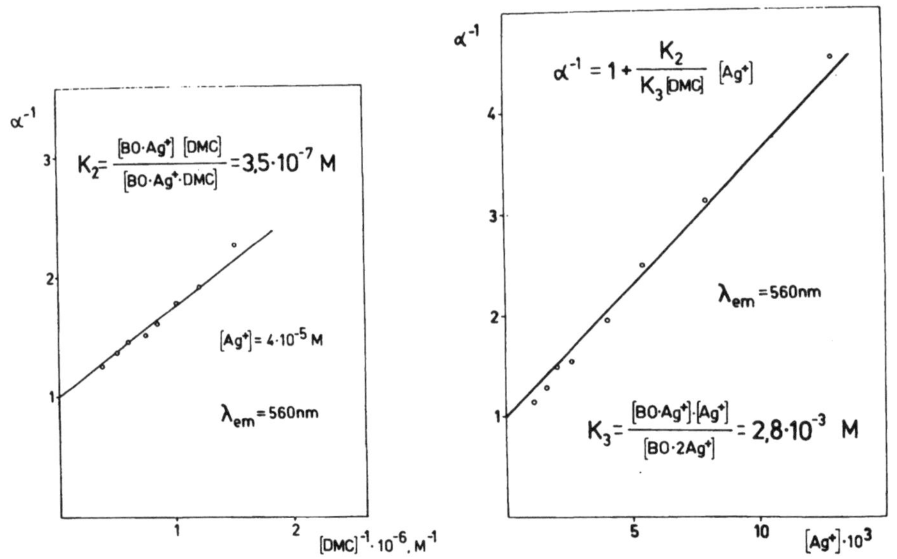

Fig. 6, 7. Determination of equilibrium constant of binding of DMCh to BO-Ag^+ complex (left). Removal of DMCh from BO by large concentrations of Ag^+-ions (right), $C_{BO}=10^{-7}$ M, MES 0,02 M, pH 6,4.

apparently induce not only the migration of DMCh from lipid to protein, but also modify the intraprotein binding site in such a way, that it can overlap the binding site of all-trans retinal. No other investigated metal ions cause such an effect. When the Ag^+ concentration is further increased, DMCh, like other retinal analogs is forced out of BO-DMCh-Ag^+ complex ($K_d = 8 \cdot 10^{-3}$ M, Fig. 7).

References

1. Oesterhelt, D., Stoeckenius, W.: Proc. Natl. Acad. Sci. U.S.A. 70, 2853 (1973).
2. Oesterhelt, D., Schuhmann, L.: FEBS Lett. 44, 262-265 (1974).
3. Schreckenbach, Th., Walckhoff, B., Oesterhelt, D.: Eur. J. Biochem. 76, 499-511 (1977).
4. Ovchinnikov, Yu.A., Shkrob, A.M., Rodionov, A.V., Mitzner, B.I.: FEBS Lett. 97, 15-19 (1978).
5. Shkrob, A.M., Rodionov, A.V.: Bioorg. Chem. 5, 376-393 (1979).

PEPTIDE FROM BEEF HEART MITOCHONDRIA INDUCING ION-SELECTIVE CHANNELS ON PLANAR BILAYER MEMBRANE

L.A. Pronevich, G.P. Mironov, G.D. Mironova
Institute of Biological Physics of the USSR Academy of Sciences, Pushchino

Introduction

Essential steps in studying the molecular mechanism of calcium transport regulation in mitochondria are isolation, identification and reconstruction of ion-transporting systems (1-6). This article is dealing with a new efficient method of isolation of beef heart mitochondrial peptide (P) which induces the selective Ca^{2+}-transport through artificial bilayer lipid membranes (BLM).

Results

1) Isolation of peptide. The mitochondrial fraction of beef heart tissue was extracted with 96% ethanol (200 ml per 1 g of mitochondrial protein; pH 7.0; 1 h under stirring; 4°C) and centrifuged at 5.000 x g for 15 min; 4°C. The pellet was resuspended in 50% ethanol and then extracted and centrifuged under the same conditions. The supernatants of two centrifugations were combined and evaporated to dryness in vacuum at 30°C, then dissolved in bidistilled water (1 ml per 2 g of wet tissue) and treated with 3-fold volume of chloroform: methanol (2:1, v/v) in order to remove lipids. The lipid-free extract was evaporated to dryness in vacuum at 30°C and dissolved in bidistilled water. 1.5 ml of this solution which corresponded to 100 mg of the dry pre-

Chemistry of Peptides and Proteins, Vol. 1

paration were applied to a Sephadex G-15 column (2.8x78 cm). The profile obtained after elution with bidistilled water (60 ml/h) is presented in Fig.1. As has been shown earlier (2), fraction 3, as identified by its ability to induce BLM conductance, corresponded to the membrane-active peptide.

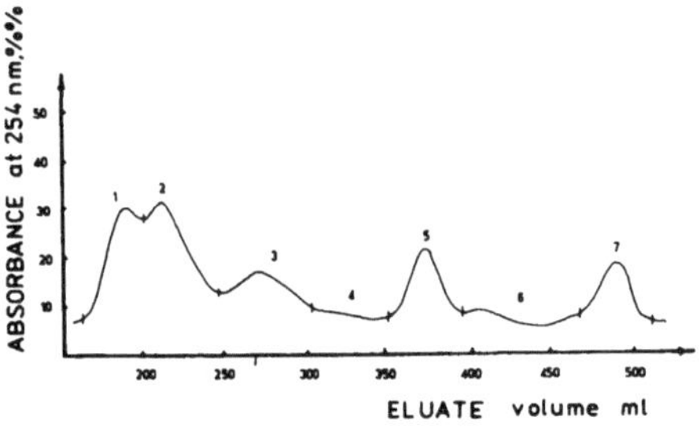

Fig.1. Elution profile of beef heart mitochondria, defatted water extract from Sephadex G-15 column (2.8x75 cm). Elution was carried out by bidistilled water at the rate of 60 ml/h.

2) Purification of P. Since we have shown earlier the presence of sulfhydryl groups in the peptide, further purification was conducted by means of affinity chromatography with reversible coupling of thiol-containing peptides. 150 mg of fraction 3 were applied onto an activated Thiol-Sepharose 4B column (4 ml).

The non-coupling components of fraction 3 were eluted with 25 ml of 0.02 M ammonium formate buffer (pH 8.0; 5 ml/h). The coupled P was eluted for 4 h with 20 mM mercaptoethanol in the same buffer. The eluate was evaporated until complete removal of mercaptoethanol and ammonium formate, then the dry preparation of P was brought onto Sephadex G-15 column (2.8x60 cm) and eluted with bidistilled water (60ml/h). The 140-170 ml fraction contained P. The homogeneity of P was analyzed by paper chromatography as described elsewhere (2).

3) Properties of P. The molecular weight of P as determined by gel filtration on Sephadex columns was about 2,000. After affinity chromatography purification the specific activity of P increases about 1000 fold. The acid hydrolysis revealed 7-8 amino acids (Fig.2). P is water- and ethanol- (70%) soluble. In water solution the Ca^{2+}-transporting activity of P decreases at room temperature in 26 h and at 100°C in 3 min. The yield of P depended on the season as in the case of glycoprotein (3) and was 2-5 µg from 100 g wet tissue.

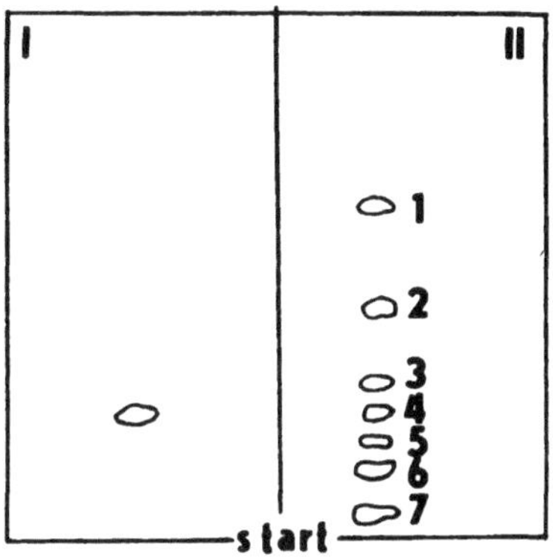

Fig.2. Ascending paper chromatography (Filtrak FN-1) in the butanol:acetic acid:water (4:1:5) system at room temperature: I - peptide before hydrolysis in 6 N HCl; II - peptide after hydrolysis in 6 N HCl: 1 - Leu; 2 - Phe; 3 - Ala; 4 - Pro; 5 - Glu,Gly; 6 - His,Lys; 7 - Cys

4) P-induced Ca^{2+}-selective conductance of BLM. Ca^{2+}-transporting activity of P was tested using BLM (beef brain lipids in n-decane, 20 mg/ml) which were formed on a hole (1 mm diameter) in a teflon wall separating two chambers each containing 2.5 ml of magnetically stirred buffer solution (20 ml Tris-HCl, pH 7.5; 5-20 mM $CaCl_2$). Adding P (final concentration1.0-5.0µg/ml) to one or two bathing solutions of BLM one could observe in 1-5 min keeping membrane voltage constant, step-wise increases of current which were interpreted as due to formation of channels. The smallest

amplitudes of these steps corresponded to an increase in BLM conductance of about 20 pS (Fig.3).

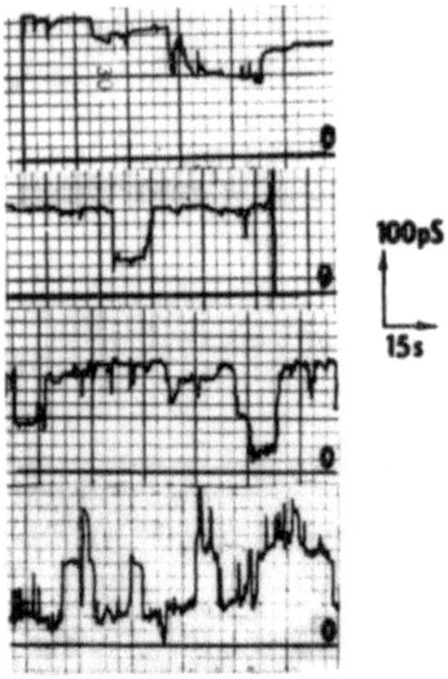

Fig.3. Current fluctuations in BLM modified with peptide P (2.0 μg/ml). Buffer solution contains 20 mM Tris-HCl and 20 mM $CaCl_2$.

Besides, large jumps of conductance up to 0.5-1.0 nS were registered which persisted for tens of minutes and resembled the increase of integral BLM conductance. The higher the concentration of P, the more probable was the opening of the channels and the longer the channels remained open. Increasing the concentration of $CaCl_2$ makes the conductance of the channels rise proportionally with the increase of the concentration of this salt. The generation of the transmembrane gradients of $CaCl_2$ gives rise to the potential difference on the membrane (21-25 mV per 10-fold gradient at pH 6.0-8.0) that corresponds to Ca^{2+}-selectivity of the channel. The selectivity for Ca^{2+} is also observed in case of large channels, which probably indicates their sieve structure, determined, in particular, by incorporation of whole P-aggregates.

5) Inhibition of Ca^{2+}-transporting properties of P by specific agents: a) sulfhydryl groups intrinsic in P proved to be

essential for its ability to induce the Ca^{2+} transport through BLM. Addition of fluorescein mercury acetate (FMA) binding sulfhydryl groups into the BLM bathing solution results in suppressing the ion-transporting properties of P; b) the specific inhibitor of Ca^{2+} transport in mitochondria ruthenium red (Rr) induces complete inhibition of the Ca^{2+}-conductance of BLM in the presence of P. When one or several Ca^{2+}-selective channels are open, the addition of Rr ($5 \cdot 10^{-6}$M) results in a 100% inhibition of the effect. The inhibition may proceed either jump-wise (abrupt closing of all or a few channels) or gradually. The inhibition rate depends on concentration of Rr.

The results obtained enable the assumption that P isolated from beef heart mitochondria is involved in transport of Ca^{2+} through the mitochondrial membrane. The proposal is also made that P in mitochondria functions in a complex with the glycoprotein.

References

1. Carafoli, E., Sottocasa G.: Dyn. Energ. Transducing membranes, Amsterdam, 455-465 (1974).
2. Mironova, G.D., Pronevich, L.A., Fedotcheva, N.I., Sirota, T.V., Trofimenko, N.V., Mironov, G.P.: Mitochondrial processes in time organization, Pushchino, 126-140 (1978).
3. Mironova, G.D., Sirota, T.V., Pronevich, L.A., Trofimenko, N.V., Mironov, G.P., Kondrashova, M.N.: Biofizika, 25, 276-280 (1980).
4. Azarashvili, T.S., Lukjanenko, A.I., Evtodienko, Yu.V.: Biochimia, 47, 1139-1142 (1978).
5. Lunevsky, V., Zherelova, O., Alexandrov, A., Vinokurov, M., Berestovsky, G.: Biofizika, 25, 685-691 (1980).
6. Sottocasa, G., Sandri, G., Panfili, E., de Bernard B., Gazzotti, P., Vasington, F., Carafoli, E.: Biochem. Biophys. Res. Commun. 47, 808-813 (1972).

FUNCTIONAL ROLE OF THE PROTEIN COMPONENT OF THE Ca^{2+}-TRANSPORTING GLYCOPROTEIN FROM BEEF HEART HOMOGENATE AND MITOCHONDRIA

T.V.Sirota, L.A.Pronevich, G.D.Mironova

Institute of Biological Physics of the Academy of Sciences of the USSR, Pushchino, Moscow Region, USSR

Introduction

Recent investigations have shown that the Ca^{2+}-transporting system of mitochondria contains substances of glycoprotein origin (1,2). A confirmation of this statement is provided by our data that mitochondria contain a glycoprotein (GP) with a molecular mass 40 kD, which, when reconstructed on a bilayer lipid membrane (BLM), activates a selective transport of calcium ions (3). There is some evidence in literature showing that Ca^{2+}-transporting activity may be due to phospholipids (4). The aim of the present work was to elucidate whether the phospholipids enter into composition of GP and what part of the GP molecule is responsible for its ionoforic properties.

Results

The attempt to identify the lipid component in the fraction of GP after electrophoresis in polyacrylamide gel (PAGE) by means of sudan black has not given positive results. At the same time our experiments have shown that the amount of inorganic phosphorus (P_i) in GP is 13.6 µg/mg of the protein whereas the preparations of GP isolated by the other authors (1,4) contain up to 300 µg of P_i per 1 mg of protein. The disparity in the results may be probably accounted for by the fact that in our experiments the partially purified preparation was treated

Chemistry of Peptides and Proteins, Vol. 1

with chloroform-methanol (2:1). After additional special treatment of GP we managed to isolate some amount of lipids using the method of thin-layer chromatography (TLC). At the first stage of extraction conducted by the method of Folch (8) negligible amount of neutral lipids was found: fatty acids, mono-, di-, triglycerides, cholesterol as well as the phospholipids phosphatidylserine and lysophosphatides. Following the second stage of GP treatment conducted by the method of Blondin (9) much less amount of the same neutral lipids and phospholipids was isolated. GP obtained after the two-stage treatment with organic solvents does not contain any lipids.
Simultaneously with determination of the lipid composition, at each stage of lipid extraction the Ca^{2+}-transporting activity of GP and the effect of ruthenium red (Rr) was studied by reconstruction on BLM.
The results are presented in Table 1. As seen from the Table, after removal of lipids from GP its Ca^{2+}-transporting activity does not change, the inhibiting effect of Rr, the specific inhibitor of calcium transport in mitochondria, being retained. The data obtained by PAGE indicated that after all stages of treatment GP contains, both, the protein and carbohydrate components. As shown by PAGE GP does not change prior and after treatment with organic solvents. This indicates that removal of lipids does not result in denaturation of the protein and that the contribution of the lipid component to the molecular mass of GP is insignificant. The data that treating with the mixture butanol:acetic acid:water does not affect the Ca^{2+}-transporting activity have been obtained earlier (3) when using the method of ascending paper chromatography in the same solvent system. The obtained results indicate that lipids are not essential for functioning of GP. It is likely that lipids, namely the phospholipids prevent a fast incorporation of GP into a model membrane, since GP from beef heart mitochondria containing minor amount of lipids starts to affect BLM conductance in 2-5 min (3), whereas GP from rat liver mitochondria containing a great amount of lipids in 40-50 min (4).

Table I. Ca^{2+}-transporting activity of GP after lipid extraction.

Preparation	GP, μg/ml	BLM conductance $(Ohm^{-1} \cdot cm^{-2})$	Rr inhibition, per cent
Control (without GP)	-	$(8.2 \pm 1.0) \cdot 10^{-9}$	-
GP initial	20	$(7.3 \pm 2.0) \cdot 10^{-7}$	100
GP after Folch procedure	20	$(6.2 \pm 2.5) \cdot 10^{-7}$	100
GP after Folch and Blonding procedures	20	$(7.1 \pm 3.1) \cdot 10^{-7}$	100

Media - tris-HCl pH 7.5 - 10 mM, $CaCl_2$ - 10 mM, Rr - $5 \cdot 10^{-6}$ M

Table 2. Proteinase treatment effect on Ca^{2+}-transporting activity of GP.

Preparation	GP μg/ml	BLM conductance $(Ohm^{-1} \cdot cm^{-2})$
Control (without GP)	-	$(6.8 \pm 1.4) \cdot 10^{-9}$
GP initial	20	$(5.5 \pm 2.0) \cdot 10^{-7}$
GP after proteinase treatment	20	$(8.8 \pm 3.2) \cdot 10^{-9}$
GP after incubation without proteinase	20	$(5.4 \pm 4.2) \cdot 10^{-7}$
Proteinase after incubation	-	$(6.5 \pm 1.5) \cdot 10^{-9}$

Media as in Table 1.

Further evidence for the essential role of the protein component in the function of GP are the experiments with proteinase (Table 2). As seen from the Table, proteinase-treated GP looses completely its Ca^{2+}-transporting activity.

Thus, it follows from the results presented in the paper that the main functional role in GP is played by its protein component. When reconstructed on BLM by means of lipid-free GP the Ca^{2+}-transporting system is inhibited by Rr. This enables the assumption that it is GP (its protein moiety) that is the main component in the system of electrogenic influx of calcium into mitochondria. The phospholipid isolated from rat liver GP whose Ca^{2+}-transporting activity is not inhibited by Rr is probably involved in the system of Ca^{2+} efflux from the mitochondria which exists in the mitochondrial membrane.

References

1. Carafoli,E., Sottocasa,G: Dyn.Energ.Transducing membranes, Amsterdam, 455-465 (1974).
2. Moore,C.: Biochem.Biophys.Res.Commun., 42, 298-305 (1971).
3. Mironova,G.D., Sirota,T.V., Pronevich,L.A., Trofimenko,N.V., Mironov,G.P., Kondrashova,M.N.: Biofizika, 25, 276-280 (1980).
4. Evtodienko,Yu.V., Medvedev,B.I., Yaguzingsky,L.S., Azarashvili,T.S., Lukjanenko,A.I., Kuzin,A.M.: DAN SSSR, 249, 1235-1238 (1979).
5. Reynafarje,B., Lehninger,A.: J.Biol.Chem., 244, 584-593 (1969).
6. Jeng,A., Ryan,E., Shamoo,A.: Proc.Nat.Acad.Sci.USA, 75, 2125-2129 (1978).
7. Tyson,C.A., Lande,A.V., Green,D.E.: J.Biol.Chem., 25, 1326-1329 (1976).
8. Folch,J., Lees,M., Stanley,J.: J.Biol.Chem., 226, 497-509 (1957).
9. Blondin,G.: Biochem. Biophys. Res. Commun., 56, 97-105 (1974).

OPTICAL SPECTROSCOPY STUDY OF SUBSTRATE BINDING BY LEUCINE SPECIFIC AND LEUCINE - ISOLEUCINE - VALINE BINDING PROTEINS FROM *E. coli*

Igor Nabiev, Sergey Trakhanov, Alexandr Surin, Tatyana Vorotyntseva, Evgenii Efremov, Vladimir Pletnev

Shemyakin Institute of Bioorganic Chemistry, USSR Academy of Sciences, Moscow, USSR

Introduction

Leucine - isoleucine - valine-binding (LIV) and leucine specific (LS) proteins are the components of high affinity transport systems for branched chain amino acids in *E. coli* K12. These proteins are located in the periplasmatic space being released on cold osmotic shock procedure (1). The polypeptide chain of the LIV protein consists of 344 amino acid residues including 3 Trp, 13 Tyr, 9 Phe and one S-S bridge (2). The homogeneous LIV protein contains tightly bound substrates as L-leucine, L-isoleucine and in less amount L-valine which might be released by reversible denaturation treatment in urea solution with subsequent urea removal. In contrast to LIV, the LS protein can be obtained in crystalline form free of L-Leu without the treatment mentioned above. The primary structure of the LS protein which is not completely determined yet (the most probable number of amino acid residues is 346, including 4 Trp, 13 Tyr and 9 Phe residues) revealed a high degree of homology with the primary structure of the LIV protein (3).

Results

Recently it has been shown that binding L-Leu, L-Ile and L-Val causes changes in optical spectra of the LIV protein (4). The positions of the maxima and fluorescence band shapes of LIV and LS proteins point to differences in polarity of Trp environment with at least one Trp buried in the protein interior. The changes of the fluorescence spectra of LIV and LS proteins upon substrate addition (Fig. 1) reflect different environments of the Trp residues in the complexed with respect to the free forms of the proteins. In case of LIV the shape of the differential spectra depends on the type of substrate (4). The ions I^- and Cs^+ quench significantly the fluorescence of the Trp residue (-s) in the Ls protein but do not influence the fluorescence of the protein. In the L-Leu complex of LS the fluorescence quenching is less pronounced. It was suggest-

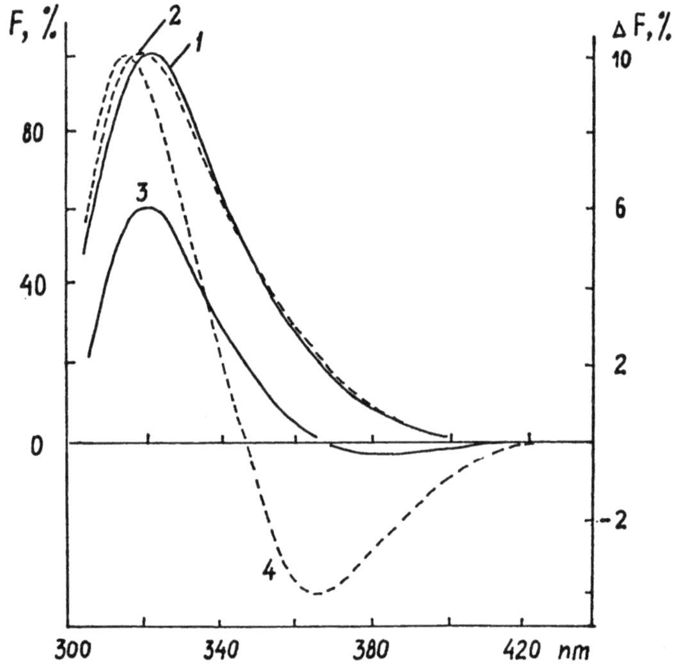

Fig. 1. Fluorescence spectra (F) LIV (1) and LS (2) and relative changes in fluorescence (ΔF) LIV (3) and LS (4) upon L-Leu complex formation. $\lambda_{exc.}$=295 nm, protein concentration 5 μM, L-Leu concentration 10 μM.

ed that a Trp residue (-s) on the surface of the LS globule becomes less accessible to the solvent environment as a result of complex formation.

Differential UV spectra obtained upon substrate binding by LIV and LS proteins reveal changes in the aromatic residue environment due to complex formation.

The common feature of differential spectra of LIV and LS proteins is the presence of bands corresponding to Tyr and Phe chromophores which might be interpreted as a transfer of the side chains of these residues into a more hydrophobic environment.

Contrary to LIV in LS protein the bands at 290 and 300 nm associated with Trp residue (-s) are clearly revealed. These bands are not observed in the LIV protein spectra. Probably the

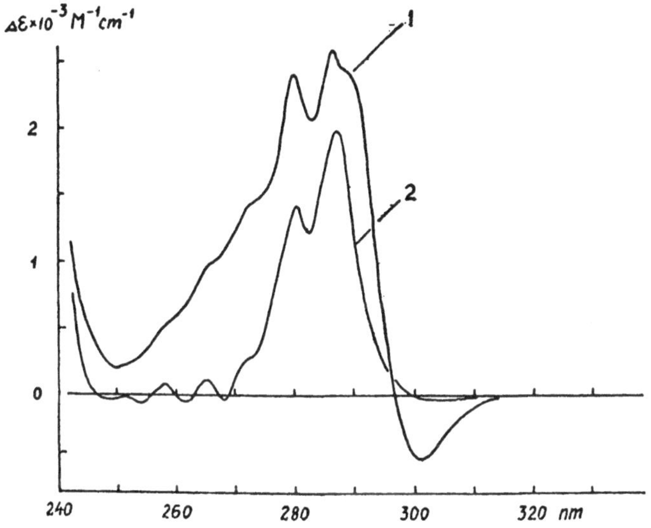

Fig. 2. Differential UV spectra of LS (1) and LIV (2) after L-Leu addition. Protein and substrate concentrations are 5 μM and 10 μM.

Table 1. The amount of Trp and Tyr residues accessible for solvent in LIV and LS proteins and in the L-Leu - protein complexes.

Protein/Perturbant	Ethylene glycol 20% v/v		Sucrose 20% w/v	
	Tyr	Trp	Tyr	Trp
LIV	7	2	7	2
LIV+L-Leu	5-6	2	5-6	2
LS	9	3		
Ls+L-Leu	5-6	1-2		

charge alteration in the vicinity of the chromophore causes the appearence of a minimum at 300 nm.

The solvent perturbation differential UV spectroscopy has been applied for quantitative estimation of the effect of substrate binding on the solvent accessibility of the aromatic amino acid residues (Table 1).

In comparison with LIV the number of accessible Trp residues of LS protein is diminished of at least one residue. In the CD and Raman spectra practically no changes are observed upon substrate binding. Complex formation does not obviously affect the secondary structure of the polypeptide chain in the LIV and LS proteins. Comparison of differential UV, fluorescence, CD and Raman spectra with the data of primary structure of these proteins enables us to suggest that Trp-18 (for the LS protein) and Tyr-18 (for the LS protein) participate in the formation of a substrate binding center. A model of the binding was also proposed (4).

The confirmation about participation of Tyr, Trp, Phe residues in complex formation has been obtained from the comparison of surface - enhanced Raman spectra of the free and complexed LS protein adsorbed on a silver electrode. The changes of the

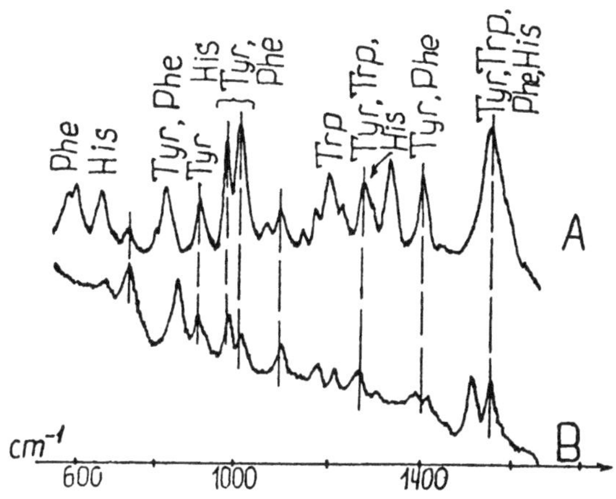

Fig. 3. Surface - enhanced Raman spectrum of LS protein (B) and LS+L-Leu (A) adsorbed on a silver electrode. Protein concentration 0.5 mg/ml, $\lambda_{exc.}$=632.8 nm.

surface - enhanced Raman spectrum on substrate binding suggest that the LS protein maintains its ability for substrate binding even after adsorption on the electrode surface (Fig. 3).

References

1. Oxender, D.L.: In Biomembranes chap. 2 (Manson, L.A., ed.), New York, Plenum Press 25-79, 1974

2. Ovchinnikov, Yu.A., Aldanova, N.A., Grinkevich, V.A., Arzamazova, N.M., Moroz, I.N.: FEBS Letters 78, 313 (1977).

3. Soldatova, L.N., Shakhparonov, M.I., Chertova, E.N., Aldanova, N.: In Abstracts of 5th All-Union Symposium on Chemistry and Physics of Peptides and Proteins. Baku, 124, 1980

4. Vorotyntseva, T.I., Surin, A.M., Trakhanov, S.D., Nabiev, I.R., Antonov, V.K.: Bioorg. Khim. 7, 45 (1981).

INVESTIGATIONS OF IMMUNOGLOBULIN STRUCTURE IN SOLUTION

Vladimir Zav'yalov, Vyacheslav Abramov, Olga Loseva, Vladimir Tishchenko, Elena Dudich, Igor Dudich

All-Union Research Institute of Applied Microbiology,
142200 Serpuckov, Moscow Region, USSR

Frantisek Franek

Institute of Molecular Genetics
Prague 6, Czechoslovakia

George Medgyesi

Institute of Haematology
H-1502 Budapest, Hungary

Introduction

The available data on the three-dimensional immunoglobulin structure have been obtained as a result of X-ray structural analysis of some human myeloma immunoglobulins G belonging to the first subclass (IgG1 Dob, Col, Mcg) and of immunoglobulin A produced by murine plasmacytoma (M 603) (1). These data give a reason to conclude only - the most general regularities of the spatial arrangement of polypeptide chains of immunoglobulins G. In particular, these data manifest the conformational mobility in the so-called "switch" and "hinge" regions. It is likely that during crystallization only one of the possible dynamic conformational states is fixed. The equilibrium in solution may be shifted towards the other conformer, and it may appear to be the decisive factor to understand the functional immunoglobulin differences and the mechanism of activation of their effector functions. Besides, it is known that even immunoglobulins G of different subclasses differ significantly in their functional properties (2). The conformatio-

Chemistry of Peptides and Proteins, Vol. 1

nal basis of these differences is still unclear,though the main differences in the primary structure of immunoglobulin G subclasses are known to be concentrated in the "hinge region" (3).One may suppose that the structure of this region can affect the functional properties of immunoglobulin G subclasses, both, by influencing the degree of relatively independent motions of Fab and Fc subunits in solution and by influencing the structure of Fc subunit and,correspondingly,the conformation of effector sites.In order to verify this supposition,we have investigated in the present paper the normal rabbit immunoglobulin G,the homogeneous rabbit antibodies to pneumococcus polysaccharides,precipitating (early) and non-precipitating (late) pig antibodies to dinitrophenyl derivatives and homogeneous human myeloma immunoglobulins G of 4 subclasses.The investigations were carried out with the help of the complex of physical methods allowing to estimate the conformational properties of proteins in solution (differential temperature- and solvent-perturbation spectrophotometry, adiabatic scanning microcalorimetry,polarization of fluorescence,spin label method of EPR,method of proton relaxation of NMR,circular dichroism) and also the electronic microscopy.

Results

1) Immunoglobulins G of different species origin and of different subclasses differ significantly not only in the character of interactions between Fab and Fc subunits (the different energy of non-covalent interactions between subunits and different degree of relatively independent motions of subunits in solution) but in the structure of Fc subunits. In addition, the differences in the structure of Fc subunits are stipulated not only by different energy of non-covalent interactions between C_H2 and C_H3 domains but by different rigidity of the structure of C-terminal part of C_H2 domains.Precisely these

parts of the Fc subunit form, according to the available data, the site of interaction with Fc receptors of different cells and the site of fixation of the first factor of the complement system (4,5).

2) Normal rabbit immunoglobulin G is characterized by the absence of the significant, to some extent, non-covalent interactions between Fab and Fc subunits and by the highest degree of relatively independent motions of Fab and Fc subunits in solution. The freshly obtained homogeneous rabbit antibodies to pneumococcus polysaccharides differ from normal rabbit immunoglobulins G in the lower stability of C-terminal part of C_H2 domains and in the less energy of non-covalent interactions between C_H2 and C_H3 domains (the expected sites of complement fixation and of interaction with cell Fc-receptors). This is stipulated by the slow kinetics of renaturation of these regions after the acid antibody elution from the affinity sorbent. The binding of bivalent antigen causes the change in the site of complement fixation and in the site of interaction with cell Fc receptors. These data are convincing evidence in the favour of the induction by an antigen of conformational changes in Fc subunit and precisely in effector sites. But the completely renaturated antibodies lack this effect. Therefore, it appears necessary to review from these positions all the previously published work with antibodies, obtained by acid elution.

3) Pig immunoglobulins G are characterized by the strongest non-covalent interactions between Fab and Fc subunits and by the lowest degree of relatively independent motions of Fab and Fc subunits in solution. Moreover, the non-precipitating (late) pig antibodies have the more rigid structure than the precipitating ones. Though the N-terminal part of Fc subunit of the non-precipitating antibodies has also the more rigid structure as compared with the precipitating ones, the energy of non-covalent interactions between C_H2 and C_H3 domains in the non-precipitating antibodies is lower than in the precipitating ones, what shows good agreement with the higher re-

sistance of this region in precipitating antibodies to pepsin proteolysis. The hapten binding induces the conformational changes in the antibody Fc subunit, bound with the destabilization of C-terminal part of C_H2 domains and also of C_H2 and C_H3 contacts. According to the data of solvent and thermal perturbation difference spectroscopy the stability of another part of C_H2 domains (probably N-terminal part) of the non-precipitating antibodies is increased as a result of hapten binding.

4) Human immunoglobulin G subclasses have different degree of relatively independent motions of Fab and Fc subunits in solution, which depends greatly on temperature, and they have also different conformation of Fc subunits. Immunoglobulins G3 are characterized by the absence of non-covalent interactions between Fab and Fc subunits and by the highest degree of mobility of these subunits to each other in solution. At the same time Fc subunit of IgG3 is characterized by the strongest non-covalent interactions between C_H2 and C_H3 domains and by the rigid structure of the C-terminal part of C_H2 domains, what explains, apparently, the highest complement fixation activity of IgG3 of human immunoglobulins G subclasses (2). The strongest non-covalent interactions between Fab and Fc subunits are observed in immunoglobulin G2. At the same time the weakest non-covalent interactions between C_H2 and C_H3 domains and the low stability of the C-terminal part of C_H2 domains are characteristic of this subclass what explains the low IgG2 activity in relation to complement fixation (2). Immunoglobulins G1 are similar to IgG3 by the energy of non-covalent interactions between C_H2 and C_H3 domains and by the stability of the C-terminal part of C_H2 domains, but the essential non-covalent interactions between Fab and Fc subunits are observed for this subclass in contrast to IgG3, what explains, apparently, the lower complement-fixating IgG1 activity (2). In immunoglobulin G4 the non-covalent interactions between C_H2 and C_H3 domains are even, to some extent, stronger and the C-terminal part of C_H2 domains

is more stable than in IgG2. Therefore, the lack of the complement-fixation activity of this subclass may be explained only by the peculiarity of the interactions between Fab and Fc subunits, as a result of the shortest "hinge region" and "rigid" practically completely polyproline sequence of this region (3).

5) The investigations carried out in the laboratories of J. Gergely (6), F. Franek (7) and V. Zav'yalov manifest that immunoglobulin Fc subunits may bind the relatively low molecular polypeptide factors, modifying Fc-subunit structure and affecting their effector functions and the other biological properties (the capacity to bind the complement, to precipitate in complex with polyvalent antigens, to form cryoprecipitates in cold). These compounds (we propose to call them as Fc-MF-Fc-modifying factors) may present the new class of regulators of immune response and take an important part in the mechanism of activation of effector functions. The unusual increase of Fc-MF may explain the changes of IgG properties as a result of some diseases. This effect was discovered earlier by us as a consequence of chronic diseases (8).

References

1. Poljak, R.: Immunoglobulins, ed. by G.W. Litman and R.A. Good, Plenum Medical Book Company, New York-London, 1978

2. Kehoe, J.M.: Immunoglobulins, ed. by G.W. Litman and R.A. Good, Plenum Medical Book Company, New York-London, 1978

3. Zav'yalov, V.P., Abramov, V.M., Ivannikov, A.I., Loseva, O.I., Dudich, E.I., Dudich, I.V., Tishchenko, V.M., Khechinashvili, N.N., Franek, F., Medgyesi, G., Zavodszky, P., Jaton, J.-C.: Haematologia (Hungary) 14, 175 (1981).

4. Calvanico, N.J., Tomasi, T.B.: Immunochemistry of Proteins, v.3, ed. by M.Z. Atassi, Plenum Publishing Corporation, New York, 1979

5. Burton, D.R., Boyd, J., Brampton, A.D., Easterbrook-Smith, S.B., Emanuel, E.J., Novotny, J., Rademacher, T.W., van Schravendijk, M.R., Sternberg, M.I.E., Dwek, R.A.: Nature 288, 338 (1980).

6. Füst, G., Csecsi-Nagy, M., Medgyesi, G.A., Kulics, J., Gergely, J.

Immunochemistry 13,793 (1976).

7. Franek,F.,Olshovska,Z.,Simek,L.:Eur.J.Immunol. 9,696, (1979).

8. Troitsky,G.V.,Selivanova,K.F.,Glushkova,O.V.,Kornienko, N.V.,Zav'yalov,V.P.: Voprosy med.chimii (USSR) 25,122 (1979).

CONFORMATIONAL LABILITY OF IMMUNOGLOBULIN M MOLECULE

Ekaterina Kaverzneva, Faina Shamkova, Yuri Khurgin
Institute of Organic Chemistry, Academy of Sciences of USSR, Moscow 117913, USSR.

Renata Kayushina
Institute of Crystallography, Academy of Sciences of USSR, Moscow 117333, USSR.

Introduction

The complement binding and some other physiological functions of immunoglobulins are antigen-dependent reactions. It was supposed that during their biological function the immunoglobulins undergo certain structural changes (1) which require rather high flexibility of the macromolecule. Up to this day the immunoglobulin flexibility is mostly evidenced by indirect ways.

It is the purpose of our study to demonstrate the structural lability of the immunoglobulin molecule as a function of surrounding media. We present here data on large-scale irreversible changes in the immunoglobulin M (IgM) structure after short-time exposition to pH lower than its isoionic point (pI 5.5). The immunoglobulin M antibodies are synthesized at earlier steps of the immune response reaction of the organism. These proteins are built up of five IgG-like subunits and altogether contain ten light (L) and ten heavy (H) chains. Each subunit contains two antigen-binding sites in their N-termini.

The protein IgM Cep was derived from the serum of a Waldenstróm disease patient. The starting solution of IgM Cep in 0.01 M tris-HCl buffer, pH 8.0 (IgM^{I}) was prepared by usual

Chemistry of Peptides and Proteins, Vol. 1

manner (2). The second sample solution (IgM^{II}) was obtained from the starting solution IgM^{I} by acidification with 1.0 M HCl down to pH 3.0 and after short time exposition (ca. 5 min) was returned up to pH 8.0 with 1 M NaOH or by exhaustive dialysis against 0.001 N HCl and then against tris-HCl buffer (0.01 M, pH 8.0).

The rate of limited proteolysis of IgM^{I} and IgM^{II} at constant pH 8.0 was measured potentiometrically with pH meter "pH 121".

The tyrosine residues were nitrated in $3.3.10^{-6}$ M IgM and $6.1.10^{-4}$ M tetranitromethane solutions. The yield of 3-nitrotyrosine was measured spectrophotometrically ($\lambda = 428$ nm, $\varepsilon = 4100\ M^{-1}cm^{-1}$).

The hydration isoterms of lyophilized samples of IgM Cep were measured by the aquametric titration method (3) using the Fisher reagent in a Sherman apparatus (4).

Results and Discussion

The limited proteolysis of both samples IgM^{I} and IgM^{II} was carried out in a buffer-free medium at pH 8.0 and 38°C with trypsin-IgM of a ratio of 1:25. During a starting period of 30 min the yield of trypsinolysis increased proportionally with time. The time of back titrations of acidified solutions up to pH 8.0 did not influence the kinetics of the trypsinolysis of IgM^{II}.

The initial trypsinolysis rate of IgM Cep in the acidic medium is 2.5-3.0 times higher than for IgM^{I}, not subjected to such treatment. The initial rate increase seems to be the result of rapid irreversible changes in the acidic medium making some of the peptide groups accessible to trypsin.

An opposite situation was noted (5) when oligosaccharides of IgM were hydrolyzed by a multienzyme glycosidase complex having a pH optimum below 5.5. Fifty per cent of the neutral monosaccharide residues were split off at pH 5.0 only after 24 days. But the same degree of hydrolysis was reached at pH 6.4 after 8 days.This effect shows a marked decrease of oligosaccharide accessibility to glacosidases below pH 5.5.

A similar "acidic" decrease in accessibility was found for tyrosine residues in IgM Cep. The initial nitration rate of acid-treated IgM^{II} at pH 8.0 and 23°C is twofold lower than that of an untreated starting sample IgM^{I}. When the reaction time was 24 hours the level of nitration of IgM^{I} and IgM^{II} was 16% respectively 10% of 190 tyrosines present in IgM Cep. These data correspond to three (IgM^{I}) respectively two (IgM^{II}) nitrated tyrosine residues in one H respectively one L chain. The irreversible "acidic" structural change seems to cause translocation of one fully exposed tyrosine residue into partially hindered interdomain space.

The aquametric titration of solid protein samples allows to estimate the total number of surface (i.e. accessible to water molecules) polar amino acid side chains (h) calculated on the basis of the BET monolayer (3). The native IgM^{I} is characterized by h = 6.13 mmoles H_2O per g protein (mmole/g) which is essentially equal to the total number of proposed hydration sites (3) i.e. to the sum of polar amino acid side chains and hydroxyl and amido groups of the oligosaccharides. The BET monolayer of an IgM sample, lyophilized from buffer-free solution at pH 3.0, has h = 4.48 mmole/g. This value is 25 % lower compared to the former one. Such a decrease of hydration capacity is an evidence for large-scale transformation of the IgM structure in acidic media. As globular proteins do not belong to the immunoglobulin group the change of the BET monolayer by acidification does not exceed 5-7 %.

The most conclusive support of large structural changes in the IgM molecules in acidic medium seems to be obtained from radii of gyration (R_g) and the dimension of the molecules (L) obtained by the small-angle X-ray scattering method. The R_g and L values were received in 0.01 M ammonium carbonate solution at pH 8.6 and 23°C.

$$IgM^{I} \quad R_g = 10.6 \text{ nm} \qquad L = 31.2 \text{ nm}$$
$$IgM^{II} \quad R_g = 8.9 \text{ nm} \qquad L = 24.3 \text{ nm}.$$

It follows from these data that the IgM Cep molecule in acidic medium is irreversibly contracted to about 20-25 %(6).

Conclusions

The short-time exposition of IgM to acidic media(pH $<$ 4.5) results in the following structural changes:

(1) A more than 20 % decrease in size,

(2) a decrease in the number of accessible hydration sites and

(3) a change in accessibility of certain groups - partial burrying of oligosaccharides and tyrosines but some lysines and arginines simultaneously become more exposed.

The compactness of IgM molecule seems to be due to the displacement of Fab regions out of the central (Fc_5) u-disc plane. The dramatic structural changes seem to be accompanied by an increase in its lability as evidenced from the fluorescence data of the group of Cathou (7). We have found rather similar changes of IgM Cep structure at 56-58°.

References

1. Feinstein, A., Munn, E.A.: Nature (London) 224, 1391-1309 (1969).

2. Lapuk, V.A., Shamkova, F.V., Kaverzneva, E.D.: Bioorganicheskaya Chimiya 1, 1134-1139 (1975).

3. Khurgin, Yu.I.: Zhurnal Vsesoyusnogo Chimicheskogo Obshchestva 1976, 684-690.

4. Sherman, F.B., Khurgin, Yu.I.: Talanta 28, 161-167 (1981).

5. Kaverzneva, E.D., Vikha, G.V., Lapuk, V.A.: Bioorganicheskaya Chimiya 1, 1379-1382 (1975).

6. Khurgin, Yu.I., Kayushina, R.L., Damaschun, G., Shmakova, F.V., Dembo, A.T., Rolbin, Yu.A., Damaschun, H., Plitz, P., Asadchikov, V.E.: Studia Biophysica 79, 143-144 (1980).

7. Holowka, D.A., Cathou, R.E.: Biochemistry 20, 3779-3790 (1976).

STRUCTURE-FUNCTIONAL STUDIES OF DNA-DEPENDENT RNA POLYMERASE FROM *E. coli*

Oleg Chertov, Vladimir Efimov, Oksana Chakhmakhcheva, Yuri Smirnov, Sergei Tsarev, Nickolai Skiba, Tatyana Marchenko, Valeri Lipkin, Eugeni Sverdlov

Shemyakin Institute of Bioorganic Chemistry, USSR Academy of Sciences, Moscow, USSR

Chemical and photochemical affinity modifications were used to find subunits of the DNA-dependent RNA polymerase responsible for binding the DNA template, substrates and RNA product.
It is known that short thymidilate polynucleotides serve as templates for RNA polymerase. Decanucleotide 5'-^{32}P-d(T_5-BrU-T_4) containing photoreactive residue of 5-bromouracil forms a stable covalent complex with RNA polymerase subunits under ultraviolet irradiation. The distribution of the radioactive label among the subunits depends on the type of enzyme (holo or core) and the presence of ATP. The irradiation of the binary holoenzyme - decanucleotide complex has led to preferential labeling of the σ-subunit. In the presence of ATP (transcribing complex) β' was labeled exclusively (1).
We have also developed an approach enabling us to detect contacts of RNA polymerase with 5'- and 3'-ends of the growing RNA chain at different stages of transcription.
To identify the enzyme contacts with the 5'-end of the nascent RNA chain we used a one-promoter fragment of a λ imm 434 DNA, γ-azidoanilide of GTP (which is according to our findings a substrate for RNA polymerase (2)), and various incomplete sets of nucleoside triphosphates, one of which was radioactively labeled (Fig. 1).
Maximum tetranucleotide can be synthesized in the presence of γ-azidoanilide of GTP and UTP, while decanucleotide in the presence of γ-azidoanilide of GTP, UTP and ATP. Obviously radio-

Chemistry of Peptides and Proteins, Vol. 1

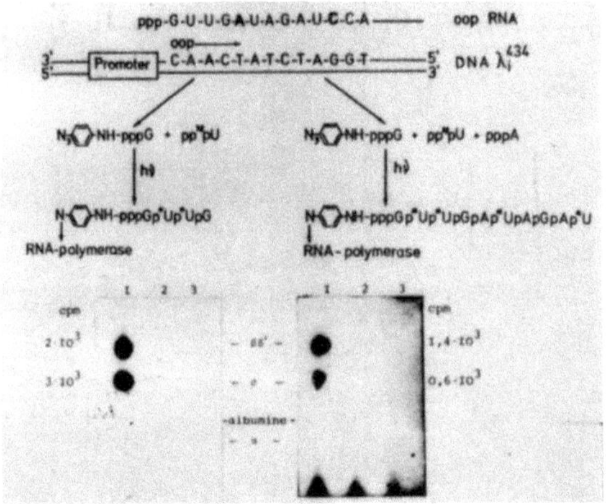

Fig. 1. Crosslinking of the RNA polymerase with photoreactive RNA in transcribing complex. (1)-after irradiation. (2)-without irradiation. (3)-the complex was destroyed before the irradiation.

active label covalently bound to RNA polymerase upon illumination at λ >290 nm may be due only to the oligonucleotides synthesized in the system, rather than to initial substrates, because the γ-azidoanilide of GTP is not radioactive, whereas the $pp^{32}pU$ is not photoreactive. After synthesis and irradiation the reaction mixtures were subjected to SDS electrophoresis. The slabs were stained with Coomassie and autoradiographed. After photoaffinity labeling (two triphosphates are present) the β,β' and σ-subunits become radioactive (Fig. 1 left-hand side), but in the case of three substrates (right-hand side) radioactivity predominates in β,β'-subunits, which presumably reflects the shift of the 5'-end of the nascent oligonucleotide within the transcribing complex when its length grows continuously up to decanucleotide.

A similar methodology was employed to establish the subunits contacting with the 3'-OH end of the growing RNA. Affinity modification was made by the oligonucleotides being synthesized in situ. They contained photoreactive 5-halogenopyrimidine residues in the vicinity of 3'-OH end. In the presence of GpU (the primer), 5-IU or 5-BrU and |α-^{32}p| GTP (substrates) and the above-mentioned λ imm DNA fragment (template) only σ-subunit incorporated the radioactive label upon UV-irradiation (3).

During primary structure determination of β-subunit (see V.M. Lipkin et al., this volume) the limited tryptic hydrolysis was made. Separation and sequencing of low molecular weight peptides together with their yields enabled to find regions of polypeptide chain easily accessible for the protease. Two regions in the middle part of the polypeptide chain (amino acid residues 528-548 and 678-758) and C-terminal part of the molecule (residues 1105-1342)are evidently on the surface of the β-subunit and possibly represent flexible connections between stable domains.

A topographic study of the RNA polymerase α-subunits was made by modifications of RNA polymerase with iodoacetate, iodoacetamide, mercury ions and $K_3|Fe(CN)_6|$. The results are summarized in the model (Fig. 2).

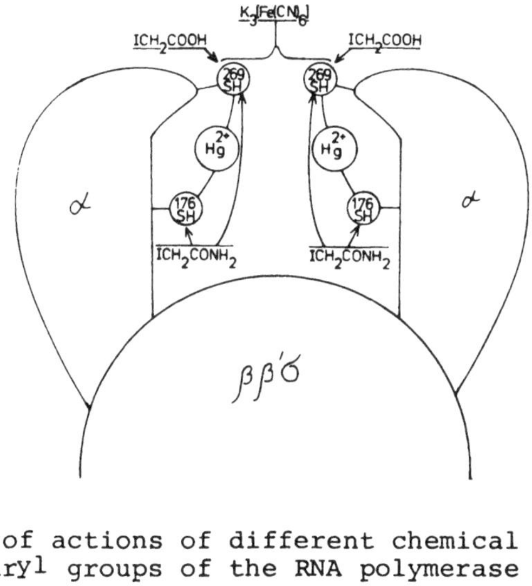

Fig. 2. Scheme of actions of different chemical agents on sulfhydryl groups of the RNA polymerase α-subunits.

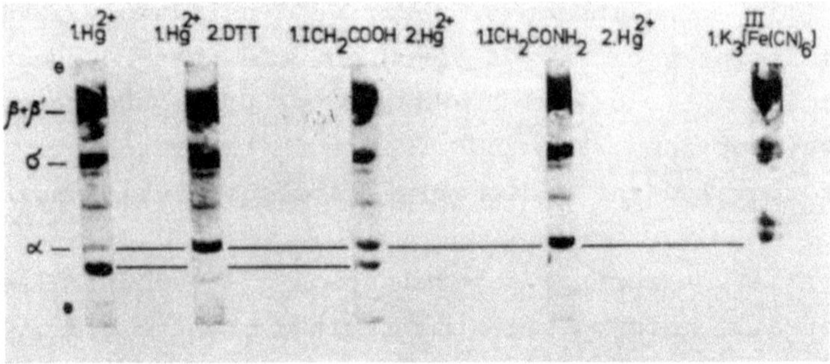

Fig. 3. SDS electrophoresis of the modified RNA polymerase. The order of modifications indicated. Buffer: 0.05 M Na-phosphate pH 7.0, 0.1% SDS.

The assignment of the alkylations to definite residues of the α-subunit polypeptide chain was made by the analysis of the α-subunit sequence (4) and electrophoretic mobilities of labeled peptides of tryptic digest on thin-layer peptide maps. Formations of intrasubunit (Hg^{2+}) and intersubunit (disulfate bridge) crosslinks were shown with the aid of SDS electrophoresis (Fig. 3).

References

1. Ovchinnikov, Yu.A., Efimov, V.A., Chakhmakhcheva, O.G., Skiba, N.P., Lipkin, V.M., Modyanov, N.N.: Bioorg. Khim. 5, 1410-1421 (1979).
2. Sverdlov, E.D., Tsarev, S.A., Kuznetsova, N.F.: FEBS Letters 112, 296-298 (1980).
3. Sverdlov, E.D., Tsarev, S.A., Begar, V.A.: FEBS Letters 114, 111-114 (1980).
4. Ovchinnikov, Yu.A., Lipkin, V.M., Modyanov, N.N., Chertov, O.Yu., Smirnov, Yu.V.: FEBS Letters 76, 108-111 (1977).

STRUCTURE OF THE PROTEOLIPID SUBUNIT OF THE ATP SYNTHASE

Jürgen Hoppe and Walter Sebald
Gesellschaft für Biotechnologische Forschung mbH.
D-3300 Braunschweig, Germany

Introduction

Membrane bound ATP synthases isolated from various organisms show common structural and functional properties [1-4]. They are composed of a membrane associated part F_1, which bears the ATPase activity and the membrane integrated part F_0, catalyzing H^+-conduction across the membrane. Both parts are necessary for energy transduction. A functional F_0 part has been isolated from the thermophilic bacterium PS-3 and from E. coli [5-8]. A minimum subunit composition of two different subunits was found in PS-3. In E. coli three different subunits were identified [6-8]. The major constituent of the F_0 part is a protein of about 8000d which most likely exists as a hexamer [9]. This protein is extractable with chloroform/methanol or with butanol and thus is referred to as a proteolipid. Beechey and cow. [10] discovered that dicyclohexylcarbodiimide (DCCD) inhibited the ATPase activity in beef heart mitochondria. The hydrophobic carbodiimide was found to be covalently bound to a proteolipid which was later identified as a subunit of the ATP synthase [1-4]. The proteolipid subunit appeared to be the target of several hydrophobic antibiotics e.g. oligomycin. But oligomycin only inhibited the activity of ATP synthases from mitochondria and some photosynthetic bacteria, as e.g. Rhodospirillum rubrum [11,13].
In recent years numerous proteolipid subunits from various organisms have been sequenced [14-20]. Amino acid substitutions which lead to inhibitor resistancy or to a nonfunctional enzyme

Chemistry of Peptides and Proteins, Vol. 1

have been determined [14,17,18,20,21].This article will therefore concentrate on the description of the primary structure of this protein, especially on the identification of an invariant carboxyl residue which seems to be intimately involved in proton conductance. A model for the conformation of the protein based on CD-measurements and secondary structure prediction will be discussed.

Results

General properties. The proteolipids can be generally purified by extraction with chloroform/methanol and subsequent chromatography on carboxymethyl-cellulose in chloroform/methanol [14-21]. All the sequences were determined by automated solid phase Edman degradation [14-21]. The proteins contain 72 - 82 amino acid residues which number is in agreement with the molecular weights determined on SDS gels. They are extremely hydrophobic containing only 16 - 25% polar residues which explain their solubility in organic solvents. Polar residues are clustered at the N-terminus and in the middle of the protein (residues 40 - 50 using the numbering of the *Neurospora crassa* sequence). Long hydrophobic stretches of about 25 amino acid residues are found both in the N-terminal as well as in the C-terminal part of the sequence.
A comparison of the sequences of these homologous proteins revealed the numerous conserved amino acid residues. These residues might be indispensable for the function or the maintainance of certain structure of the protein. Extensive homologies among the sequences of the proteolipid subunit were found. In the middle of the N-terminal hydrophobic stretch (residues 27 - 33) four glycine residues are conserved. Gly_{42}, arg_{45}, asn/gln_{46}, pro_{47} localized in the hydrophilic segment were also found to be invariant. In this segment the single conserved basic amino acid residue (arg_{45}) occurs. The only invariant acidic amino acid residue is observed in the middle of the C-terminal hydrophobic stretch (position 65). With one exception

	10 ... 20
Neurospora crassa	Tyr-Ser-Ser-Glu-Ile-Ala-Gln-Ala-Met-Val-Glu-Val-Ser-Lys-Asn-Leu-Gly-Met-Gly-Ser-Ala-Ala-Ile-Gly-Leu-
Bovine	Asp-Ile-Asp-Thr-Ala-Ala-Lys-Phe-Ile-Gly-Ala-Gly-Ala-Ala-Thr-Val-Gly-Val-
Saccharomyces cerevisiae	f-Met-Gln-Leu-Val-Leu-Ala-Ala-Lys-Tyr-Ile-Gly-Ala-Gly-Ile-Ser-Thr-Ile-Gly-Leu-
Rhodospirillum rubrum	f-Met-Asp-Ala-Glu-Ala-Ala-Lys-Met-Ile-Gly-Ala-Gly-Leu-Ala-Ala-Ile-Gly-Met-
Spinach	f-Met-Asn-Pro-Leu-Ile-Ala-Ala-Ala-Ser-Val-Ile-Ala-Ala-Gly-Leu-Ala-Val-Gly-Leu-Ala-Ser-
Mastigocladus laminosus	f-Met-Asp-Pro-Leu-Ile-Ser-Ala-Ala-Ser-Val-Leu-Ala-Ala-Ala-Leu-Ala-Ile-Gly-Leu-Ala-Ala-
E. coli	f-Met-Glu-Asn-Leu-Asn-Met-Asp-Leu-Leu-Tyr-Met-Ala-Ala-Ala-Val-Met-Met-Gly-Leu-Ala-Ala-
Acido caldarius	f-Met-Gln-Leu-Asp-Met-Val-Lys-Ala-Ile-Tyr-Asp-Ile-Ala-Val-Ala-Leu-Leu-Leu-Gly-Leu-Ala-Ala-
PS-3	f-Met-Ser-Leu-Gly-Val-Leu-Ala-Ala-Ala-Ile-Ala-Val-Gly-Leu-Gly-Ala-

	30 ... 40 ... 50
Neurospora crassa	Tyr-Gly-Ala-Gly-Ile-Gly-Ile-Gly-Leu-Val-Phe-Ala-Ala-Leu-Leu-Asn-Gly-Val-Ala-Arg-Asn-Pro-Ala-Leu-Arg-
Bovine	Ala-Gly-Ser-Gly-Ala-Gly-Ile-Gly-Thr-Val-Phe-Gly-Ser-Leu-Ile-Ile-Gly-Tyr-Ala-Arg-Asn-Pro-Ser-Leu-Lys-
Saccharomyces cerevisiae	Leu-Gly-Ala-Gly-Ile-Gly-Ile-Ala-Ile-Val-Phe-Ala-Ala-Leu-Ile-Asn-Gly-Val-Ser-Arg-Asn-Pro-Ser-Ile-Lys-
Rhodospirillum rubrum	Ile-Gly-Ser-Gly-Ile-Gly-Val-Gly-Asn-Ile-Trp-Ala-Asn-Leu-Ile-Ser-Thr-Val-Gly-Arg-Asn-Pro-Ala-Ala-Lys-
Spinach	Ile-Gly-Pro-Gly-Val-Gly-Gln-Gly-Thr-Ala-Ala-Gly-Gln-Ala-Val-Glu-Gly-Ile-Ala-Arg-Gln-Pro-Glu-Ala-Glu-
Mastigocladus laminosus	Ile-Gly-Pro-Gly-Ile-Gly-Gln-Gly-Asn-Ala-Ala-Gly-Gln-Ala-Val-Glu-Gly-Ile-Ala-Arg-Gln-Pro-Glu-Ala-Glu-
E. coli	Ile-Gly-Ala-Ala-Ile-Gly-Ile-Gly-Ile-Leu-Gly-Gly-Lys-Phe-Leu-Glu-Gly-Ala-Ala-Arg-Gln-Pro-Asp-Leu-Ile-
Acido caldarius	Val-Gly-Ser-Gly-Val-Gly-Asp-Gly-Met-Val-Met-Ser-Lys-Tyr-Val-Glu-Gly-Val-Ala-Arg-Gln-Pro-Glu-Ala-Arg-
PS-3	Leu-Gly-Ala-Gly-Ile-Gly-Asn-Gly-Leu-Ile-Val-Ser-Arg-Thr-Ile-Glu-Gly-Ile-Ala-Arg-Gln-Pro-Glu-Leu-Arg-

	60 70
Neurospora crassa	Gly-Gln-Leu-Phe-Ser-Tyr-Ala-Ile-Leu-Gly-Phe-Ala-Phe-Val-Glu-Ala-Ile-Gly-Leu-Phe-Asp-Leu-Met-Val-Ala-
Bovine	Gln-Gln-Leu-Phe-Ser-Tyr-Ala-Ile-Leu-Gly-Phe-Ala-Leu-Ser-Glu-Ala-Met-Gly-Leu-Phe-Cys-Leu-Met-Val-Ala-
Saccharomyces cerevisiae	Asp-Thr-Val-Phe-Pro-Met-Ala-Ile-Leu-Gly-Phe-Ala-Leu-Ser-Glu-Ala-Thr-Gly-Leu-Phe-Cys-Leu-Met-Val-Ser-
Rhodospirillum rubrum	Ser-Thr-Val-Glu-Leu-Tyr-Gly-Trp-Ile-Gly-Phe-Ala-Val-Thr-Glu-Ala-Ile-Ala-Leu-Phe-Ala-Leu-Val-Val-Ala-
Spinach	Gly-Lys-Ile-Arg-Gly-Thr-Leu-Leu-Leu-Ser-Leu-Ala-Phe-Met-Glu-Ala-Leu-Thr-Ile-Tyr-Gly-Leu-Val-Val-Ala-
Mastigocladus laminosus	Gly-Lys-Ile-Arg-Gly-Thr-Leu-Leu-Leu-Thr-Leu-Ala-Phe-Met-Glu-Ser-Leu-Thr-Ile-Tyr-Gly-Leu-Val-Ile-Ala-
E. coli	Pro-Leu-Leu-Arg-Thr-Gln-Phe-Phe-Ile-Val-Met-Gly-Leu-Val-Asp-Ala-Ile-Pro-Met-Ile-Ala-Val-Gly-Leu-Gly-
Acido caldarius	Gly-Ser-Ile-Phe-Gly-Ser-Ala-Leu-Leu-Gly-Val-Ala-Leu-Val-Glu-Ala-Phe-Pro-Val-Ile-Ala-Leu-Ala-Phe-Gly-
PS-3	Pro-Val-Leu-Gln-Thr-Thr-Met-Phe-Ile-Gly-Val-Ala-Leu-Val-Glu-Ala-Leu-Pro-Ile-Ile-Gly-Val-Val-Phe-Ser-

	80
Neurospora crassa	Leu-Met-Ala-Lys-Phe-Thr
Bovine	Phe-Leu-Ile-Leu-Phe-Ala-Met
Saccharomyces cerevisiae	Phe-Leu-Leu-Leu-Phe-Gly-Val
Rhodospirillum rubrum	Leu-Ile-Leu-Leu-Phe-Ala-Ala
Spinach	Leu-Ala-Leu-Leu-Phe-Ala-Asn-Pro-Phe-Val
Mastigocladus laminosus	Leu-Val-Leu-Leu-Phe-Ala-Asn-Pro-Phe-Ser
E. coli	Leu-Tyr-Val-Met-Phe-Ala-Val-Ala
Acido caldarius	Ile-Ile-Ile-Leu-Phe-Thr-Lys-Gly-Ala-Phe
PS-3	Phe-Ile-Tyr-Leu-Gly-Arg

Fig. 1

Amino acid sequence of the proteolipid subunit of various ATP synthases.

Solid lines indicate conserved amino acid residues, dashed lines amino acids specific for a group of organisms.

(aspartic acid in E. coli), this amino acid is a glutamic acid. This acidic residue is flanked by two conserved alanyl residues (position 62 and 66). At the C-terminus (position 80) a conserved phenylalanine was found.

DCCD binding. DCCD binds covalently to the proteolipid subunit. The use of ^{14}C DCCD allowed the identification of the target amino acid [14-16,19]. In all analyzed proteolipids the covalently bound ^{14}C-labelled inhibitor could be traced back to a single fragment produced by CNBr and/or N-bromosuccinimide cleavage [14-16,19]. Automated solid-phase Edman degradation of the modified fragment lead to the liberation of bound radioactivity during one degradation step only. In all organisms the invariant acid amino acid residue 65 was exclusively labelled adding further evidence for the importance of this residue. Interestingly, labelling of only a part of the proteolipid oligomer with DCCD completely abolishes the enzymatic activity of the ATP synthase [9,15] indicating that the proteolipid monomers act in cooperative fashion.

DCCD resistant mutants from E. coli have been isolated. Two classes according to their decreased sensitivity towards DCCD were identified [18]. In both cases the isoleucine residue 32 is substituted, which is located in a highly conserved region of the polypeptide. Class one contains a valine residue at this position leading to a moderate resistancy while class two is strongly resistant. Here a threonine is found instead of the isoleucine. It is thus a distinct possibility that this part of the sequence provides a noncovalent binding site for DCCD.

Mutants with defective F_0. It was shown above that the invariant acidic amino acid 65 can be covalently modified by DCCD. But covalent modification or just noncovalent binding could induce allosteric conformational changes leading to a nonactive ATP synthase complex. Mutants defective in H^+ conduction were selected for nonbinding of DCCD and analyzed for amino acid exchanges in the proteolipid subunit. Mutant DG 7/1 contains a glycine instead the aspartic acid [19]. Diploid strains containing both wild type and mutant alleles were analyzed for H^+ conduction [24]. A strong negative complementation was found indicating that the mutant protein could substitute the wild type one. Thus, most likely the mutant protein exhibits the same conformation *in vivo* than the wild type protein. All experimental evidence suggests that the aspartic acid 65 is located in the middle of the lipid bilayer. Firstly, it occurs in the middle of a hydrophobic stretch in the amino acid sequence; secondly, only hydrophobic carbodiimide can interact with this residue.

Secondary structure. At the present time no direct measurements on the secondary and tertiary structure of the membrane-bound part of the ATP synthase are available. One attempt is to measure the secondary structure composition of the isolated proteolipid in solvents which provoke different effects on the secondary structure. This will yield information on preferential structural properties of the protein. These measurements may be compared with secondary structure predictions to locate the respective secondary structure in the sequence of the protein. CD-spectra of the *E. coli* proteolipid subunit were recorded in trifluoroethanol and chloroethanol. These solvents promote the formation of α-helical structures [24]. Thus a high fraction of α-helix was calculated from the spectrum. Long stretches of about 25 residues in α-helices were found. But both solvents could not shift the conformation entirely to 100% α-helix as in [24], indicating that parts of the protein do not form α-helices under any conditions. Addition of water to the solution gradually decreases the α-helical content. At 30% trifluoroethanol a drastic change in the spectrum occurred (Fig. 2B). The proteolipid remains as a clear solution down to 2% of trifluoroethanol or chloroethanol without significant change in the CD-spectrum. Still a high fraction of α-helix is observed ($\sim$65%). But the length of the helix-segments significantly decreased from about 25 amino acids in 100% trifluoroethanol (A) to about 6 amino acids (B).

Complexes of the proteolipid subunit and vesicles were prepared. Vesicles containing phosphatidylethanolamine and phospatidyl-

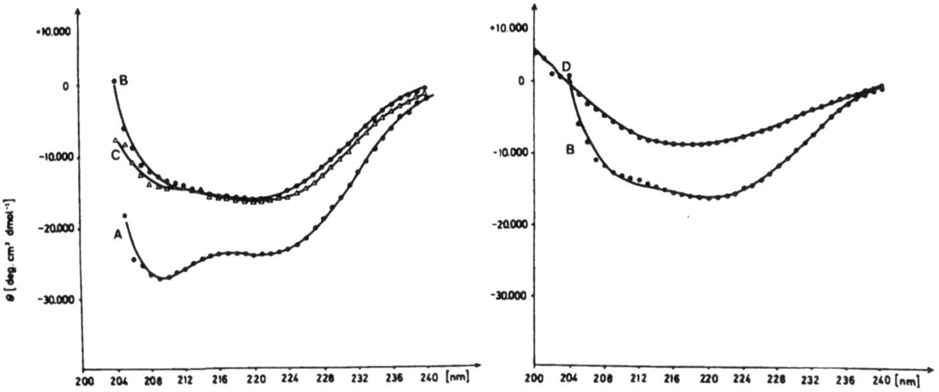

	A	B	C	D
		%		
α -helix	72.5	67.9	65.9	37
β -bond	19.6	31	25	37.4
turns	4.1	1.0	9.9	5.8
coil	3.8	-	0.8	19.8
mean α -helix length	24.5	5.8	6.7	5.6

Fig. 2.

CD-spectra of the E. coli proteolipid and the large CNBr-fragment B6.

The points represent the measured spectra. The solid lines are the curves fitted by a computer program using a non-linear leastsquare fit method described by Chen et al. 25 . Reference data were kindly provided by Dr. J.T. Yang.
A) proteolipid in 100% trifluoroethanol B) in 10% chloroethanol
C) in vesicles D) fragment B6 in 10% chloroethanol.

choline at a ratio of 3/2 were prepared by sonication. Protein either in pure chloroethanol or in a 10% solution was added to the vesicle solution at a ratio of 1/5. The residual organic solvent was removed by dialysis. In both cases similar CD-spectra were observed (C) which closely resemble spectrum B recorded in 10% chloroethanol. In all systems a high content of α-helix was observed. This indicates that the protein may be largely composed of α-helical structures in the native membrane. In the most thoroughly investigated membrane protein, the bacteriorhodopsin from Halobacterium halobium a similar content of α-helix was found. Seven stretches of each 25 hydrophobic amino acids were found to traverse the membrane in α-helical conformation [26,27].

Secondary structure prediction methods according to Chou and Fasman [28], Robson and Suzuki [29], and Nagano [30] were used to locate the α-helical segments in the proteolipid subunit. Fig. 3 shows the result of the calculations for the <u>E. coli</u> proteolipid. α-helical segments were predicted for the N-terminal sequence and at the C-terminus. In all prediction models the amino acids in the conserved hydrophilic segment containing the invariant arginine residue are located in a turn or loop. A high probability for the occurrance of a turn was also found in the glycine-rich segment close to the position where mutations were found leading to DCCD resistance. To substantiate this model the CD spectrum of the large cyanogenbromide fragment B6 from <u>E. coli</u> was measured. In this peptide (residue 22 to 61 in Fig. 1) large parts of the N-terminus and the C-terminus are missing. As shown in Fig. 2D a lower fraction of α-helix is observed for this peptide compared with the whole protein, suggesting that indeed the α-helical segments are located at both ends of the protein.

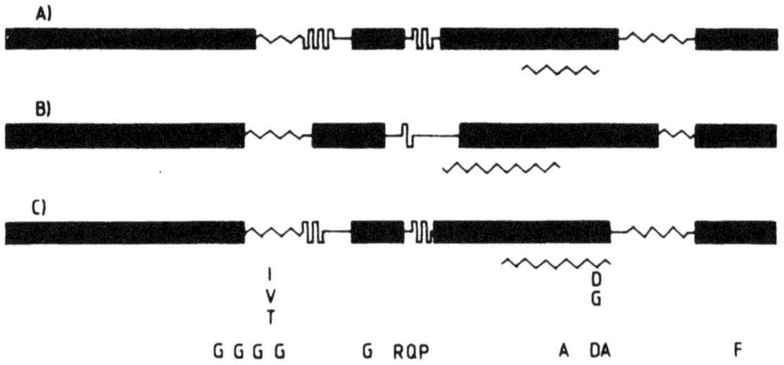

Fig. 3.

Secondary structure predictions.

The methods of A) Robson [29], B) Chou and Fasman [28], and C) Nagano [30] were applied.

▬ α - helix, ∿ ß-bend, ⊓⊓ turns, — coil conformation

The localization of the invariant amino acids and the amino acid exchanges in mutant protein are indicated by the one letter code for amino acids.

Conclusions

To construct models how the proteolipid may be orientated in the lipid bilayer the following points seem to be important:

1. The essential carboxyl group (glu_{65}) is probably located in the bilayer, as labelling is only possible with hydrophobic carbodiimides.
2. The amino acid segment arround ile 32 may provide a binding site for DCCD. This would bring this part of the protein into the proximity of the invariant glu_{65}.
3. The long hydrophobic stretches at both ends of the protein may traverse the membrane similar as in the bacteriorhodopsin in Halobacterium halobium.
4. The hydrophilic segment containing the invariant sequence arg/asn/gln-pro most likely forms a loop which according to its polar character should be located in the water phase.

References

1. Sone, N., Yoshida, M., Hirata, H., Kagawa, Y.: J. Biol. Chem. 250, 7917-7923 (1975).
2. Friedl, P., Friedl, C., Schairer, H.U.: Eur. J. Biochem. 100, 175-180 (1979).
3. Pick, U., Racker, E.: J. Biol. Chem. 254, 2793-2799 (1979).
4. Foster, D.L., Fillingame, R.H.: J. Biol. Chem. 254, 8230-8236 (1979).
5. Sone, N., Yoshida, M., Hirata, H. and Kagawa, Y.: Proc. Natl. Acad. Sci. U.S.A. 75, 4219-4223 (1978).
6. Negrin, R.S., Foster, D.L., Fillingame, R.H.: J. Biol. Chem. 255, 5643-5648 (1980).
7. Schneider, E., Altendorf, K.: FEBS Lett. 116, 173-176 (1980).
8. Friedl, P., Schairer, H.U.: FEBS Lett. in press (1981).
9. Sebald, W., Graf, T., Lukins, H.B.: Eur. J. Biochem. 93, 587-599 (1979).
10. Beechey, A.M.: Biochem. Biophys. Res. Commun. 23, 75-80 (1966).
11. Lardy, H.A., Johnsen, D., Murray, W.C.: Arch. Biochem. Biophys. 78, 587-597 (1958).
12. Glehn, M., Norrestam, R., Kierkegaard, P., Marun, L., Ernster, L.: FEBS Lett. 20, 267-269 (1972).

13. Schneider, E., Schwulêra, U., Müller, H.W., Dose, K.: FEBS Lett. 87, 257-260 (1978).

14. Sebald, W., Hoppe, J., Wachter, E.: in Functional and Molecular Aspects of Biomembrane Transport(E. Quagliariello et al. eds) pp. 63-74, Elsevier/North-Holland Biomedical Press (1979).

15. Hoppe, J., Sebald, W.: Eur. J. Biochem. 107, 57-65 (1980).

16. Sebald, W., Machleidt, W., Wachter, E.: Proc. Natl. Acad. Sci. U.S.A. 77, 785-789 (1980).

17. Hoppe, J., Schairer, H.U., Sebald, W.: FEBS Lett. 109, 107-111 (1980).

18. Hoppe, J., Schairer, H.U., Sebald, W.: Eur. J. Biochem. 122, 17-24 (1980).

19. Sebald, W., Wachter, E.: FEBS Lett. 122, 307-311 (1980).

20. Sebald, W., Wachter, E., Tzagoloff, A.: Eur. J. Biochem. 100, 599-607 (1979).

21. Wachter, E., Schmid, R., Deckers, G., Altendorf, K.: FEBS Lett. 113, 265-270 (1980).

22. Friedl, P., Friedl, C., Schairer, H.U.: FEBS Lett. 119, 254-256 (1980).

23. Cattel, K.J., Londop, C.R., Knight, I.G., Beechey, R.B.: Biochem. J. 125, 169-177 (1971).

24. Cockle, S.A., Epand, R.M., Boggs, J.M., Moscarello, M.A.: Biochemistry 17, 624-629 (1978).

25. Chen, Y.-H., Yang, J.T., Chan, K.H.: Biochemistry 13, 3350-3359 (1974).

26. Unwin, P.N.T., Henderson, R.: J. Mol. Biol. 94, 425-440 (1975).

27. Engelman, B.M., Henderson, R. McLachlan, A.D., Wallace, B. A.: Proc. Natl. Acad. Sci. U.S.A. 77, 2033-2027 (1980).

28. Chou, P.Y., Fasman, G.D.: in Advances in Enzymology (A. Meister ed.) pp. 45-148, John Wiley, New York (1978).

29. Robson, B., Suzuki, E.: J. Mol. Biol. 107, 327-356 (1976).

30. Nagano, K.: J. Mol. Biol. 109, 251-274 (1977).

QUATERNARY STRUCTURE AND RECONSTITUTION OF ACETYLCHOLINE RECEPTOR FROM *Torpedo californica*

W. Schiebler, L. Lauffer, G. Bandini, and F. Hucho
Freie Universität Berlin, Institut für Biochemie,
Fabeckstr. 34, 1000 Berlin 33

Introduction

Nerve cells communicate via synapses. Receptor proteins located in the postsynaptic membrane play a key role in intercellular communication. They recognize specific signals (emitted from other cells) and convert them into intracellular effects. The acetylcholine receptor is a typical postsynaptic "signal converter". Nicotinic cholinergic synapses are widespread in the periphery and form the neuromuscular junctions (endplates) of voluntary muscles. Their presence in the central nervous system has been proven but it is not clear whether nAChR from brain is similar or identical with the peripheral protein.

But although pure receptor protein is readily available in large quantities most of the basic problems of receptor functional mechanism are unsolved. The AChR regulates the ion permeability of the postsynaptic membrane. This presentation is focused on the structural basis of the interaction between signal recognition (transmitter binding) and permeability change (ion channel). Elucidation of the quaternary structure of the receptor protein is a prerequisit for solving this problem.

In this report we discuss evidence that in the case of the nAChR the bifunctional receptor complex is indeed a protein complex consisting of several different subunits. Polypeptide

Chemistry of Peptides and Proteins, Vol. 1

chains of different size are present and have different functions. Furthermore we present evidence based on reconstitution experiments that the binding site and the ion channel may be located on different subunits of the protein complex which can be separated by biochemical methods.

Results

Quaternary structure of AChR from *Torpedo californica*

The experimental model system unusually rich in nAChR are the electric organs of electric fish. Especially from the electric ray *Torpedo* and the electric eel *Electrophorus electricus* hundreds of milligram of receptor protein can be easily extracted and most of the biochemical investigations so far have been performed with this material.

nAChR from *Torpedo californica* is a pentameric protein of molecular weight 27o,ooo. This structure was deduced from SDS gel electrophoresis patterns of cross-linked nAChR (Fig. 1) and was recently proven by amino acid and endgroup analysis (1).

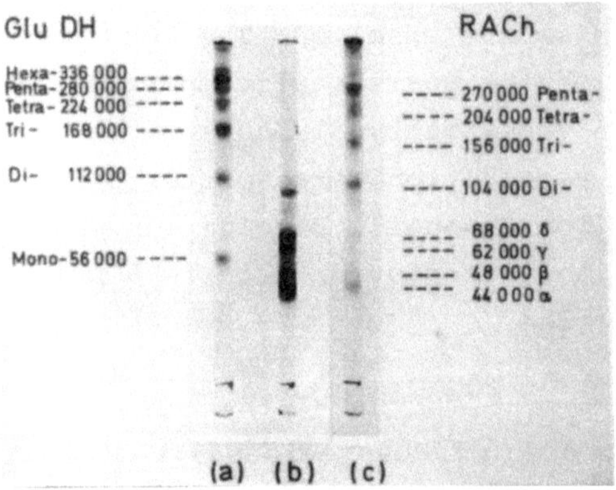

Fig. 1: Dodecylsulfate polyacrylamide gel electrophoresis of AChR cross-linked with suberimidate. (a) calibration with glutamate dehydrogenase; (b) uncross-linked AChR; (c) cross-linked AChR (2).

Two of the five polypeptide chains are identical; the other three possess partly homologous primary structures (in the first 56 N-terminal amino acids sequenced so far) but are considerably longer. The quaternary structure thus appears to be $\alpha_2 \beta \gamma \delta$, a structure first proposed by Karlin (3). All polypeptide chains contain oligosaccharides of unknown size and function. Several other quaternary structures have been proposed and considerable confusion existed until recently as to the true structure (4). It now appears clear that this was caused by proteolytic activity in the respective receptor preparations.

$\alpha_2 \beta \gamma \delta$ are the predominant polypeptide chains of receptor-rich membrane fragments isolated from electric tissue and purified by density gradient centrifugation. They comprise more than 60% of the dry weight and more than 80% of the protein in these membranes (2). Membrane bound receptor can be covalently cross-linked <u>in situ</u> and subsequently analysed by SDS gel electrophoresis. By this technique it can be shown that the four polypeptide chains α β γ and δ are part of a protein complex embedded in the membrane (2). Two of such complexes appear to form a dimer (5) covalently linked togehter by a disulfide bridge between two δ-chains(6).

Electron microscopy of receptor-rich membranes revealed ring structures ("doughnuts") believed to represent receptor molecules. Each molecule appears to have a diameter of 80 Å with a central pit of 20 Å and it protrudes about 50 Å from the lipid matrix into the extracellular space (Fig. 2 A). Although the 20 Å resolution obtained by electron microscopy does not allow to locate single polypeptide chains of the protein complex the obvious asymmetry of the image is in agreement with the $\alpha_2 \beta \gamma \delta$ subunit structure. A symmetrical complex cannot be constructed from these subunits when they all have to be accomodated facing the extracèllular space.

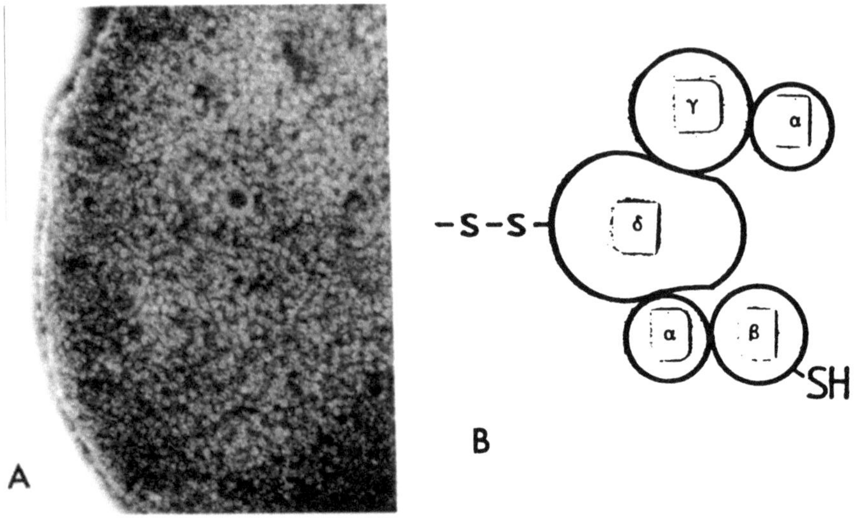

Fig. 2: nAChR from Torpedo californica. A electron micrograph of negatively stained receptor-rich membrane. B model of the quaternary structure.

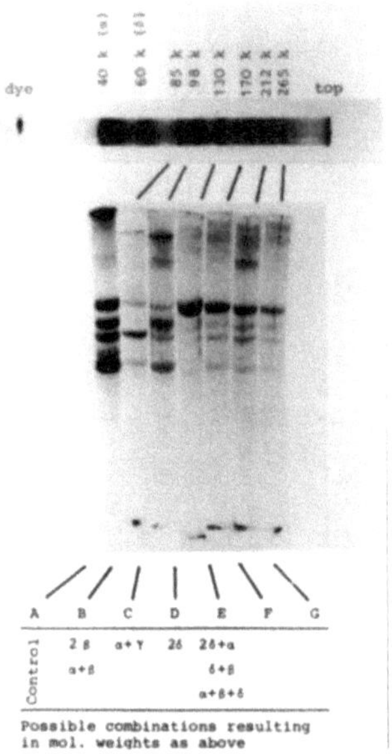

Fig. 3: SDS polyacrylamide gel-electrophoresis of nAChR cross-linked with the cleavable cross-linker 3,3'dithiopropionimidate. Top gel (horizontal) : first dimension. Vertical gels: second dimension after reductive cleavage of the cross-links.

This must be the case because all four types of polypeptide chains are glycoproteins.

Further information about the subunit arrangement can be obtained by nearest neighbour analysis using cleavable cross-linkers and twodimensional SDS gel-electrophoresis. After cross-linking neighbouring polypeptide chains of the AChR complex with 3,3-dithiopropionimidate denaturing electrophoresis allows to seperate cross-linked subunits (Fig. 3). After cleavage of the cross-link SDS electrophoresis in a second dimension shows which were the polypeptide chains contained in a cross-linked fragment of the complex. By this method it can be shown that in the membrane two δ-chains are located next to each other, also two β-chains and an α-chain and a γ-chain. The two α-chains never could be linked to each other indicating that they are either too remote to be bridged by the bifunctional reagent or reactive side chains are not accessible. A photoaffinity label bound to the α-chain also reacted with the δ-chain suggestingα and δ being close neighbours. Taking the available information together a model of the AChR complex can be drawn (Fig. 2 B).

Subunit function unknown:

The functional significance of the α_2 β γ δ-subunit structure is largely unknown (Fig. 4). Only the α-chains have been identified unambiguously as the binding proteins for cholinergic ligands (7). The alkylating agonist bromoacetylcholine labels selectively the α-chain of reduced AChR (8). In native AChR α and β react with the photoaffinity antagonist 4-azido 2-nitrobenzyltrimethylammonium fluoroborate (9), but only the α-chain is protected against this label by α-bungarotoxin, the curare-like snake venom peptide toxin.

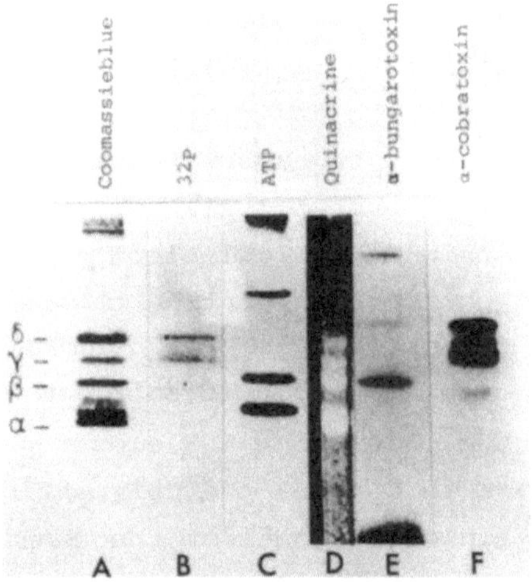

Fig. 4 : Affinity labeling of receptor rich membranes. A, control, coomassieblue stained SDS gel. B, ^{32}P, phosphorylation by ATP-kinase, C, ATP-binding, D, quinacrine mustard, E, α-bungarotoxin, F, α-cobratoxin. B, C, E, F are autoradiograms, D shows quinacrine fluorescence. The molecular weights obtained from bands of the autoradiograms E and F must be corrected for the molecular weights of the toxin (8,000) which is attached to the membrane protein.

Photoaffinity derivatives of α-bungarotoxin also react mainly with the α-chains again proving their role as the binding protein for competitive cholinergic effectors (10, 11). Interestingly, with photoaffinity labels an overspill to other chains can be observed, again showing their close proximity within the membrane-bound complex and possibly hinting at their participation in the early steps of transmitter action.

Stoichiometry of ligand binding was a matter of controvers, but now there appears to be general agreement that there are two agonist or antagonist binding sites (including α-bungarotoxin) on each AChR complex of MW 270,000.

The functions of β, γ, and δremain to be determined. δ and to a lesser degree γ are the target of an ATP-dependent phosphorylation catalysed by a membrane-bound protein kinase. A role of this reaction in the regulation of AChR function or metabolism has been discussed (4) but no experimental evidence is available so far. (Seven serin-OH bound phosphate groups have been found, distributed over the four different polypeptide chains.) Neither β, γ, or δ appear to be the protein kinase (or the phosphatase, reversing the ATP-dependent phosphorylation). This has been shown by labeling the membrane-bound AChR with a photoaffinity derivative of ATP: Two polypeptide chains incorporated the label but α β γ and δ were not among them (12).

Hybridisation-reconstitution:

From structural investigations alone the relation between acetylcholine binding site (the receptor in the true sense of the word) and ion channel cannot be deduced. We therefore attempted to tackle the problem by reconstitution experiments. In this report we present evidence that in the case of the nicotinic acetylcholine receptor from the electric organ of *Torpedo californica* transmitter binding site and ion channel are not necessarily located on the same protein molecule. By hybridisation of receptor complexes in which the transmitter binding site is blocked with complexes in which the ion channel is irreversibly inhibited we were able to reconstitute active acetylcholine receptor complexes. The reconstituted system was similar to the native receptor in its ability to regulate the ion permeability of lipid vesicles in response to nicotinic cholinergic effectors.

The hybridisation experiment was performed with membrane fragments enriched in acetylcholine receptor prepared from the electric organs of *Torpedo californica* according to (13) with the slight modifications described (14). One sample of the receptor-enriched membranes was treated with α-bungarotoxin, a strong and virtually irreversible competitive inhibitor of the transmitter binding site in nicotinic acetylcholine receptors. Another sample of this preparation was treated with quinacrine mustard. Quinacrine has been shown to act upon the postsynaptic membrane of nicotinic cholinergic synapses as a local anaesthetic (15), characterised by its blocking effect on the ion translocation through the membrane without competitively inhibiting transmitter binding. We have shown that the alkylating derivative quinacrine mustard irreversibly blocks the ion permeability response to agonists in receptor enriched membrane microsacs (16). These modified membranes still contained at least fifty percent of their high-affinity binding sites for ^{3}H-acetylcholine (Fig.5).

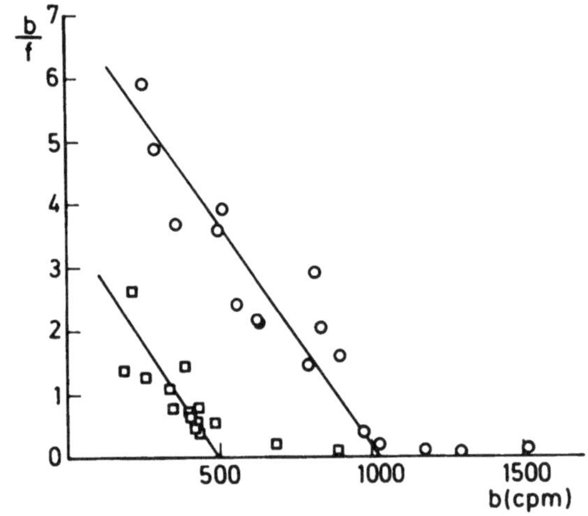

Fig. 5: Scatchard plot of acetylcholine binding to untreated (O-O-O) and quinacrine mustard modified (□-□-□) receptor-rich membranes (16).

Both these modified samples were dissolved in detergent supplemented with lecithine, first separately, for the controls, later jointly, for the hybridisation experiment (Fig. 6). Reconstitution of the detergent solubilised membranes to functional liposomal receptor-membrane systems was performed by column chromatography with Sephadex G 50 as described before

(14). Receptor function was measured by determining the ^{22}Na efflux from reconstituted vesicles by the Millipore filtration assay (17). Maximum efflux was determined by adding gramicidin to the vesicles (open squares in figure 7). The first control experiment was aimed to show that under the conditions of the reconstitution experiment the bungarotoxin block is virtually irreversible (top frame of fig. 7 , labeled "T", for toxin).

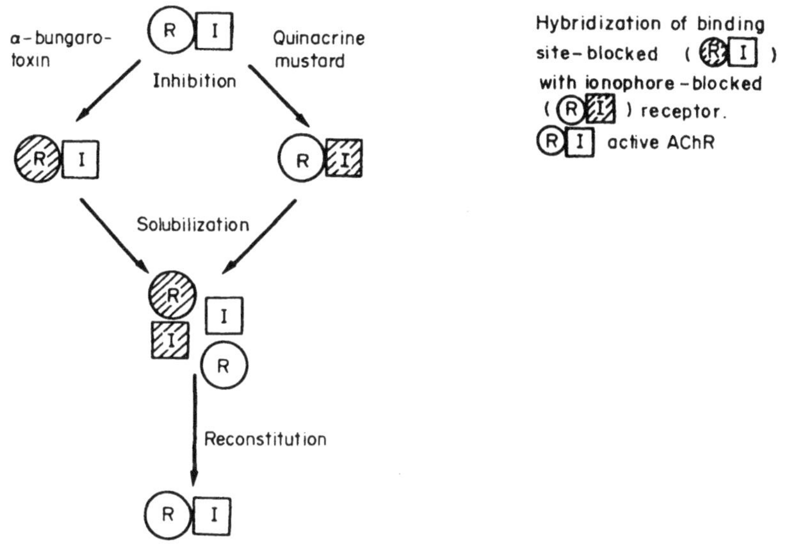

Fig. 6: Scheme of the hybridisation-reconstitution experiment.

The time dependent ^{22}Na efflux from vesicles reconstituted with toxin-blocked receptor could not be stimulated by o.1 mM carbamoylcholine although considerable amounts of radioactivity could be released with gramicidin. A similar result was obtained in the second control experiment with quinacrine mustard blocked receptor (middle frame of the figure, labeled "Q"): the ^{22}Na efflux from vesicles reconstituted with mustard-blocked receptor could also not be stimulated by o.1 mM carbamoyl choline. (In this experiment unstimulated "leakage" of ^{22}Na was higher as compared to the "T"-experiment. This increased leakage was consistently observed with quinacrine modified membranes (16) and may be due to the increase of lipid fluidity caused by the local anaesthetic).

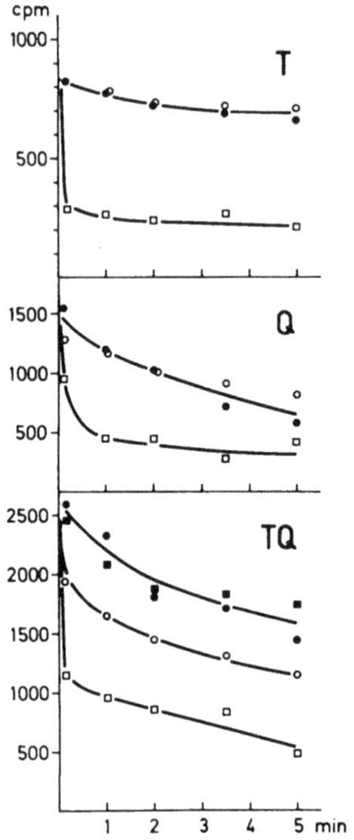

Fig. 7: Hybridisation of binding site-blocked (T) and ion channel blocked (Q) AChR yielding functionally active AChR (TQ). ^{22}Na efflux from reconstituted AChR-lipid vesicles in the absence (● - ● - ●) and presence (o - o - o) of 10^{-4} M carbamoylcholine. Inhibition of carbamoylcholine stimulation by α-neurotoxin from Naja naja siamensis (■ - ■ - ■). ^{22}Na efflux in the presence of o.1 mg/ml gramicidin A (□ - □ - □).

Reconstitution of a one to one mixture of toxin-blocked and and quinacrine mustard-blocked receptor-rich membranes resulted in vesicles in which the ^{22}Na efflux could be stimulated by o.1 mM carbamoyl choline (bottom frame, labeled "TQ"). This stimulation (open circles) was prevented by preincubation of the reconstituted "TQ"-vesicles with Naja naja siamensis toxin, another competitive inhibitor of nicotinic acetylcholine receptors (closed squares).

The simplest explanation of this result would be based on the following model: The nicotinic acetylcholine receptor of the Torpedo membrane is composed of two types of subunits, one carrying the transmitter binding site (being the receptor "sensu stricto"), the other translocating the cations (being the ion channel, ionophore).

Another possibility would be a complementation of two identical polypeptide chains forming together the active AChR complex, each containing both a ligand binding site and an ionophor. In our reconstitution experiment polypeptides with unblocked receptor sites would complement with chains containing unblocked ion channels. Both models, the first assuming distinct proteins for receptor and channel, the second postulating only one type of polypeptide chain, would imply that receptor and channel are not necessarily located on the same polypeptide chain and that in our reconstitution experiment exchange of chains, identical or not, took place. Several alternative explanations were investigated and ruled out by appropriate control experiments (not shown).

References

1. Raftery, M.A., Hunkapiller, M.W., Strader, C.D., and Hood, L.E.: Science 208, 1454-1457 (1980).
2. Hucho, F., Bandini, G., and Suarez-Isla, B.A.: Eur.J.Biochem. 83, 335-340 (1978).
3. Reynolds, J., and Karlin, A.: Biochemistry 17, 2035-2038 (1978).
4. Heidmann, T., and Changeux, J.-P.: Ann.Rev.Biochem.47, 317-357 (1978).
5. Chang, H.W., and Bock, E.: Biochemistry 16, 4513-4520 (1977).
6. Suárez-Isla,B.A., and Hucho, F.: FEBS LETTERS 75, 65-69 (1977).

7. Reiter, M.J., Cowburn, D.A., Prives, J.M., Karlin, A.: Proc.Natl.Acad.Sci. USA 69, 1168-1172 (1972).

8. Damle, V.N., McLaughlin, M., and Karlin, A.: Biochem. Biophys.Res.Commun. 84, 845-851 (1978).

9. Hucho, F., Layer, P., Kiefer, H.R., and Bandini, G.: Proc. Natl.Acad.Sci. USA 73, 2624-2628 (1976).

10. Witzemann, V., and Raftery, M.A.: Biochemistry 16, 5862-5868 (1977).

11. Hucho, F., FEBS LETTERS 103, 27-32 (1979).

12. Gordon, A.S., Guillory, R.J., Diamond, I., and Hucho, F.: FEBS LETTERS 108, 37-39 (1979).

13. Duguid, J.R., and Raftery, M.A.: Biochemistry 12, 3593-3597 (1973).

14. Schiebler, W., Hucho, F.: Eur.J.Biochem. 85, 55-63 (1978).

15. Grünhagen, H.H., Iwatsubo, M. and Changeux, J.-P.: Eur. J.Biochem.80, 225-242 (1977).

16. Lauffer, L., Weber, K.-H., Hucho, F., Biochem.et Biophys. Acta 587, 42-48 (1979).

17. Popot, J.L., Sugiyama, H., and Changeux, J.P.: J.Mol.Biol. 106, 469-483 (1976).

AUTHOR INDEX

SUBJECT INDEX